Contamination Control and Cleanrooms

Contamination Control and Cleanrooms

Problems, Engineering Solutions, and Applications

Alvin Lieberman

Library of Congress Catalog Card Number 91-44921
ISBN 0-442-00574-1

Manufactured in the United States of America

Published by Van Nostrand Reinhold
115 Fifth Avenue
New York, New York 10003

Chapman and Hall
2-6 Boundary Row
London, SE1 8HN, England

Thomas Nelson Australia
102 Dodds Street
South Melbourne 3205
Victoria, Australia

Nelson Canada
1120 Birchmount Road
Scarborough, Ontario M1K 5G4, Canada

16 15 14 13 12 11 10 9 8 7 6 5 4 3 2 1

Library of Congress Cataloging-in-Publication Data

Lieberman, Alvin, 1921–
Contamination control and cleanrooms : problems, engineering solutions, and applications / Alvin Lieberman.
p. cm.
ISBN 0-442-00574-1
1. Clean rooms. 2. Decontamination (from gases, chemicals, etc.)
I. Title.
TH7694.L54 1992
620.8′6—dc20

91-44921
CIP

To the engineers and the scientists who worked in the early stages of contamination control and to those who are carrying on the development of new technology at this time. This appreciation includes the technicians, laboratory workers, and engineers who developed the early white room technology by trial and error, and by applying basic engineering to solve the problems of cleaning and maintaining cleanliness in those areas. It also includes the scientists and engineers of today who are developing and applying the needed and technically realistic models for the phenomena causing contamination in our present, more critical world and the procedures needed for control.

A special appreciation to Andrew R. Gutacker of ARGOT, Inc. He is one of the pioneers in this field who has moved with the times to add the latest scientific approaches to a wealth of practical experience. A further thanks to Andy for permission to use many of the illustrations shown in this book.

Contents

Contamination Control and Cleanrooms

1

Contamination Control Overview

Many high-technology products currently manufactured are affected adversely if contamination is deposited in or on the product during manufacture or use. *Contamination* can be defined as any condition, material, process, or effect that can degrade quality, performance, or yield for any product. The product can be an electronic device, an optical or mechanical assembly, a pharmaceutical product or device, or essentially any modern "high-technology" product. The product can also be a simple system, such as a food item or plant that can be affected adversely by airborne bacteria. A contaminant can be a particulate or a gaseous material; it can be a harmful electric field or an excessive vibration level. It can be the condition of excessive moisture (relative humidity) in an assembly area, or it can be a high light flux at harmful wavelengths in a lithographic operation. It can be an excess of lubricant in a precise mechanical device, or it can be a condition of inadequate lubrication. An examination of an ultralarge-scale integrated (ULSI) semiconductor device manufacturing requirement might result in the conclusion that the entire manufacturing operation must use ultraclean technology (Ohmi 1988). As a strong example, regarding the design and operation of present-day logic and memory devices used in computers, it can be concluded that contamination control is a vital necessity, without which no computer or any of the many microprocessor-controlled devices in our modern world would be operating today. The dimensions of the circuitry used in these devices simply does not allow operation if the particulate contaminants found in "normal" air were to deposit upon these devices during the manufacturing process. The manufacturer must not only control particulate and vapor contamination but also control exposure to high-energy conditions and excessively variable thermal conditions, ensure freedom from vibration, and eliminate excessive process variations. These precautions are required during the manufacturing operation, during storage, and, in many cases, during product operation.

A few of the product-fabrication processes that require contamination control measures are mentioned here. Modern electronic products that would not be possible without contamination control include semiconductors, printed circuit boards, rotating memory devices, and switching elements. Optical components include precision imaging devices, fiber-optical signal transmission systems, compact disks, photographic components, and reprographic equipment. Pharmaceutical and medical devices include parenteral and ingested liquids, implants and medical devices, and surgical theaters. Food product systems whose quality is improved by effective contamination control measures include beverages, foods, animal growth and holding areas, and packaged food preparations that will be stored at room temperature without degradation. It has been shown that preparation of many milk products in contamination-controlled areas permits long shelf life without the use of chemical preservatives (Bruderer and Schicht 1989). Power and energy system operations include fluids used for power transmission, lubrication, and cooling; fuel injection components; and nuclear energy system controls.

Contamination can result from personnel errors: an assembler is incorrectly garbed or inadequately trained for a sensitive procedure and carries out procedures that may be acceptable in normal manufacturing areas but can destroy extremely sensitive devices. Contamination can also result from normal manufacturing operations. A tool or a machine needed for production can also generate a variety of contaminants in normal operation that can degrade some products unless they are kept free from contamination. A common example of this problem is the presence of metal chips and cutting oil residues from normal machining operations. In all these situations, procedures are required whereby contamination can be controlled and the product yield and quality brought to a usable and profitable level.

Contamination control (CC) describes the technology to reduce hazards from any contamination source. The technology is only one of the tools needed in any well-controlled manufacturing operation. Although CC deals with materials and processes that may differ from what is considered a normal manufacturing environment, it is an essential tool for efficient production. The tool is based on the same kind of technology that controls and defines any manufacturing process. However, it also includes a need for the operators, their supervisors, and *their* management to realize, understand, and control all of the processes that may cause contamination.

CC technology consists of cleanrooms, clean procedures, and devices for ensuring cleanliness. The most important component, however, is personnel who understand their part in the success or failure of the contamination control program. They must understand the CC tool operation, how it works, what it can and cannot do, and how to keep it operating efficiently under a variety of conditions. This discussion is designed to introduce some of the basics of CC, to explain both the capabilities and limitations of this necessary tool, and to help to extend its use by appropriate actions. CC is not an automatic process that operates without personnel interaction. Personnel activities in the CC environment can make or break the process, as well as simply affecting costs and profitability.

CC operations are established for a variety of reasons, sometimes based on good planning and other times simply to meet competitive claims. Figure 1-1 (Gutacker 1987) describes how application of effective CC practices can result in a more reliable product when the right CC technology is used at the right time and place. As shown, a very important time for CC consideration is in the design of the product. This stage includes definition of product performance, components, and physical and mechan-

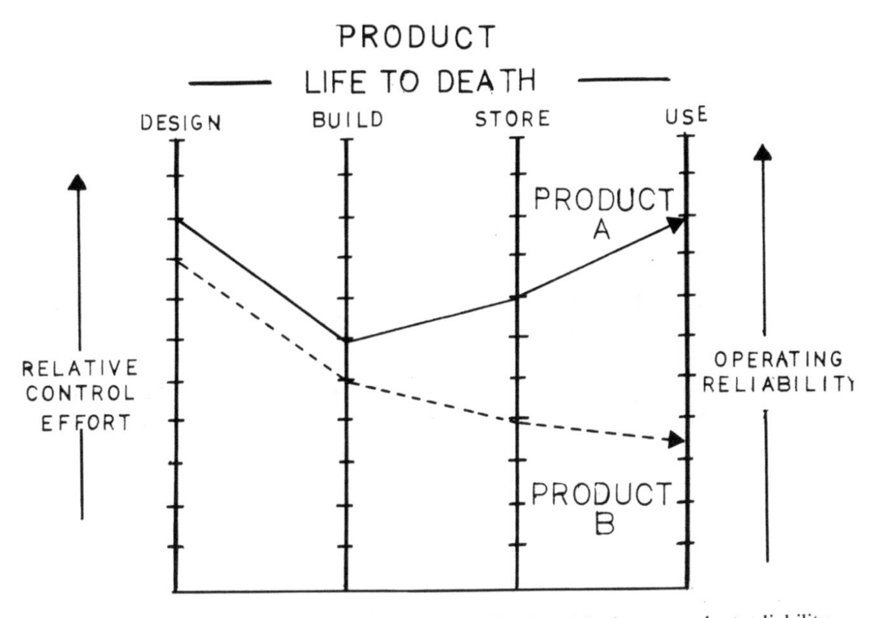

FIGURE 1-1. Product quality control effort distribution. Maximum product reliability requires well-planned contamination control during all design, production, testing, and packaging stages. (Courtesy A. Gutacker, ARGOT, Inc.)

ical constraints, as well as assembly and sealing features. Wherever possible, contamination limits must be included in the product design parameters. If the product and its components are well designed, then the manufacturing procedure requirements may not be as severe as those for a less carefully planned system. A good CC program is effective in producing higher real throughput and better yields of products at lower total costs. Many high-technology products could not be manufactured at all without effective contamination control. As part of such a program, transfer of information on problems, solutions, and operating procedures to both supervisors and operators is critical in improving the real efficiency of the process.

The CC problems and solutions discussed here are those of this time. Knowledge is being developed continuously. This information is presented at technical meetings, described in specific technical journals, and based on fundamental scientific studies. The major technical meetings where CC is a concern are the meetings of the Institute of Environmental Science (IES), the Microcontamination meetings, the Parenteral Drug Association (PDA) meetings, the Pure Water Conferences, and the biyearly International Contamination Control Committees Society (ICCCS). Additional fundamental and application studies have been discussed at the meetings of the American Association for Aerosol Research and of the Fine Particle Society; these two groups tend to discuss the physics of particulate contamination. Some papers appear in their journals, *Aerosol Science & Technology* and *Particulate Science & Technology*. Some of the trade journals published in the United States that contain a good part of the CC literature are *Microcontamination, Journal of Environmental Sciences, CleanRooms, Semiconductor International, Pharmaceutical Technology, Solid State Technology*, and *Ultrapure Water*.

CC developmental work in the United States is carried out at the University of Arizona, Clarkson University, the University of Minnesota, Rochester Institute of Technology, Stanford University, Research Triangle Institute, and other research organizations. For the most part, CC research and development efforts are carried out in departments of chemical, civil, electrical, and mechanical engineering. Much of this work is supported by consortia, mainly from the electronic and pharmaceutical industries.

CC standards and specifications have been developed mainly by voluntary standards organizations, such as the American Society for Testing and Materials (ASTM), the Institute of Environmental Sciences (IES), the Parenteral Drug Association (PDA), Semiconductor Equipment and Materials International (SEMI), and the National Fluid Power Association (NFPA). Many U.S. government specifications have been derived directly from these sources. A large number of specifications were developed by government agencies, both military and civilian. Many of the NASA procedures for defining environmental and product cleanliness levels were developed internally by that agency rather than adopted from a separate source. The major exception is, of course, U.S. Federal Standard 209 for cleanroom air quality. This standard was developed first by a task group directed by a government agency and is being updated by the IES. It is of interest that the personnel involved in most voluntary standards groups are employed both by private industry and by government agencies. Note that most of the agencies, universities, and associations listed here are located in the United States. At the same time, extensive work in contamination control is carried out in parallel European and Asian organizations. Around 1988, information on contamination control work from the Soviet Union, East Germany, the Peoples Republic of China, and other countries also became available. As of 1990, the requirement for standards that can be used internationally is obvious to all standards preparation agencies. Cooperative work on standards preparation is expanding rapidly throughout the world.

A brief historical review of CC is useful in helping the reader understand better the sources of the technology, how it developed to the present state, and the sources of many of the procedures in use. The first meaningful use of CC can be traced to nineteenth-century bacterial control efforts in some hospital surgeries. The connection between dirt and infection frequency pointed out the need for cleanliness of surfaces, furnishings, clothing, and the overall hospital environment. Shortly after that time, food preparation and long-term storage needs indicated why cleanliness and sterilization methods were important in that area. As part of this development, some "sterile" fluid and material handling methods were investigated that are still in use. Many of the terminal sterilization methods now used in pharmaceutical manufacturing, as well as work area gas and liquid phase sterilization, follow from work of that time. The gowning procedures still used in many CC areas are based on the surgical gowning techniques used for many years in operating areas.

In the late 1930s, a major effort was carried out in the United States, Great Britain, and Germany to develop respiratory protection devices because of fear of possible wartime use of chemical and biological weapons. This work resulted in initial development of high-efficiency particulate air (HEPA) filter media for truly effective particle filtration and of activated carbon adsorbents for gas cleaning. These materials were used in gas masks and for building air cleaning systems for removal of chemical and biological warfare agents. Development was begun at about that time on design of the photometer and the optical particle counter now used in all

CC areas for testing of HEPA systems. Much of this work was carried out under the control of the U.S. Army at the Chemical Research and Development Laboratories. Similar activities were in progress at the British military facilities at Porton.

As part of offensive military development in the United States, the air arm began work on the Norden optomechanical bomb sight. It soon found that reliability and operability of this device were improved significantly when it was assembled under clean conditions, rather than those found in typical machine shops of that time. Small-scale "clean" cabinets were used for critical assembly operations. At the close of World War II, gyroscopic guidance devices were also being developed for many air and sea navigation systems. The importance of cleanliness in all parts of these devices was understood early in their design stage. Military system needs resulted in development of the first family of mechanical assembly area "white rooms," which were probably equivalent to a class 500,000 area, using our present terminology.

In the early 1960s, work on extra-atmospheric vehicles showed the need for extreme cleanliness in their mechanical and fluid control systems, as well as for guidance devices. Concern for possible contamination of extraterrestrial areas by terrestrial bacteria resulted in rigid bacterial controls on vehicle components that might impact on extraterrestrial objects whose biological nature was unknown. Many CC procedures were developed and used during the period when the aerospace industry was the primary area for new CC applications. Because of time constraints on production, most of the technology was empirical. Manufacturers were desperate to produce products for their contractual obligations. Some extreme CC concepts were explored at that time. At one electronic fabrication company, the white room was constructed with stainless-steel panel walls that sloped outward from the ceiling to the floor. This costly construction was justified on the assumption that particles settle vertically (in a mixed flow room!) and could not deposit on a wall surface that sloped outward from the vertical. Neither the proponents of the concept nor the engineers who approved construction considered particle physics or fluid dynamics.

In the 1960s, the U.S. Food and Drug Administration (FDA) became seriously concerned about possible contamination of drug products. Both cross-contamination of one product with airborne powders or fumes from another and bacterial contamination from improperly cleaned areas became concerns. At about that time, the first semiconductor devices were being developed and the need for effective CC became paramount. The effects of contaminants other than particles and gases also became a concern with these products. From the viewpoint of product yield effects and potential profit loss, the value of the CC tool in production has been realized by the electronics industry almost more than by any other source. The needs of the semiconductor and pharmaceutical industries have led to the application of solid science and engineering to CC.

A brief review of many of the papers that appear in the CC literature in the 1980s shows that most of the development and design descriptions consider the fundamental physics and engineering that deal with fluid flow and particle transport. One of the first examples of this activity is the work of Whitfield and his associates at the Sandia Laboratory that led to the development of the large-area unidirectional filtered airflow system, which has been known as "laminar flow" clean systems (Whitfield 1963). As Whitfield has frequently stated, the Reynolds number in these areas cannot be used to define laminar flow, but the airflow streamlines are close enough to being parallel so that very little reentrainment of particles or other contamination can occur. Work is in progress now on physics and hydraulics of applied cleaning technology to remove very small particles from semiconductor materials. A reasonable understanding of the van der Waals and intermolecular forces, along with the energy fields that can be applied to these materials, should lead to significant improvements in cleanrooms in the near future. A summary of anticipated cleanroom developments was presented recently (McIlvaine 1988); a number of new application areas and estimated market sizes for a number of cleanroom components were summarized. In 1990, a market report was released by Frost and Sullivan, 106 Fulton Street, New York, NY 10038. This report, *The U.S. Market for Cleanrooms and Clean-room Supplies and Equipment* (#12242), states that the cleanroom industry will be larger than $3.3 billion a year by 1993. Cleanroom structures represented $1.8 billion in volume in 1989. Supplies were $229 million, and monitors $74 million, totaling $2.1 billion. The 1989 cleanroom area of 3.6 million square feet will exceed 5.4 million by 1993. Semiconductor producers represent nearly half of the total market for cleanroom area and exceed the remaining industries for cleanroom costs.

The discussions that follow cover the fundamentals, engineering, and applications of contamination control. The specific points discussed include description and definitions of particulate, gaseous, and energy field contaminants, along with their effects on products and on manufacturing operations. It includes some description of contaminant sources and the mechanisms that generate and transport contamination within and from these sources. In particular, the forces that affect particle contamination are mentioned because many technically trained personnel are not very familiar with particle physics. Methods used to protect products are summarized, and some discussion is given of CC methods and CC results. Environmental control such as gas, surface, and liquid cleaning methods are covered. The standards and specifications that are important for cleanroom and clean process design and operation are discussed. Clean area control specifications are described, as well as those for the systems and devices to test CC systems. Cleanroom designs and cleanroom materials, processes, and procedures to minimize contamination are discussed for optimization of CC operations, including those aspects of CC operation from heating, ventilating, and air conditioning (HVAC) and filtration systems to layout of cleanroom ancillary and support areas, as well as maintenance and cleaning considerations. Personnel problems are discussed, including suggestions for selection and training in optimal work procedures and disciplines. Usual cleanroom personnel requirements, such as clothing, wipers, and gloves, are discussed. Air sampling procedures for maximum sampling efficiency are described for cleanroom verification, routine monitoring, and filter testing. The operation and problem areas that can be expected when using optical particle counters are described in some detail. Means of improving counter reliability and establishing correct operation are pointed out. Because much of the CC work incorporates use of clean liquids, the discussion also includes some of the problem areas in cleaning, using, and verifying those liquids.

References

Bruderer, J., & Schicht, H. H., 1989. Laminar Flow Protection for the Sterile Filling of Yoghurt and Other Milk Products. *Swiss Contamination Control* 2(1):41–45.

Gutacker, A., 1987. *Fundamentals of Contamination Control Handbook*, ARGOT, Inc., Webster, NY.

McIlvaine, R. W., 1988. An Industry Overview: Current Status of the Cleanroom Industry. *CleanRooms* 2(2):20–21.

Ohmi, T., 1988. Ultraclean Technology: ULSI Processing's Crucial Factor. *Microcontamination* 6(10):49–58.

Whitfield, W. J., 1963. State of the Art (Contamination Control) and Laminar Air Flow Concept. Conference on Clean Room Specifications, SCR-652, Office of Technical Services, Dept. of Commerce, Washington, DC.

2

Particulate Contaminant Descriptions and Definitions

Particulate contaminants can be either solid or liquid. Many of these materials were originally suspended in air or in a process fluid; others derive from nearby sources, such as activities of personnel working in the cleanroom or operations of process equipment. Until particles are about to contact a product or already have been deposited upon that product, some schools of thought consider that any particles remote from critical areas are not really contaminants, but simply environmental components. Even so, because the fluid in which they are suspended will probably contact the product sooner or later, it is necessary to keep that fluid as clean as possible.

Whether contamination is defined as potential or actual, product quality is more easily maintained if manipulations of the product are minimized. If contaminants are present in process fluids or on tool surfaces, then they must be removed before there is any chance of transfer to products. If particles are present on surfaces anywhere in the work area, then those particles can be entrained, migrate, or otherwise move to a location where they may affect the product. For these reasons, *any* undesired particulate material in the work area should be controlled.

Contaminant particles can range in size from approximately 0.001 to 100 micrometers (μm) in diameter. Size is defined by a physical dimension of the particle. To place these numbers in context, the smallest object visible to normal, unaided eyes in bright light with good contrast is about 30 μm in diameter; human hair is from 50 to 150 μm in diameter. Although particles are three-dimensional objects and are normally irregular in shape, a single value is used to define the size of the particle. Because most particles are not spherical, it is necessary to define size specifically. There are many definitions of *particle size* used in studies of particle phenomena. Some definitions are based on the particle size measurement procedure in use; for example, the term *aerodynamic diameter* is used to define the size of a particle that has the same gravitational settling velocity as a unit density sphere of that diameter. This term is derived from use of a measuring system for characterizing particle size as a function of time to accommodate to an airstream moving at a velocity different from that of the particle that is introduced into that air. Small particles accommodate to such changes faster than larger ones.

Figure 2-1 shows the definitions commonly used for particle size characterization in the contamination control area and how they are applied to an irregular particle. These definitions are also developed from the operation of the particle-measuring devices used in contamination control work. When either an optical or an electron microscope is used to observe collected particles, the "longest dimension" size is frequently used to define the particle size, and the diameter of an "equal area circle" is the dimension associated with use of an optical particle counter. Needless to say, sizes obtained by use of the two methods seldom agree for the same particle, unless that particle is a sphere. Differences in reported particle size for the same particle can vary by more than an order of magnitude,

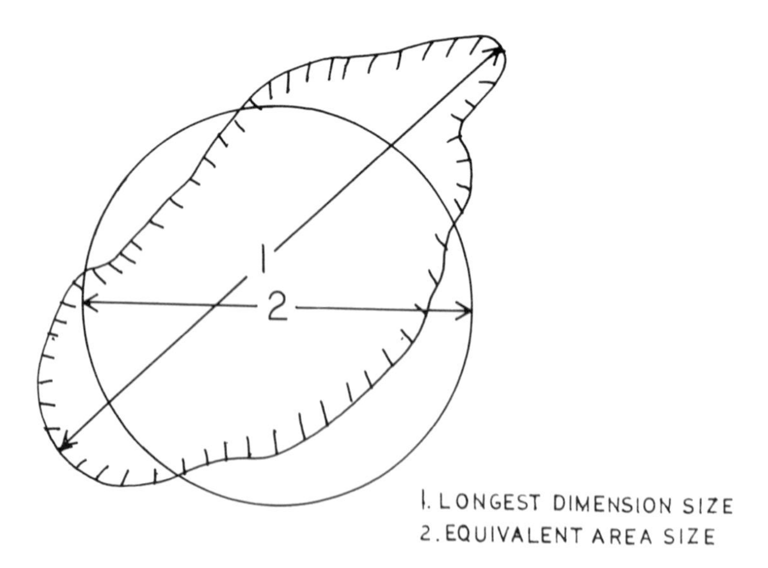

FIGURE 2-1. Particle size definitions used in particle science. Only the equivalent area diameter, used for instrumental measurements, or the longest dimension diameter, used for microscopic measurement, is used in contamination control definitions. Other dimensions are used in describing particle behavior during motion.

depending on which sizing method is used, the accuracy of the equipment, and the skill of the technologist. For this reason, it is necessary to state the measurement method in characterizing particles.

Many other particle size definition terms are used in other particle physics areas that are not part of the CC measurement lexicon. Even though they are not used in defining reported contaminant particle size, they can be important in understanding particle motion and transport within the cleanroom. There are more than a dozen different definitions of *particle size* associated with a wide variety of technologies. These terms are usually associated with the measurement procedure used. Sieve sizes are used when particles are sized by screening; sedimentation systems provide a particle size measurement based on the rate at which a particle settles in a particular fluid. Even though these terms are seldom used in contamination control areas, their importance must be accepted. For example, Air Cleaner Fine Test Dust, sometimes referred to as *Arizona road dust,* is a popular coarse filter test dust. It is characterized in terms of fraction by weight of particles in specific size ranges. These fractions are determined by separation in an air elutriation system, using a match between gravitational settling rate and buoyancy in a rising airstream to define particle size.

Some fundamental particle shapes that may occur in cleanrooms and cleanroom processes are shown in Figure 2-2. Many particle shapes are

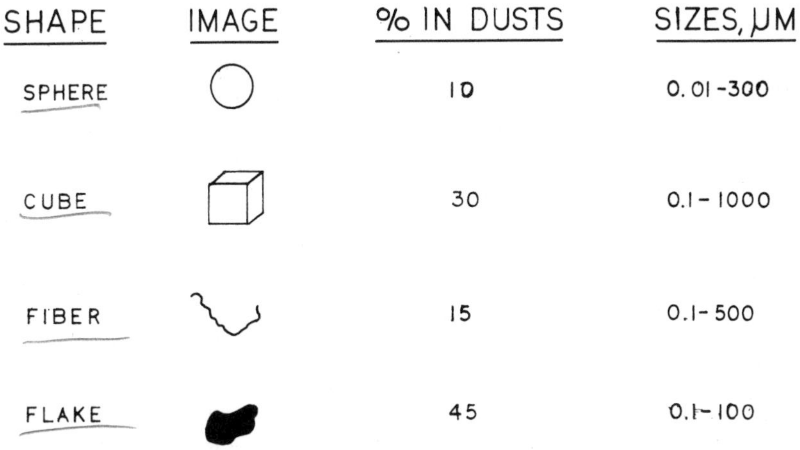

SHAPE	IMAGE	% IN DUSTS	SIZES, μM
SPHERE		10	0.01-300
CUBE		30	0.1-1000
FIBER		15	0.1-500
FLAKE		45	0.1-100

FIGURE 2-2. Some particle shapes commonly seen in cleanrooms. The spherical particles are produced from liquid condensation and may be "frozen" as the liquid cools. The cubes represent crystalline fragments produced by chemical reaction. Fibers come from fabric degradation. Flakes are produced from wear of metal or plastic or from personnel emissions.

source-related. Spheres usually result from droplet formation. Spray formation, as may occur from liquid stream impact on a surface or excess lubricant release from rapidly rotating bearing surfaces, produces spherical droplets. Liquid droplets frequently result from condensation of vapors to form mists or fogs; solid spheres can result from condensation and rapid freezing of a high-boiling material. Industrial smog consists mainly of droplets produced by chemical reactions in the gas phase. Some vegetation emits nearly spherical particles, such as spores or pollens, in sizes ranging from 1 μm to nearly 100 μm. Irregular slabs can result from a chemical reaction that produces specific crystal structures. Soil particles in the ambient atmosphere are typical of this particle form. These may also be produced by erosion or by frictional grinding mechanisms; flakes are also generated as debris from organic sources. Fibers are also usually fragments stripped from surfaces or fabrics; some chainlike particles may be flocculated fume particles generated by combustion processes. These are commonly seen in the vicinity of some high-temperature metal-processing operations. Viable particles are frequently found in air systems, even in cleanrooms. Individual bacteria, usually in sizes just over a micrometer, can be nearly spherical to rod-shaped, depending on bacterial variety. Most bacteria are attached to dust particles. There is a wide variety of microbiological organisms in the air, such as bacteria, fungi, protozoa, and viruses (Morey and Feeley 1988). These materials exist in any environment where living things may be present, including cleanrooms if adequate control is not maintained.

The surface nature and composition of the particles are very important in terms of both how the particles affect a surface and how the particles may move to that surface. The surface composition of most droplet spheres is mainly liquid, either oleophobic or hydrophobic. If the droplet strikes a surface that is easily wetted, then a film forms; surface tension spreads the liquid drop over the surface upon which it has deposited. Optical surface integrity can be severely deteriorated, whereas mechanical devices may not be affected at all. If the droplet is composed of a conductive liquid, then an electronic component may be damaged, whereas an inert, nonconductive liquid droplet may be of little concern in this area. Most irregularly shaped solid particles are as hard or harder than the surfaces upon which they may deposit. As is described later, small particles that deposit upon a surface are retained by that surface by strong electrical and/or molecular forces. If the product contains moving parts or if abrasive material can cause damage, then the irregular particle is a severe hazard to that product. In most semiconductor manufacture, any particle of significant size deposited upon a product surface will cause damage sooner or later.

A significant source of all particle contaminant types comes from activity of personnel in the work area. Droplets are produced by coughing, sneezing, or even agitated speech. Skin flakes are emitted continuously; if any cosmetics are used, then that material also flakes off. Dirt or skin particle transmission from and through normal clothing—sometimes from cleanroom garb, as well—is always a potential problem in cleanrooms. Fiber emission from hair and clothing must also be controlled or contained at all times.

Another source of particulate (as well as other types) contamination that has been recently recognized is the operation of process equipment in the cleanroom. Older devices have been designed and delivered with surfaces and finishes that can emit contaminants, have moving parts that abrade, provide physical entrapment for contaminant particles, are not cleanable and are delivered with inadequate cleaning, or have product handling components that are not adequate for protection of the product or that may even harm the product by erosive or abrasive action.

Concentration of particles in the atmosphere is expressed in terms of number or weight of particles per unit volume of air or per unit area of product surface. Cleanroom particle concentrations are defined in terms of number of particles equal to and greater than a stated diameter per unit volume of fluid. Environmental pollution contamination levels are usually defined in terms of mass of particles per unit volume of fluid. Particle size in that case is usually associated with the measurement method; in atmospheric environmental pollution standards, particle size may be defined in terms of inhalable or respirable particles.

Typical cleanroom particle concentrations are in the range of one to 100,000 particles ≥ 0.5 μm per cubic foot of air. Compare this concentration with values for clean rural air of 10 micrograms of total suspended particulate matter per cubic meter. That weight loading is equivalent to about 10 million 0.5 μm particles per cubic foot. Concentrations in urban environments are significantly higher and tend to have a rather broad particle size distribution. Urban aerosol particles contain coarse materials that derive from entrained soil or building surfaces, along with fine aerosols derived from vehicular sources (Bullin et al. 1985). Most sources of air pollution generally produce gaseous pollutants, except for direct particle sources, such as fossil fuel power plants, cement kilns, or metal ore processing operations. It must be remembered that many fine particulate atmospheric pollutants are the result of gas phase reactions of gaseous materials.

Normal atmospheric particles have a size distribution with many more small than large particles. Figure 2-3 shows a cumulative particle size distribution by number for a stable, well-aged atmospheric aerosol. Stud-

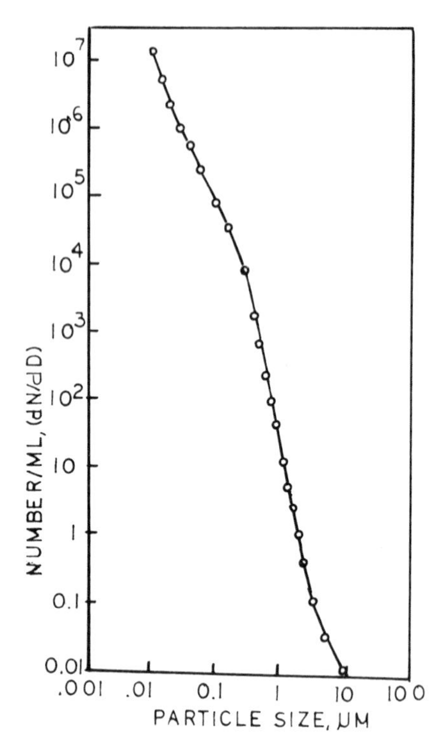

FIGURE 2-3. Distribution of stable, aged atmospheric aerosol particle sizes in urban areas. There are many more small than large particles until the size decreases to the point where the small particles are lost rapidly because of their rapid diffusion and deposition on surfaces and particles. (From Whitby, Husar, and Liu, The Aerosol Size Distribution of Los Angeles Smog. In *Aerosols and Atmospheric Chemistry*, ed. G. M. Hidy, 1972. Academic Press, New York.)

ies in various parts of the world show that this type of particle size distribution occurs widely (Junge 1963). Some variation in concentration is found, but the size distribution function is very similar in all areas. The particle size distribution of most aged atmospheric aerosols results from physical processes, such as gas phase reactions, to form particles, agglomeration of small particles, deposition losses, and the like, which are part of the atmospheric reactions occurring at all locations and not associated with the particle generation source.

Closer examination of the atmospheric aerosol size distribution by volume shows that there are three groups of particle sizes present (Whitby, Husar and Liu 1972). The smallest size, known as the *nuclei mode,* is produced mainly by gas phase reactions in the atmosphere and from combustion. The largest number of particles is in the nuclei mode. Most of these particles are in the 0.01–0.03 μm size range and are the product of combustion or other chemical reactions. Because of the larger diffusion constant of these small particles, if they are present in sufficient concentration they tend to coagulate rapidly to form larger particles. These materials can persist in the atmosphere for very long times unless they are removed by deposition upon nearby surfaces; washout by rain, snow, or fog; or agglomeration to form larger particles (Andreas 1983). The *midrange accumulation mode,* at approximately 0.3 μm, is produced primarily by agglomeration of the smaller nuclei mode particles. Although the number of the accumulation mode particles per unit volume of air is smaller than that of the nuclei mode materials, the total mass of particles in this size range is much greater than that of the nuclei mode materials. The largest size range at 5–10 μm and greater, known as the *coarse particle mode,* is generated by a number of processes.

Liquid drops form by vapor condensation, liquid atomization, or bubble bursting. Solid particles are formed by grinding of materials, by erosion, or by impact of hard, airborne particles onto softer or friable substrate surfaces. Particles in this mode are aerosolized by normal air movement, resulting in pickup of soil fragments and plant materials and of large anthropogenic particle debris. These relatively large particles (2–50 μm) tend to fall out of the air relatively fast and close to their point of generation. Atmospheric aerosols may penetrate into clean areas and must be removed by filtration. If any filter penetration occurs, some materials will be present in the clean area; further, as more materials must be removed, the filter life is affected. Thus, the size distribution of this material is of importance. Figure 2-4 shows the three modes and the particle sizes frequently seen in a stable atmospheric aerosol.

Most products sensitive to contamination are manufactured in industrial areas. If cleanrooms are located in rural areas, employee living areas and service components, as well as process and product material transportation facilities, are still not too distant. The contamination control filter system will be challenged by some fraction of the particle loading present in the ambient air. Even though the plant ventilation system removes most of the gross contamination and the cleanroom final filter removes more than 99.9% of the particles from air emitted from the prefilter and the plant ventilation system, the overall system always affects only a percentage of the challenge load. As the ambient air concentration changes, the filter system life will be affected by that load and also the penetration of some nuclei and accumulation mode particles will change.

Measurements in the ambient air have shown that the particle concentration varies with the local climate, season, and time of day. These variations affect cleanroom air quality in a way similar to the way in which

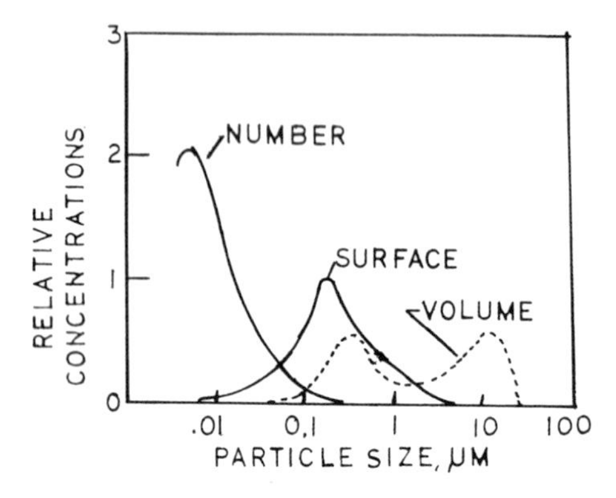

FIGURE 2-4. Atmospheric aerosol particle size modes. Smaller particles are produced mainly as a result of gas phase reactions; midrange particles result from agglomeration of small particles and larger ones from abrasion of natural surfaces. (From Whitby, Husar, and Liu, The Aerosol Size Distribution of Los Angeles Smog. In *Aerosols and Atmospheric Chemistry*, ed. G. M. Hidy, 1972. Academic Press, New York.)

changes in a municipal water supply system affect the quality of deionized water. Climatic effects are especially important in the temperate zones, as electric power plant emissions from fossil fuel power plants vary with load requirements, and industrial process emissions are transported by local winds. Seasonal effects arise from both natural and anthropogenic causes. Changes in ambient relative humidity, winds, and temperature affect emissions from vegetation, as well as the dust entrainment rate. In urban areas, industrial activity, motor vehicle use, and traffic patterns result in diurnal variations in ambient atmosphere particle concentrations. These variations are superimposed on seasonal effects. Measurements made over a period of weeks inside and outside a light manufacturing plant in a suburban area show the effects of the diurnal changes, atmospheric effects, and the relationship between the internal and external particle concentrations (Lieberman 1971). These changes are shown in Figure 2-5. Internal particle concentrations tend to follow external levels, as might be expected when the filtration system is simply that of an air conditioning system. Even so, the effects of transient internal local activities in the plant sometimes override the effects of variations in the external ambient air particle loading.

Chemical properties of particulate contaminants make a difference not so much as to whether a particle will cause damage but in the amount of damage that any particle can cause to a sensitive product. Even relatively "inert" particles can degrade the integrity of a lithographic image in semiconductor fabrication, affect clarity of beverages or pharmaceutical products, or cause data dropout in recording systems. However, some

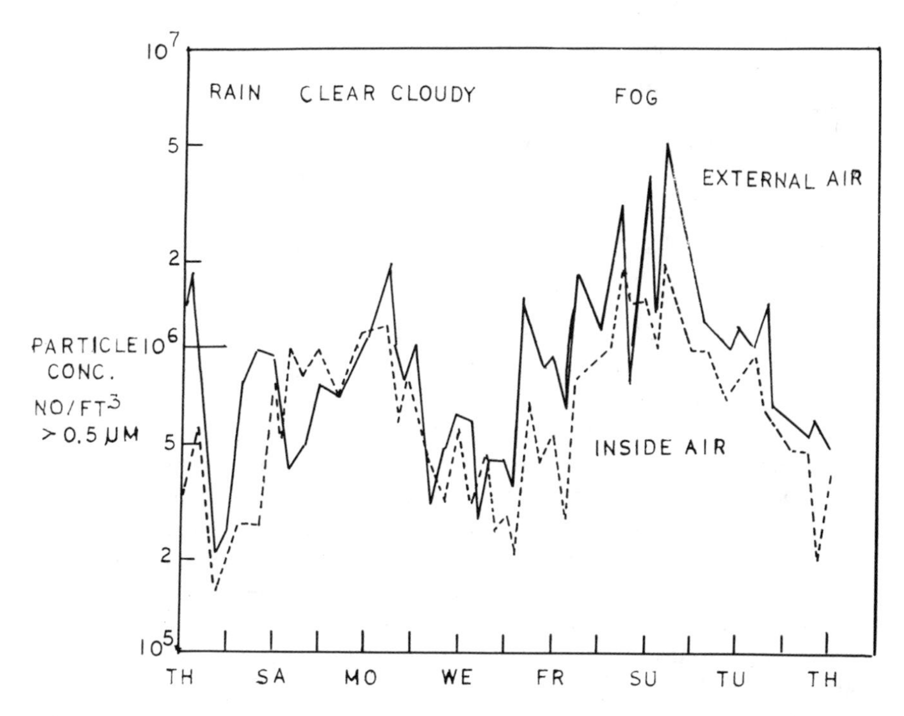

FIGURE 2-5. Variation of aerosol concentration with time. Note the similarity in trends between indoor and outdoor air concentration changes. The entering air was treated only with normal air conditioning filters. Only larger particles are controlled by this type of filter. The cleanroom filtration system must handle these and internally generated materials.

materials can cause even more damage because of their reactivity with product component materials. Semiconductor component functions are often based on the actions of trace-quantity addition of specific compounds to form desired circuit elements. Many metal particles cause excessive yield loss if they occur on a wafer surface during the manufacturing processes. Preparation of the basic ultrapure single-crystal silicon wafers, chemical vapor or plasma vapor deposition of desired materials, wet or dry pattern etching, ion implantation, and sputtering are some of the processes that require close control of trace quantities of materials. The presence of undesired materials during any of these processes can result in unacceptable products.

In addition to reactivity with product components, particle chemical properties are of concern for their potential effects on personnel and on the environment. Many of the materials used in manufacturing contaminant-sensitive products may be toxic, carcinogenic, flammable, corrosive, or otherwise hazardous. These materials can be hazardous to plant personnel. In addition, disposal of waste materials can be a serious problem to the environment. It is important that the chemical properties of materials be known to the production manager so that any necessary safety precautions can be implemented. These precautions can include ensuring availability of needed protective devices, such as respirators or separate air supplies. Special air handling and purifying systems for hazardous material generation areas may be needed; special treatment before disposal may also be required.

The relationship between particle shape and source has been pointed out. Similarly, the composition of the particles often depends on their source. Thus, if a problem in defining particle sources occurs in a clean area, sometimes it can be solved by defining the composition and morphology of the contaminant particles so as to relate these factors to the source of the problem. As examples, skin flakes obviously arise from inadequate or incorrect gowning and personnel procedures, lead or bronze flakes may come from a poorly lubricated bearing, fiber fragments can usually be identified by fiber material morphology to determine their source, and irregular silica particles are probably brought into the area from external rooms by personnel or on materials. Some of the materials that are seen in urban air are shown in Table 2-1. Obviously, if any of them are detected in any measurable quantity in a cleanroom, it is necessary to inspect the HVAC and filter systems for leaks and to inspect product raw material and work element transport systems. Most of these contaminant materials can be recognized easily. They are often present in a quantity sufficient to be easily detected with low-power visual light microscopy.

The characteristics of the particle surface are important because the

TABLE 2-1 Urban Air Pollutant Particle Compositions, Sources, and Typical Size Ranges

POLLUTANT	SOURCE	SIZE RANGE, µm
Oil Smoke	Combustion	0.03-1.0
Coal Fly Ash	Power Plant	1-200
Tobacco Smoke	People	0.01-2.0
Coal Dust	Power Plant	1-100
Cement Dust	Construction	3-100
Insecticide Dust	Agriculture	0.5-10
Plant Spores	Plant Life	10-30
Pollens	Plant Life	10-100
Bacteria	Varied	0.1-20
Soils	Varied	0.5-100

Source: (Courtesy A. Gutacker, ARGOT, Inc.)

way in which a particle affects its target is controlled by the surface. Electrical charge on particles is concentrated on the surface, even though the charge level depends on the particle composition. If the particle surface is smooth or rough, wet or dry, the effects on interparticle reactions, deposition velocity, or surface retention differ. The particles may be rigid or plastic in nature, depending on the particle material and generation mechanism. All these properties are source related, just as is particle shape and overall composition; all of these properties are related to the way in which particles affect the product.

When the quantity of material is small, then sophisticated analytical methods must be used for accurate analysis. For analysis of airborne particles, they must be first collected on a suitable substrate; then several analytical techniques can be used. Particles are usually collected with membrane filters, by impaction, or by gravitational or diffusional settling. Once the particles are on a suitable substrate, then analysis can be carried out. A complete discussion of all of the analytical techniques that can be used to identify each and every contaminant is beyond the scope of this discussion; however, some of the more common methods are mentioned here.

A summary of methodology for identification of typical cleanroom contaminants was recently presented (Vander Wood and Rebstock 1988). It describes the capabilities and limitations of several visual and electron microscopic methods for analysis of individual particles in sizes of concern in most cleanrooms. Raman microprobe analysis can provide information on the molecular structure of both inorganic and organic particles as small as 1 μm in size (Muggli and Andersen 1985). Secondary ion mass spectroscopy is very sensitive for many inorganic elements (Gavrilovic 1988). For analysis of materials on product surfaces, a number of methods can be used. Methods for observation of contaminants on semiconductor devices have been summarized (Koellen and Saxon 1985). Another summary was presented on use of infrared spectroscopy for aerosol particle analysis (Allen and Palen 1989). Single-particle spectroscopy, infrared microscopy, attenuated total reflection spectroscopy, photoacoustic spectroscopy, and infrared spectroscopy were reviewed. The use of single particle spectroscopy for levitated single particles was shown effective for identification of some materials in quantities as small as 10^{-12} g. The application of transmission electron microscopy for detection of surface deposits of metals was used to identify contamination sources (Augustus 1985). The use of a scanning electron microscope to give morphological and composition information on contaminants has been invaluable in identifying their sources and means of control (Brar and Narayan 1988). Identification is based on crystal structure, refractive index, and appearance. These techniques can be used for either collected airborne particles or those removed from product surfaces.

When the product is a liquid, as is common in the pharmaceutical industry, similar analytical procedures can be used with some care. Techniques for inspection, counting, and identification of particles in liquids, elucidating their sources and formation mechanisms, are well known in that industry. Common analytical methods include light and electron microscopy, atomic and molecular spectroscopic methods, and chromatography. An excellent review of these methods, especially in terms of interest to the pharmaceutical industries, has been presented (Borchert 1986).

The last important particle definition of concern in CC results from the fact that particles are not uniform or identical. In any consideration of particles, individual particles are seldom seen. Generally, particles are present in some number per unit volume of fluid or per unit area of surface. The particles present in any environment have a range of particle sizes, shapes, and compositions, and they are not uniformly distributed in space. These parameters are a result of the processes that first generated the particles and those that affect their motion and interactions in the fluid. These factors are of importance in classifying and monitoring cleanrooms. Essentially all of the chemical and physical properties of contaminants are important. The degree and kind of damage produced depend on these properties and how the cleanroom product may be affected. In mechanical device operation, the hardness and/or friability of particles determines whether the particles are broken up or the device is damaged. The electrical properties affect both particle deposition rates and the potential for damage to electronic components. Diffusion rates and the resulting particle motion depend on both the contaminant and the substrate materials. If a particle can absorb water vapor and grow, the initial small, dry particle may not be immediately hazardous, but the corrosive salt solution droplet that grows from that particle can cause problems at a later time. The initial small particle can be then considered as a delayed-action contaminant.

The term *particle size distribution* was used previously to show the relationship between size and number of particles. In a stable atmosphere where normal phenomena such as particle growth, agglomeration, and deposition on surfaces have occurred, these are generally particles in an aerosol that has stabilized so that the atmospheric aerosol size distribution is fairly constant with both time and location. In fact, this stabilization process has resulted in the generation of the term *self-preserving size distribution* (Friedlander 1977) for many urban aerosols. In a cleanroom, however, aerosol stability seldom exists for the time necessary for the aerosol to stabilize. Air is recirculated rapidly through the cleanroom with the HVAC system in operation to control air temperature, relative humidity, and cleanliness. Most of the larger and the smallest particles are removed by the HEPA filters. The major number of particles present in a cleanroom are those that are recently generated within the room by local processes. Thus the particle size distributions of Figures 2-3 and 2-4 are seldom found in cleanroom air. Data obtained in a number of cleanrooms have shown that a logarithmic distribution of particle sizes often exists in the cleanroom air. The smallest particles are removed by the HEPA filters

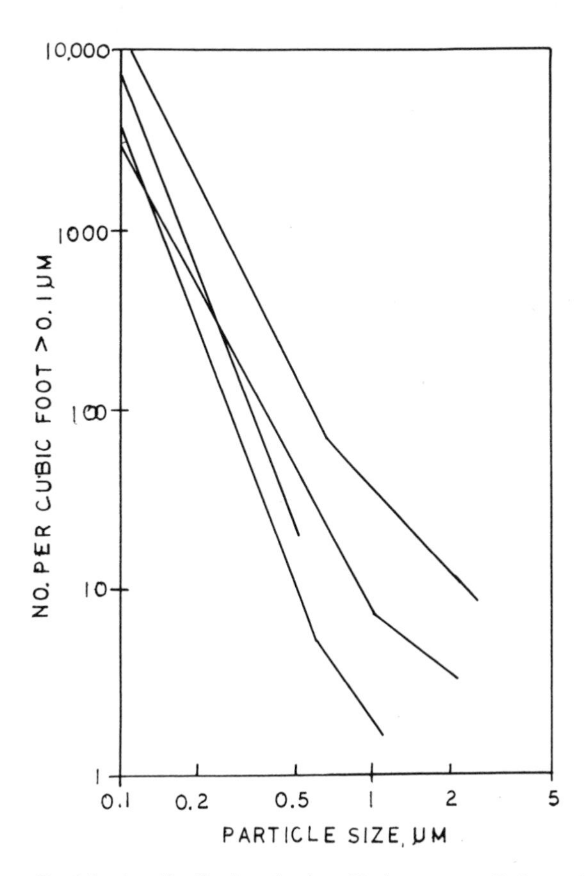

FIGURE 2-6. Particle size distributions in class 10 cleanrooms. Only particles larger than 0.1 μm were measured. Particles larger than 0.5 μm or so are generated within the cleanroom and have not yet been treated by the cleanroom filtration system. The straight line represents the distribution for a federal standard 209 class. (From Knollenberg, R. G., 1987. Sizing Particles at High Sensitivity in High Molecular Scattering Environments. Proceedings of the 34th Institute of Environmental Sciences Annual Technical Meeting, May 1987, King of Prussia, PA.)

with very high efficiencies, and the larger particles are usually generated within the cleanroom by specific processing operations. Size distribution data for contaminant particles larger than 0.1 μm found in some very clean areas are shown in Figure 2-6 (Knollenberg 1988). Note how the large particle data do not fit well within the anticipated values expected by extrapolating the small particle data. This difference occurs because these larger particles were not affected by the same phenomena that affected the overall particle size distribution. Although the major particle population is composed of materials that were probably suspended in the air system and circulated through the filter and HVAC system for at least some time, the

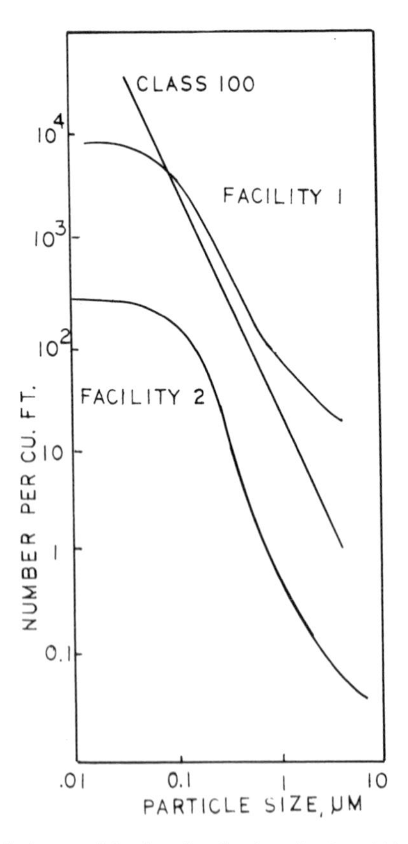

FIGURE 2-7. Cumulative particle size distributions in class 100 cleanrooms. Particles larger than approximately 0.01 μm were measured. It can be seen that many particles smaller than 0.1 μm are recirculated in the airstream and are being effectively removed by the cleanroom filters. (From Ensor, Donovan, and Locke, Particle Size Distributions in Clean Rooms, *Journal of Environmental Sciences* 29 (6):44–49.)

larger particles were probably generated very recently by local activity close to the point of measurement.

When the entire particle population is to be defined, the number of very small particles decreases markedly. This phenomenon is due to the extremely high removal efficiency of HEPA or ultralow penetration air (ULPA) filters for particles smaller than the size of maximum penetration and the high air recirculation rate through the filters that is typical of clean unidirectional flow systems. In addition, the usual gas phase reactions that generate submicrometer particles are inhibited by the removal of most of the reactive gases by the gas purification components of the HVAC system. Figure 2-7 shows the submicrometer particle size distribution in a clean area with normal air flows (Ensor, Donovan and Locke 1987). This distribution is associated with an aerosol that is affected mainly by repeated passage through a filter system. Note that few particles are smaller than 0.1 μm in this air sample. If a point source of such particles exists, some particles from that source will remain in the air until that air parcel has passed through the filters at least once or twice.

References

Allen, D. T., & Palen, E., 1989. Recent Advances in Aerosol Analysis by Infrared Spectroscopy. *Journal of Aerosol Science* 20(4):441–455.

Andreas, M. O., 1983. Soot Carbon and Excess Fine Potassium: Long-Range Transport of Combustion-Derived Aerosols. *Science* 220(4602):1148–1151.

Augustus, P., 1985. Detection of Fe and Ni Surface Precipitates by TEM. *Semiconductor International* 8(11):88–91.

Borchert, S. J., 1986. Particulate Matter in Parenteral Products: A Review. *Journal of Parenteral Science and Technology* 40(5):212–241.

Brar, A. S., & Narayan, P. B., 1988. Quality Control Analysis Reduces Disk Drive Failures. *Research and Development* 30(5):84–86.

Bullin, J. A., et al., 1985. Aerosols near Urban Street Intersections. *Journal of the Air Pollution Control Association* 35(4):355–358.

Ensor, D. S., Donovan, R. P., & Locke, B. R., 1987. Particle Size Distributions in Clean Rooms. *Journal of Environmental Science* 29(6):44–49.

Friedlander, S. K., 1977. *Smoke, Dust and Haze,* New York: J. Wiley & Sons.

Gavrilovic, J., 1988. Secondary Ion Mass Spectroscopy for Bulk and Surface Analysis of Particulate Contaminants. *Solid State Technology* 28(4):299–302.

Junge, C. E., 1963. *Air Chemistry and Radioactivity,* New York: Academic Press.

Knollenberg, R. G., 1988. Sizing Particles at High Sensitivity in High Molecular Scattering Environments. Proceedings of the 3rd Institute of Environmental Science Annual Technical Meeting, pp. 428–435, May 1987, San Jose, CA.

Koellen, D. S., & Saxon, D. I., 1985. Application of Surface Analysis: Detection and Identification of Contamination on Semiconductor Devices. *Microcontamination* 3(7):47–55.

Lieberman, A., 1971. Comparison of Continuous Measurement of Interior and Exterior Aerosol Levels. Paper read at American Industrial Hygiene Association Conference, May 24, 1971, Toronto.
Morey, P. R., & Feeley, J. C., 1988. Microbiological Aerosols Indoors in Workplaces, Schools and Homes. *ASTM Standardization News* 16(12):54–58.
Muggli, R. Z., & Andersen, M. E., 1985. Raman Micro-Analysis of Integrated Circuit Contamination. *Solid State Technology* 28(4):287–291.
Vander Wood, T. B., & Rebstock, J. M., 1988. Using Automated Analysis to Identify Cleanroom Particulate Contaminants. *Microcontamination* 6(2):24–27.
Whitby, K. T., Husar, R. B., & Liu, B. Y. H., 1972. The Aerosol Size Distribution of Los Angeles Smog. In *Aerosols and Atmospheric Chemistry*, ed. G. M. Hidy, pp. 237–264. New York: Academic Press.

3

Particle Effects on Products

Although many mechanical, optical, and electrical products that are presently being manufactured do not require cleanroom facilities, the need is increasing. The number and type of precise mechanical devices whose operation is degraded if particles are present to interfere with the rate and/or extent of movement increase almost daily. These devices include electronic memory systems, machine tool controls, fluid flow control systems, and even inexpensive ballpoint pens. Two examples of areas where increasing contamination control requirements appear are entertainment and transportation. Musical reproduction systems based on long-playing records or cassette tapes can be produced with only reasonable care to avoid gross contamination. However, high-quality digital audio tape and compact disk systems must be kept much cleaner for optimum performance. Automobile manufacturing is carried out mainly in normal factory areas. However, paint facilities must be kept very clean because the ingression of atmospheric debris can damage the appearance and life of modern glossy painted surfaces. Many vehicle transmissions that may have to operate without needing new fluids replacement for many months are now built in clean areas. This occurs particularly in large, off-road hydraulic machines and in control systems. The precision positioning of components used in these machines means that the hydraulic fluids and their control assemblies must be very clean. As automobile carburetors are being replaced by fuel injection systems, the small-dimension injection nozzles used in these systems require fabrication in clean areas; the dimensions of the nozzles approach the size of particles that may be present in factory areas that would have been suitable for carburetor manufacture. Engine performance is degraded if the nozzle orifices are partially blocked.

If the injector valve solenoids capture particulate debris that can interfere with normal cycling, then uncontrolled engine speedup can result. Similarly, computer-controlled antilocking brake systems rely on cleanliness in the hydraulically operated brake systems to prevent any brake operation malfunction.

A wide variety of military systems must be built, tested, and maintained in clean areas because the components are precise and mechanical operation control can be degraded by particles that might be ingested at any stage in the system life cycle. This problem is special because most military devices must be operated in areas that are very heavily contaminated with dust and smoke. These systems must be very carefully sealed during operation, yet still allow access for maintenance. Many navigational guidance systems operate with precision high-speed gear assemblies. Obviously, the normal operating environment for these systems means that they must be sealed after any work is carried out on interior components. Hydraulic guidance components in many military vehicles are operated at pressures up to 650 bars, with the hydraulic fluids moving through valve and actuator assemblies with clearances in the 2–5 μm range. Abrasion of these precise passageways can occur easily if any hard particles are present to be forced through the components. The result is fluid leakage with imprecise system operation. Partial or slow valve operation in fuel lines can result in vehicle speed or attitude errors. This requirement for precise control also exists for the very large valve assemblies used in space vehicle fuel and oxidizer flow control systems.

Many optical components can be damaged easily by particles. High-resolution film and optical elements are damaged by particles in the micrometer size range. Image values are degraded, and light transmission efficiency can be reduced, particularly in fiber-optic connectors. With the advent of high-power laser systems, any particles deposited on light transmission or amplifier elements may absorb sufficient light to cause excessive heating of any surface upon which the particles are deposited. Work in laser fusion studies has shown that the amplifier elements must be kept totally free of particles in the micrometer size range, or the high-energy beams can cause rapid heating and nearly explosive evaporation of any particles on the amplifier surface, with severe damage to that surface. Figure 3-1 (Gutacker 1987) shows how even inert transparent particles can cause problems in lens and film performance. Add to direct optical interference effects the possibility that small reactive particles can nucleate sensitive film to cause unwanted image formation in areas much larger than the particle cross-section. The importance of maintaining cleanliness in optical systems becomes apparent.

The particle content of pharmaceutical products *must* be controlled,

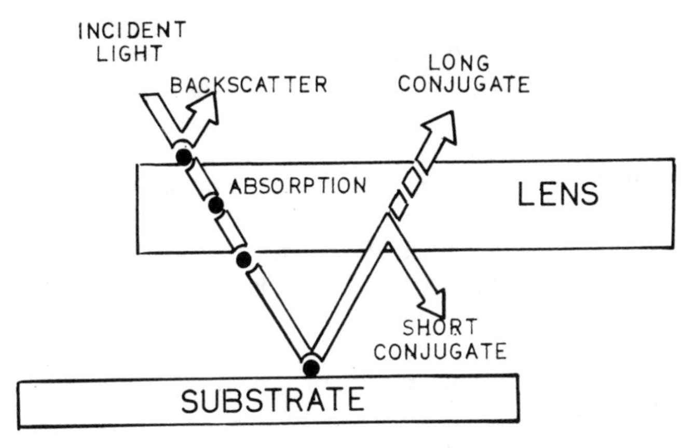

FIGURE 3-1. Contaminant particle effects in optical paths. Each particle can cause scattering and/or absorption of the direct beam with resulting interference to the desired optical paths. (Courtesy A. Gutacker, ARGOT, Inc.)

whether particles are inert or viable. The potential harm from ingested viable particles, such as bacteria, fungi, spores, or viruses, is intuitively apparent even to the layman. Inert particles can affect the appearance of the pharmaceutical product and interfere with bodily functions if ingested with a supposedly clean parenteral liquid. For these reasons alone, the need for cleanliness of pharmaceutical products is apparent. Governmental agencies specify cleanliness requirements for pharmaceutical products in most areas of the world. In the United States, the *United States Pharmacopeia* (U.S. Pharmacopeia 1985) defines standards for maximum particle contents of parenteral products, and the U.S. Food and Drug Administration (FDA) enforces the standards promulgated by the USP. If an FDA audit shows unsatisfactory product quality for any of several reasons, including excessive particle content, an entire product batch may have to be destroyed. For inert particles, the FDA allows up to 50 particles $\geq$ 10 μm and 5 particles $\geq$ 25 μm in 100 ml of large-volume parenteral liquid products. For small-volume injectable products (< 100 ml), the FDA maximum population is 10,000 particles $\geq$ 10 μm and 1,000 particles $\geq$ 25 μm per small-volume container.

Production of dry products in the pharmaceutical industry usually involves powder handling. There is some potential for dust escaping either the processing or packing areas and cross-contaminating different material. Medical devices that are fabricated with potential dust ingestion possibilities must be kept clean during manufacture, packaging, and use to avoid potential patient harm. Measurements of commercially available sterile needles and syringes have shown that thousands of particles as large as 20–30 μm per syringe with needle can be present (Taylor 1982). Even though the devices were sterile, the inert particles can cause problems in use. Some blood vessel capillaries are approximately 10 μm in diameter, and inert particles can cause irritation upon being deposited on sensitive tissues in the patient's body.

Essentially the same problems occur with food products and with pharmaceutical materials. The difference is that pharmaceutical materials can be injected as well as ingested, whereas food products are *only* ingested. Problems that can occur from particles in foods are connected mainly with product appearance and sensory effects, unless the foods are contaminated with harmful bacteria. When food products may have collected harmful viable particles, the effects are often noted by the appearance or odor of the food product. This is not always the case; infection of food products by some bacteria does not always result in easily noticed effects until after the food is eaten.

Production of some specialty foods requires cleanroom processing in preparation and packaging. These are mainly food products that are not refrigerated or heat-sterilized after packaging. If not handled in particle-controlled environments, the packaged food almost always contains bacterial contaminants that reduce shelf life. As the use of microwavable convenience foods that do not require low-temperature storage becomes more widespread for main course foods, the use of polymer containers that cannot be heat-sterilized results in a need for sterile packaging environments.

Perhaps the largest area of concern for the effects of particles on products is electronic products. Both the basic components and the assembled products can be adversely affected by particles. Semiconductor components, for example, are made by etching and/or depositing very fine, carefully controlled patterns of specific compounds that may contain trace materials on silicon wafers. If the patterns are inadvertently interrupted or bridged by interfering particles, then the semiconductor device will be defective. Particle contamination can lead to device failure through both mechanical and chemical interferences. Mechanical effects are most commonly reported. Chemical interactions may be more difficult to control. For example, particles in the 0.1 μm size range can degrade metal-oxide semiconductor (MOS) performance through incorporation of defects into gate oxides (Duffalo and Monkowski 1984). If the particles contain ionic impurities, they can react with the growing oxide and incorporate impurities into it (Monkowski 1984). These effects can occur during pattern establishment, deposition, or etching operations to produce short or open circuits or component value shifts. The time during which a wafer can be exposed to cleanroom air during fabrication may be as long as 40 hours.

Table 3-1 illustrates time of exposure for typical processes during processing (Larrabee 1982). During this time, even in a class 10 area, five or more particles larger than 0.5 μm can settle on the wafer. Many present-day components are affected by "killer defects" no larger than 0.1 μm. As many as 175 of these can deposit on the wafer during the 40-hour period. Added to processing and manufacturing contaminant deposition potential, occurrence of low-yield levels can be more easily understood. Recent studies showing that particle-induced defects can cause more than 50% of total yield loss in modern high-performance semiconductor manufacturing indicate the pressing need to control particle contamination during this process (Castel and Moslehi 1986). Consideration of anticipated directions in semiconductor designs and performance shows that product yields will decrease drastically as device dimensions increase while feature dimensions on the devices decrease (Osborn et al. 1988). Similar work in Japan has yielded essentially the same conclusions (Hattori 1988).

Contamination particles are not all deposited from the air surrounding the devices. A major source of particulate contamination is the processing chemicals that are used to fabricate the devices. These materials may be a greater contaminant source than most of the other sources considered a problem in cleanroom operations. For example, measurements of particle content of the ubiquitous photoresist used in semiconductor fabrication has shown deposition of up to several hundred particles per wafer, with significant lot-to-lot variations in particle levels in the products (Hecht and Reardon 1986). The need for care in use of this material is evident.

Many computer memory systems consist of rotating disks or moving

TABLE 3-1 Typical Operating Time for Some Semiconductor Component Manufacturing Processes

CONTAMINATION CONTROL IN SEMICONDUCTOR MANUFACTURING
TIME A WAFER "SEES" CLEAN ROOM AIR

PROCESS		HOURS
LITHOGRAPHY	11 MASKS	11.0
PLASMA ETCH	17 ETCHES	17.0
IMPLANT	6 IMPLANTS	4.0
FURNACE	32 OPERATIONS	5.3
CLEANUP	34 OPERATIONS	5.6
ASHING	10 OPERATIONS	0.8
METALLIZATION	4 OPERATIONS	0.7
TOTAL		40.4

Source: (Courtesy G. Larrabee, Texas Instruments.)

tapes with magnetic coatings. The disk rotation may be at rates of several thousand revolutions per minute. A magnetic head moves over the surface of the disk or tape to write or detect written information on the magnetic surface. Examples (Gutacker 1987) of this process are shown in Figure 3-2 for disk memories and in Figure 3-3 for tape systems. The "flying height" of heads in some disk systems is 10–20 microinches or less, with the head suspended by a combination of mechanical and aerodynamic force. As the disk rotates, it pumps air by centrifugal force outward along the disk surface. Disk systems usually include recirculating air filters to remove any particles that may be generated within the system or enter from the environment. If particulate material enters the space between the head and the disk surface, then erroneous information may be read. In extreme cases, the disk or even the head surfaces can be damaged. Figure 3-4 shows some particle data within a disk drive enclosure when operating normally. Note that small changes in particle concentration occur because of the movement of the head actuator system. If excessive particle content occurs in the enclosure, however, then damage ranging from data loss to catastrophe can occur. Figure 3-5 shows the changes in particle concentration in an operating system when damage occurs (Lieberman and Ogle 1971). Excessive levels of damaging particles were deliberately introduced into the head-disk area, and these particles generated additional particles by erosion of the disk surface. This process causes a snowballing effect until head-disk interference or "head crash"occurs. Manufacturing and design requirements for disk drive systems include procedures that will ensure that particulate contamination will not be included in the system

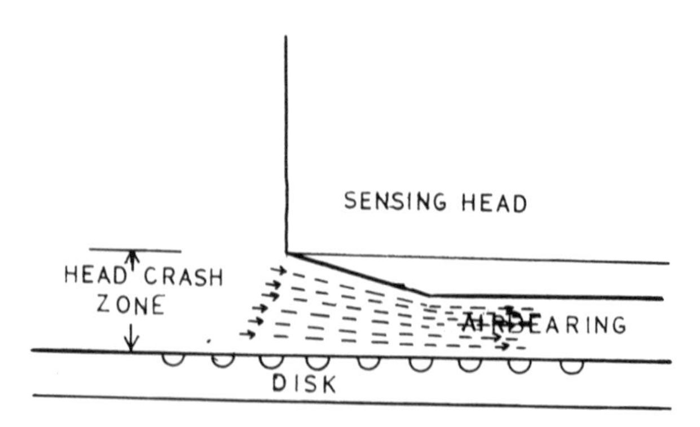

FIGURE 3-2. Disk drive operation patterns showing small clearances and potential contamination collection points. If any particles deposit within the microinch spacing between the disk and the sensing head, data dropouts, abrasive damage, or head crash can occur. (Courtesy A. Gutacker, ARGOT, Inc.)

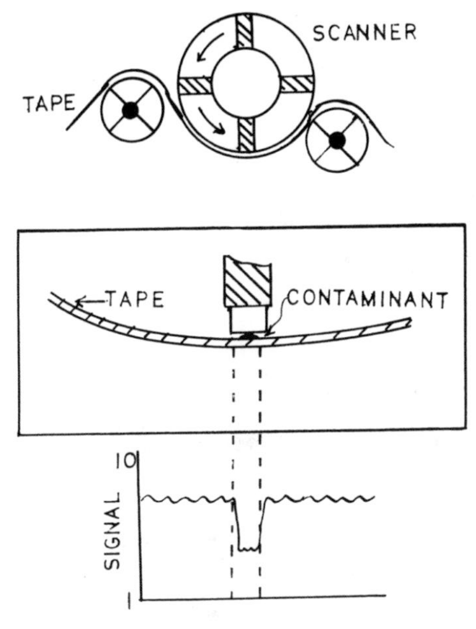

FIGURE 3-3. Tape drive operation and data dropouts as may occur because of contamination collection in the system. Even though the head-to-tape clearances are normally larger than those in disk drive operations, contaminants can change the distance between the scanner head and the magnetic fields on the tape, causing data losses. (Courtesy A. Gutacker, ARGOT, Inc.)

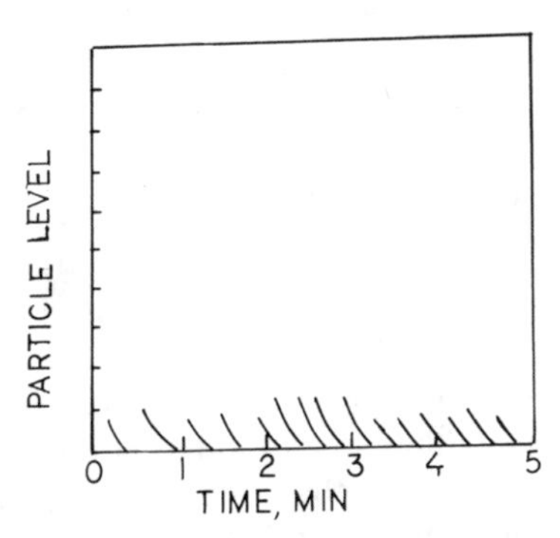

FIGURE 3-4. Normal particle concentrations in operating disk drive memory. Each time the head is moved, some particulate contamination is generated, but the particles are removed quickly and no problems occur in normal operation.

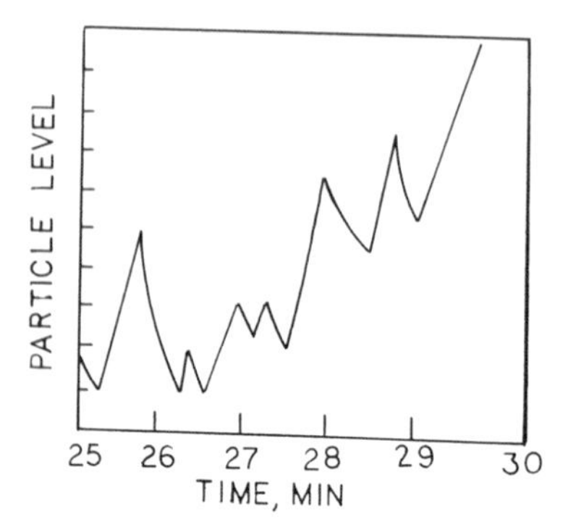

FIGURE 3-5. Particle concentrations in an operating disk drive during head crash. Many additional particles are being generated by extreme frictional wear. The wear particles cause a cascading abrasion effect that causes rapid destruction of the disk surface and/or the head.

during the manufacturing process and that critical elements in the system will be safeguarded from hazardous materials within the system itself. In particular, both permanently magnetic and magnetizable particles must be eliminated from any components used in the device (Viswanadham 1987).

Experience has shown that contaminant particles can be damaging to the reliability of precision products. Those products and devices that are manufactured or operated in clean areas operate more reliably and accurately than those subjected to contaminant particles during either assembly or operation. Such devices include bearings, timers, gyroscopes, pneumatic controllers, mechanical actuators, power transmission and conversion systems, fluid flow metering and controlling devices, high-speed electrical switches, electronic data recording and reading systems, and optical beam controllers. Products of concern in this area include pharmaceutical and food products, as well as many device components and subassemblies. They can include fuels and hydraulic fluids for many vehicles, as well as the surfaces of storage vessels for these liquids. For effective contamination control, a better understanding of how contaminant particles actually affect these items is helpful in planning corrective action. The major problems with particulate contamination arise from mechanical, electrical, or optical interferences. In addition to these physical interference effects, acceptance of many commercial products is controlled by specifications defining maximum contaminant particle content. The effects of these interference phenomena are discussed next.

Mechanical effects of contaminant particles involve interference with motion within a precise device. Motion may be uniform, accelerating, or intermittent and can be sliding or rolling. Contaminants can affect the end position of components or can interfere with fluid flow. The mechanisms by which contaminants cause problems are many and varied. Dimensional changes in parts can result from wear over a sufficient time period. Even though minute, these changes can cause secondary problems of component degradation caused by galling, seizing, or slipstick motion. An extreme situation can arise if solid particle interference causes increased friction in moving parts with small clearances to the extent that local heating can occur. Thermal expansion may result in subsequent component seizures. In high-speed parts, structural failure can be expected; in slow-moving or intermittent-motion devices, particularly those where large forces are applied, undesired welding can also occur.

When particles deposit on a surface that may be heated, then chemical reaction can occur with some materials. This can result in formation of a compound or alloy with physical and/or chemical properties different from those of the original substrate material. Local failure of the substrate surface or interior caused by changes in tensile strength, creep rate, brittleness, or conductivity change can occur later. Ingestion of particles into sensitive components can cause either immediate or delayed damage. A common delayed effect is seen when hard particles deposit within low-clearance systems and cause additional wear particle generation, which accelerates wear effects in the system. The same delayed effect occurs in rotating disk memory systems. Particle ingestion into the head-disk interface area can cause erosion of the disk surface or of the read/write head. The eroded particles cause additional erosion, resulting in an increased particle concentration and subsequent head crash, as seen in Figure 3-5.

Control of mechanical effects of contaminant particles can be simplified by taking into consideration the device affected, its operation, the observed and potential failure modes, and the contaminant types and their properties that will affect that device. One of the first action plans that used a sequence similar to this was first prepared in 1962 as part of a study on contamination control for the Sandia Corporation (Lieberman 1962). A contaminant effects summary allowed characterization of device operation. It lists probable macroscopic factors that might cause device failure, as well as contaminant particle characteristics and potential particle sources to aid in identifying those sources that might produce specific particles of concern. The listing was not intended for application to any particular system or device but rather was designed so that almost any device could be considered by applying appropriate parameters. This table is included as an appendix to this chapter.

Most contaminant particles in the atmosphere are composed of materials of low conductivity, such as soil particles, organic fragments, and plant and polymer fibers. In addition, there are also many conductive metal fragments, ionic liquid droplets, and inorganic crystals. Because many contaminant-sensitive devices are electrical in their operation, the effects of either high- or low-conductivity contaminant particles can be strong. Semiconductor devices that have collected conductive particles during assembly or storage can be exposed to electrostatic fields that can cause electrical discharge through a conductive contaminant particle. Usually the discharge is strong enough to destroy many components in the semiconductor. If a conductive particle falls across traces on a wafer during device fabrication, then the die where that particle exists will be defective. If a nonconductive particle falls onto a trace, then a conductive path is interrupted or is replaced by a resistive path; again, the die can be defective.

In relays or switching devices, problems can arise in conducting electrical energy from one component of a device to another. For maximum electrical energy transfer efficiency, intimate contacts should be established over as much area as is practical. The presence of contaminant particles on one or another of the contact points can cause high-resistance paths; in some cases, arcs can exist where the contact points are prevented from contact but are close enough to allow overcoming atmospheric resistivity. The severity of contaminant problems in this situation is inversely proportional to the force used in obtaining contact, to the available voltage, and to the contact area. For many high-speed switching devices, the available contact force and system voltages may be so low so that particles larger than a few micrometers in diameter can cause intermittent or erratic switching operation.

Optical systems are used for a variety of operations. Image formation and graphics storage devices are the heart of conventional photographic and reprographic optical systems. High-resolution direct-viewing devices, such as microscopes and telescopes, require carefully controlled optical systems. The use of a wide range of laser systems for information processing and for analytical devices increases continually. Development work on laser fusion systems has shown the need for contaminant-free components in all parts of the system. Optical fibers have become widely used for data transmission with high efficiency and high data density. In all these devices, the operating optical system requires an emitting surface, a transmission medium, a collecting surface, and a light detector to convert collected light to either analog information or digital data. These requirements apply to transmission of electromagnetic radiation at any wavelength.

When a well-defined beam is required, the contour of the energy-emitting surface and the emitter-medium and medium-collector interfaces must be free of contaminants. If high-energy fluxes are of concern, the radiation absorption within a contaminant particle surface with a real refractive index and/or an absorption coefficient different from that of the optical element can result in extreme levels of heat at the location of the contaminant particle, causing potential damage to the underlying optical substrate. When low-level changes in light transmission are of concern, as in information transfer situations, then even small losses in light by scattering or absorption from contaminant particles in the system can be of concern. These effects are aggravated because many contaminant particles are of nearly the same dimension as the wavelength of the illumination. Figure 3-1 showed some of these effects.

As pointed out elsewhere, many products and systems cannot be accepted by a purchaser or sold commercially if the particle content exceeds specific limits. Contamination control specifications are discussed later; because imposition of these specifications affects the disposition of the products, however, the importance of these specifications is also pointed out here. An example of one very important specification is the limit on particle concentration for both large- and small-volume parenteral liquids defined in the *United States Pharmacopeia,* volume 22, chapter 788. Without specifying numerical values here, FDA auditing procedures allow entry to an operating pharmaceutical manufacturing plant for inspection of processes and records. If satisfactory records and data are not present, then the auditors may require that the plant discard an entire product batch if the limits are exceeded. In the same way, U.S. Department of Defense inspectors can require extensive cleaning and/or rework on certain weapons system components if particle concentration in a hydraulic oil system exceeds a specified level, as abstracted from the tables in MIL-STD-1246B (U.S. Army 1972). Assembly of liquid-fueled rocket components requires that the fuel and oxidizer tank interior surface cleanliness meet a (different) level of MIL-STD-1246B. Although the use of specifications does not affect the way in which particles can physically affect sensitive products, requiring their use can decide whether the product can be manufactured in a particular facility and used afterwards.

References

Castel, E. D., & Moslehi, B., 1986. Reducing Defects in Photomasking Operations. *Microcontamination* 7(2):29–36.

Duffalo, J. M., & Monkowski, J. R., 1984. Particulate Contamination and Device Performance. *Solid State Technology* 27(3):109–114.

Gutacker, A. R., 1987. *Fundamentals of Contamination Control Handbook,* ARGOT, Inc., Webster, NY.

Hattori, T., 1988. Total Contamination Control for ULSI Wafer Processing. *Microelectronic Manufacturing and Testing* 11(5):31–35.

Hecht, J. K., & Reardon, E. J., 1986. Particulate Analysis and Control in Photoresist Manufacturing. *Microcontamination* 4(10):45–50.

Larrabee, G., 1982. Private communication.

Lieberman, A., 1962. Contamination Effects Study. ARF 3216-5, Armour Research Foundation Report to Sandia Corp.

Lieberman, A., & Ogle, H. M., 1971. Particle Generation in Computer Memory Systems. Proceedings of the International Powder Technology and Bulk Solids Conference, pp. 99–102, London.

Monkowski, J. R., 1984. Significance of Submicrometer Particles to Device Performance. Proceedings of the 30th Annual Technical Meeting, Institute of Environmental Sciences, pp. 81–84, May 1, 1984, Orlando.

Osburn, C. M., et al., 1988. The Effects of Contamination on Semiconductor Manufacturing Yield. *Journal of Environmental Sciences* 31(2):45–57.

Taylor, S. A., 1982. Particulate Contamination of Sterile Syringes and Needles. *Journal of Pharmacy and Pharmacology* 34:493–495.

U.S. Army, 1972. MIL-STD-1246B: *Product Cleanliness Levels and Contamination Control Program,* Philadelphia: U.S. Naval and Publication Forms Center.

U.S. Pharmacopeia, 1985. Particulate Matter in Injections. In *United States Pharmacopeia,* 21st ed., p. 1257, Easton, PA: Mack Publishing Co.

Viswanadham, P., 1987. Contamination Control in Disk-Drive Manufacturing: A Quality and Reliability Perspective. *Microcontamination* 5(4):41–45.

Appendix

Contaminant Effects on Mechanical Parameters Summary

I. Functional operation of device
   A. Device type examples
      1. Low-torque sliding contacts (wafer switches)
      2. Small electric motors (bearings, shafts, seals)
      3. Relay contacts
      4. Fluid metering devices (orifices, turbines, etc.)
      5. Gyroscope mechanisms
      6. Near-contact memory systems (disks, tapes)
      7. Optical linkages
   B. Hardware considerations
      1. Materials of construction
      2. Design and geometry
         a. Shape
         b. Clearances and finish
         c. Linkages
   C. Operation considerations
      1. Accuracy tolerance
      2. Required precision
         a. Intended use
         b. Environments (storage and use)
      3. Required operating life
         a. Temporal duration
         b. Number and frequency of operational cycles
         c. Obsolescence factors (modify or replace)
      4. Mode of operation
         a. Continuous

b. Intermittent
c. One shot
5. Shelf life
a. Storage before operation
b. Storage between intermittent operations

II. Implementation of device operation
A. Position control considerations
1. Rate of motion
a. Uniform or accelerating motion
b. Rate control
2. Linear or angular motion
a. Accuracy of position
b. Tolerance limits
3. Intracomponent motion with and without finite clearances
a. Sliding motion
(1). Unidirectional
(2). Reciprocating
(3). Intermittent
(4). Rotational
b. Rolling motion
(1). Unidirectional
(2). Reciprocating
(3). Intermittent
c. Nonuniform motion
(1). Velocity considerations
(2). Acceleration effects
(3). Rotational effects
d. Constraints on motion
(1). Fixed component stops
(2). Journal or sleeve container restraint
(3). Antifriction materials
(4). Elastic materials
(5). Fluid flow containment
e. Mechanical loading considerations
(1). Force transmission methods
(2). Deformation—strain
(3). Stresses
f. Energy transmission
(1). Mechanical systems
(a). Positive drive systems
(b). Frictional drive systems
(c). Magnetic connections

- (2). Power transmission
  - (a). Hydraulic and pneumatic systems
    - i. Flow and pressure constraints
    - ii. Fluid system sizing, leaks, and reservoirs needs
  - (b). Thermal energy transmission
    - i. Radiation
    - ii. Conduction
    - iii. Convection
  - (c). Electrical energy transmission
    - i. Radiation
    - ii. Conduction
    - iii. Interferences

III. Device failure modes and mechanisms
- A. Catastrophic failures
  - 1. Gross design errors
  - 2. Manufacturing errors
  - 3. Material selection error
  - 4. Assembly errors
  - 5. Functional misapplication
  - 6. Component malfunction
- B. Use or wear failures (predictable)
  - 1. Dimensional changes
    - a. Linear or irregular
    - b. Local or general change
  - 2. Friction and wear considerations
    - a. Loading variations
    - b. Stress-strain considerations
    - c. Materials interactions
      - (1). Galling, seizing, stickslip movements
      - (2). Temperature effects, overall or local
  - 3. Material property variations
    - a. Surface finish variations
      - (1). Changes in specified values
      - (2). Effects on device performance
    - b. Material diffusion effects
      - (1). Acceleration by fatigue
      - (2). Creep rate change
        - (a). Plastic deformation
        - (b). Embrittlement

IV. Contaminant properties
- A. Delayed contaminant effects on substrates

1. Gaseous material penetration and reaction
2. Particle effects
    a. Diffusion into materials to produce delayed weakness
    b. Slow erosion by repeated impact
    c. Surface migration to sensitive locations
    d. Particle growth to hazardous size by agglomeration
    e. Secondary abrasion by fragments
3. Energy gradient effects
    a. Electrostatic discharges
    b. Material degradation at high temperature
    c. Bacterial growth

B. Particle contaminant prompt effects
1. Hard particle wear, physical interference, etc.
2. Soft particle deposition and resulting dimensional changes, weight imbalance, motion interference, etc.

C. Particle-substrate energy transfer
1. Electromagnetic energy
2. Kinetic energy
3. Thermal energy

D. Pertinent physical and chemical effects
1. Nucleation from gases
2. Adhesion due to electrostatic forces or from chemical effects
3. Effects of reaction products

V. Contaminant sources

A. Component manufacturing
1. Carryover from previous machine tool operations
2. Casting and forming inclusions
3. Deposition from processing materials

B. Component assembly
1. Fragments from attachment processing (solder, flux, adhesives, etc.)
2. Grinding or other fitting operations
3. Tool deposits or deformation products

C. Storage conditions
1. Deposition from environment or tool interior
    a. Deposition mechanisms (sedimentation, electrophoretic or thermophoretic forces, impaction, diffusion)
    b. Air contamination source considerations
        (1). Air-handling system
        (2). Reentrainment from tools, surfaces
        (3). Personnel activities

   2. Transfer of material from container
   3. Secondary corrosion product movements
- D. Cleaning operations, wet and dry
- E. Normal system operation
   1. Wear, overstress of mechanical devices
   2. Overheating of mechanical or electrical devices
   3. Deposition from coolant/power transfer fluids

4

Gaseous, Chemical, and Other Contaminant Descriptions

Most contamination control technology considers generalized and often unidentified particulate material as the major contaminant, but there are many situations in which gases, chemical films, microbiological materials, environment physical conditions, and unusual ambient energy levels are hazardous to the well-being of the products. This variety of contaminants can degrade product quality at least as seriously as can particulate contaminants. In fact, as particle removal and control methods improve, contaminating gases and vapors become a major problem area. However, their sources can be different than the particle sources, their transport and deposition mechanisms are different, their interaction with products and their control and/or removal procedures are different, and their spatial and temporal distributions in the manufacturing and processing environment are usually different from that of particulate contamination.

Some gases and vapors are normally present in all areas, but if the concentration exceeds the limits for a particular product, then the gas or vapor can degrade product quality. Gases and vapors have high diffusion constants, and most of these materials tend to disperse widely and uniformly in air. Although most gases and vapors have densities similar to that of air, some may be so much more or less dense that they concentrate at low or high points in an enclosure unless airflow is sufficiently turbulent to keep the materials well dispersed or to remove them from the enclosure completely. Many organic solvent vapors have specific gravity greater than that of air, whereas the specific gravity of hydrogen or helium is much less than that of air. Most gases and vapors that may cause contamination problems in the cleanroom rapidly become well mixed in the atmosphere.

Both organic and inorganic gases can cause problems in cleanrooms.

The needs and restrictions of the particular product determine whether a particular material is a problem in any application. For example, even low levels of certain organic vapors can cause great problems in semiconductor manufacturing, but the same materials may have no effects in a pharmaceutical manufacturing area. Airborne ions can cause specific component damage in semiconductor manufacturing operations, whereas the same concentrations in a pharmaceutical manufacturing area are only a minor problem in terms of modified particle deposition in the area.

A variety of gases, such as water vapor, carbon dioxide, and many organic vapors, are normal components of the atmosphere. The major difference between dirty urban air and "unpolluted" rural air is not the difference in materials in the two air systems as much as the concentration of materials present in both air parcels. Some industrial and natural pollution generation sources observed in the 1960–1980 time frame emitted both particulate and gaseous materials that were detected hundreds of kilometers from their generation points. The natural sources included such phenomena as volcanic eruptions and desert sandstorms. The concentration of pollutants may be reduced by a variety of removal processes, such as chemical reactions that produce relatively innocuous materials, or by simple dilution as the emission cloud diffuses into the atmosphere. Even so, the large number of pollutant sources results in the presence of some contaminants in the air throughout the world. This is a special problem in industrial areas where atmospheric conditions allow pollutant gases to remain in the area for long periods of time.

As an example of a material that is not normally considered to be a pollutant, water vapor in a cleanroom must be controlled to an optimum range other than that in most air parcels. The cleanroom relative humidity may need to be controlled to avoid water condensation during temperature changes or to make sure that excessive electric charge does not build up as a result of low humidity levels. Minor differences in CO_2 are seldom a problem in most manufacturing processes, but if excessive quantities of other organic vapors are present, then the cleanroom air may require scrubbing to reduce the organic vapor pollutant levels to an acceptable range. Air cleanliness requirements in the working cleanroom arise from both the potential of damage to products and from the need to keep the concentration of noxious materials below toxic and hazardous levels in the cleanroom and in the air exhausted to the environment.

In most urban areas, industrial activity results in production of atmospheric gas and vapor components other than water. These materials include sulfur compounds emitted from power plants burning fossil fuels, a wide variety of organic vapors from petroleum refinery operations, inorganic vapors from chemical manufacturing or metal ore processing, and acid gases from wood or pulp processing plants. Agricultural activities can also cause release of potentially harmful gases. Ammonia vapors from fertilizer application or from animal feed lots and organic fumes from some crop-dusting operations are only a few of the gaseous contaminants from farming operations. Emissions of organic materials from plant life include organic vapors as well as the normally anticipated pollens and spores.

Problems can arise from emissions from industrial operations that may be several miles distant from the cleanroom facility. Industrial emissions are normally controlled by state and local environmental pollution control agencies to levels that are not health hazards. Even if safe for people, the gas levels that may be present in the air entering the cleanroom HVAC system may still be excessive for products that require cleanroom manufacturing conditions. For this reason, the production supervisor should be aware of the possible effects of gas or vapor contaminants on the product and should follow the local pollution agency reports on levels of the specific gases or vapors that may affect his operation. Knowing the kinds of industrial activity in the nearby areas and the local wind patterns is important. If any significant changes in levels of hazardous vapors may occur, then special control procedures or devices for cleaning the incoming air may have to be installed. These may take the form of additional air scrubbing or adsorption systems for the HVAC system or for specific manufacturing areas. Work has also been reported on reduction of some contaminant gases by interaction with complex electrical fields (Frey 1986). Reductions up to 49% were reported for formaldehyde vapors present in ppm ranges. No data are reported for effects of this method for control of trace concentrations of this material as may be seen in a controlled operating cleanroom.

Harmful vapors may also be released from an internal processing operation within the cleanroom that uses chemical gas or vapor treatment. Local concentrations of vapors can be much higher than that expected by infiltration or dilution in the working area. The contaminant effect can be twofold. The vapors can affect operations in the cleanroom, or the vapors may be a hazard from an industrial hygiene and air pollution standpoint. The latter hazards may not affect production, but the relatively low-concentration, high-volume material must still be removed to protect personnel and products in the area as well as the overall environment. Vapors can be produced by organic solvent or reactive chemical materials used in processing. Process chemical materials can be extremely corrosive. Semiconductor manufacturing operations involve more and more gas phase reaction processes. Some of the materials used in deposition procedures include such noxious materials as ammonia, arsine, chlorine, diborane, dichlorosilane, hydrogen, hydrogen chloride, hydrogen fluoride, ozone, phosphine, silane, and tungsten hexafluoride (McMahon 1989). Etching procedures use a variety of halogenated organic materials, acids, and oxidants. Many of these materials may deposit ionic contaminants that have been dissolved from containers or collected during the material preparation or storage conditions (Burkman et al. 1988). Several process operations were examined and various metallic ion deposition sources were traced to processes of reverse plating or adsorption from process chemicals used in manufacturing.

Many operations in semiconductor manufacturing require intermediate cleaning operations between steps in the production process. For example, photolithographic processing requires that unreacted photoresist be removed from the wafer surface after illumination and etching operations have been completed. Chlorinated fluorocarbon solvents have been used for cleaning operations for many years in this area. Vented exhausts from process lines must be cleaned before release to the atmosphere. This problem area is especially important if halogenated fluorocarbon solvents are used in a process because these materials can cause degradation of the stratospheric ozone layers. Even though replacement of these materials by others that should not affect the high-altitude ozone layer is progressing, those operations that still use them should minimize their release to the atmosphere as much as possible.

Essentially all problem gases can be removed from exhausts by wet scrubbing before release. The gas to be cleaned is passed through a sprayed suspension of absorbent liquid droplets in a turbulent flow field. The absorbent liquid is selected for the specific gas to be cleaned. Some gases are cleaned with ionic solution liquids, and others with organic liquids. The droplets are collected after the gas or vapor has been absorbed into the droplets. A variety of scrubbing systems are now in use that may involve countercurrent or cocurrent gas and droplet streams with means of inducing intimate gas-liquid contact. Turbulent mixing systems are usually used for this purpose. Careful disposition of scrubber waste liquid is required so that the air contamination problem is not replaced by a waste disposal problem. A detailed discussion of scrubbing options for a wide range of materials was given (Librizzi and Manna 1983) in 1983. This discussion was directed toward semiconductor manufacturing cleanroom applications, although gas scrubbing has been used for decades in many industrial applications.

A number of chemical contaminants come from normal cleanroom operations, as shown in Table 4-1. Light ends from mineral lubricants vaporize if a motor or bearing is overheated. Fluxes and vapors from inadequately contained soldering operations can escape into the cleanroom air during processing if on-the-spot repairs must be carried out.

TABLE 4-1 Organic Contaminants in Cleanrooms

ORGANIC CLEANROOM CONTAMINANTS
Motor Oils
Bearing Lubricants
Fluxes and Vapors
Solvent Vapors
Plastics
DOP (from filter testing)
Skin Oils
Cosmetics
Vacuum Pump Exhausts
Plasticizer Vapors

Solvent vapors can be present where cleaning operations are carried out. DOP (dioctyl phthalate) mist is widely used as a HEPA filter challenge material; among other properties, DOP has a very low vapor pressure. However, that pressure is not zero. After filter testing for efficiency or for leak determinations, sufficient DOP may be collected in the filter so that normal air passage can result in some level of vapor emission. In addition to possible release from tested filters, DOP vapors may be released from polyvinyl chloride plastic devices, which may have been molded with DOP used as a plasticizer. Skin oils, cosmetics vapors and particles, and other material are emitted from personnel in the cleanroom, requiring careful control of personnel selection, activity, and clothing. In modern, well-controlled cleanrooms, vacuum pump exhaust is not a problem. The vacuum pumps are removed from the room itself, or oil vapor pumps are not used.

Many gases and vapors are produced by outgassing from materials used in cleanroom components; therefore, as shown in Table 4-2, cleanroom materials should be carefully selected. The materials shown in Table 4-2 outgas at a low rate. Other polymers that should be avoided include those formulated with plasticizers. These materials emit plasticizer vapors for long periods. Plasticized polyvinyl chloride is widely used for flexible tubing or for area separators. If products in the cleanroom are not affected by minor organic film deposition, then this material is acceptable. However, many precise optical systems and some pharmaceutical materials are adversely affected by deposition of a film resulting from condensation of plasticizer vapor from this material. Other potential hazards from

TABLE 4-2 Outgassing Rates from Cleanroom Components at Normal Work Area Temperatures; They Are Not Considered Excessive

MATERIAL	OUTGASSING RATE (micro-Torr litre/sec cm2)
Viton	0.03
Polyimide (Kapton-Vespel)	0.6
Polymonochlorotrifluoroethylene (Kel F)	1.7
Teflon	7.5
Polyethylene	11
Polyvinylchloride	30-80
Polycarbonate (Lexan)	40
Mylar V-200 (after two hours of pumping)	40-80
Polyamide (Nylon)	120

vapors include component damage as a result of undesired reactivity with substrate materials and hazard to personnel or facilities because of toxicity or flammability. Adsorption of vapors on powder surfaces in pharmaceutical manufacturing areas may result in degradation of product properties.

In addition to the gaseous contaminants released from cleanroom manufacturing materials, many gaseous contaminants are released by handling previously clean process gases in storage and delivery systems that degrade the purity of the process material. For this reason, many newer semiconductor fabrication facilities find it necessary to monitor process gas acceptability almost continuously. It was estimated in 1984 that particulate contamination in process gases comprise no more than 10% of the total semiconductor particle problem (Harper and Bailey 1984). At that time, the need to control gaseous contaminants in process material gases was the more serious problem area. As semiconductor features decrease in dimension while the device itself grows, the problem ratio between particle and gas contamination effects may change, but gaseous contaminants will still be a serious problem in terms of contamination control, especially in this area.

Viable particles are a particular problem in many areas where one would not normally expect to find microbes or bacteria. The possibility of bacterial contamination of food or pharmaceutical products or of contamination in a surgical operating room is easily understood. The growth of microorganism colonies in portions of deionized water systems where flow may be restricted can also be expected. In addition, many microorganisms can thrive in mineral oils, fuels, and lubricants, and some survive even in corrosive acids. Fluid flow restriction caused by growth of microbial colonies in a flow control system can cause serious problems; deposited microorganisms can destroy the integrity of precision optical systems both by simple optical interference and, in some cases, by etching of the precision surfaces. Microorganisms and their waste products can deposit on wafer surfaces during semiconductor manufacturing processes and cause the same product damage as can inert particles.

Airborne bacteria can be brought into the clean area on personnel skin surfaces or by personnel emissions (sneezing, coughing, exhalations, perspiration or other body fluids). Bacteria can also be carried on the surfaces of airborne dust particles or released from organic material surfaces that may serve as nutrients. Bacteria seldom exist as individual entities in most environments. They are normally found on or in a larger dust particle or water droplet. Most individual bacteria in cleanrooms are in the size range from 1 to 3 μm in diameter, whereas the particles carrying bacteria may be as large as 15–20 μm (Whyte 1968). Bacterial contamination of products in clean areas occurs by contact from contaminated liquid systems, by contact from contaminated surfaces, and from the air; the last source may involve bacteria dispersed from personnel (Whyte 1986).

In the normal (50–60%) cleanroom relative humidity, an individual organism dehydrates rapidly. Some bacteria, however, can enter into a dormant spore phase and survive under rigorous conditions of both temperature and relative humidity. During this stage, the bacteria do not multiply. When environmental conditions are such as to stimulate the spore to return to the vegetative cell phase, then these bacteria multiply and form colonies as long as sufficient nutrients and moisture are present. Many cleanroom environments are ideal for the vegetative phase. These environments include the surfaces of heating and ventilating system components, organic material surfaces, and the interiors of many liquid storage and transport systems. For this reason, sterilization of cleanroom surfaces is a common maintenance operation in most food and pharmaceutical manufacturing areas, as well as in many hospital areas. The sterilization process can include exposure of all surfaces and components to a bactericidal vapor or wet swabbing with a suitable bactericide.

Infectious bacteria in surgical operating rooms can be a serious problem, particularly if the surgery may require large wound area exposure and extended time exposure to the environment, as occurs in organ replacement surgery. Airborne bacteria are almost always present in hospital facilities. The major organism responsible for postoperative infections is hemolytic *Staphylococcus aureus*. This organism causes most of the infections in hospitals. It is approximately 0.5 μm in dimension so that it can remain suspended in air for a long time. Individual "staph" bacteria tend to be quite viable in the warm, humid hospital air supply. For these reasons, many hospital operating rooms where major surgery occurs are fitted with horizontal or vertical flow filtered air systems that should be capable of removing all airborne particles in that size range.

Electrostatic charges exist on all surfaces. The charge level may vary, and the charge polarity can be either positive or negative. On some surfaces, a distribution of both charge levels and polarities can exist. Charge on surfaces varies depending on the conductivity and structure of the surface, proximity of other conductive surfaces, and the nature of any electric field that may be present. Once a charge has accumulated on a surface, then changes depend on whether additional charge is transferred to or from the local surface. Charged surfaces include relatively massive solid substrate surfaces and smaller isolated element surfaces. Examples of the former include pipelines, polymer or metal containers, room walls, and floor covers; examples of the latter include airborne particles or individual ions in the air, liquid droplets, tools, and personnel.

Several charge generation and transport mechanisms exist. Air is normally ionized to some extent as a result of cosmic ray ionization and the presence of natural or artificial radioactive materials in the environment. It has been estimated that normal ionization intensity in the atmosphere is some 10 ion pairs per cm^3 per second (Liu et al. 1987). This value may change in cleanrooms where ionization sources are used to reduce electrostatic charge effects. Charge accumulation on surfaces results from contact and separation from other surfaces (triboelectric charging) or from charged fluid streams. Surface potentials of 3–5 kV have been observed on semiconductor wafers during normal processing (Blitshteyn and Martinez 1986). Although some charge loss might be expected with time, short-term accelerated deposition of particles during the period when charge is high can also be expected. Control of electrostatic discharge is particularly important in electronic system manufacturing because of semiconductor device susceptibility to electrostatic discharge. Normal operations in a semiconductor fabrication area can result in very high electrostatic charge voltage levels. These charges can arise from ion deposition or from normal handling and processing operations in wafer manufacturing. A study on charge levels produced as a result of triboelectric effects (Turner 1983) showed voltage levels in the kilovolt range. As might be expected, handling operations in which any frictional effects might occur produce the greatest voltages on the components. In addition, insulating surfaces and those with low moisture adsorption characteristics developed and retained the highest voltage levels.

Electrostatic discharge is a major problem in electronic device manufacturing but is not as important a problem in many other production areas. In electronic device production, conductive trace integrity can be destroyed by electrical discharges. In other areas, the major problem from electrical discharge is the possibility of fire when combustible materials are present. However, when surfaces in any production area have acquired an electric charge, any nearby particles that may be present with an opposite charge are attracted to those surfaces and result in an aggravated contamination problem because of charged particle deposition. Particle deposition on critical surfaces caused by electrostatic forces has been reported as being more significant than deposition from almost any other force (Cooper et al. 1988).

In areas where relative humidity is low, electrostatic charge problems range from a minor personnel annoyance to a major fire hazard. The latter situation can arise when combustible low-conductivity liquids or powders are transferred with containers and lines inadequately grounded.

The presence of both electrostatic and magnetic fields can also cause problems in precise fabrication devices and in measurement instruments used in semiconductor production. Magnetic field fluctuations can cause operational variability in any tool using electron beams, including both electron microscope observation and analysis instruments, as well as in electron beam lithographic systems (Ohmi, Inaba, and Takenama 1989). Electron beam paths are strongly affected by magnetic fields. External electromagnetic fields can be detected by very sensitive instruments, with the result that the external, uncontrolled field signal interferes with the signal that is being measured. Shielding with materials having high magnetic permeability, such as iron or permalloy, is sometimes required for effective operation of some devices. In some conditions, isolation of sensitive devices may be the only control method. Removal of high-power alternating current motors or heater controls to another location may be required in some cases.

Radioactivity in normal manufacturing areas is a minor contaminant because levels are usually quite low. Radioactive particle emission can be in the form of high- and/or low-energy particles, with either positive or negative charge. Radioactive particles may be alpha particles, which are strongly ionizing but lose energy rapidly, seldom traveling more than an inch or so in air. Negatively charged beta particles are lower-energy-level ionizers and may move several times further than the alpha particles. Gamma rays are similar to x-ray emission but have much greater energy levels. They are seldom seen in normal manufacturing areas. Some radiation can be emitted from low-level radioactive materials, which can include such structural materials as cement or cinder block in some areas. When buildings are very well sealed for conservation of energy, radon emitted from such building materials can accumulate in those areas where air movement is minimal. In most modern cleanroom facilities, radiation from this type of source is not a concern; radon levels are kept low by the high air circulation rates in cleanrooms.

Some radiation can also be emitted from analytical devices or manufacturing tools, such as x-ray spectrometers and electron microscopes; it can be in the form of electromagnetic interference from poorly shielded electronic devices. Visible light emission at unacceptable wavelengths is a well-known problem in film manufacturing and photolithography processing. Radiation emission levels are usually low in a well-controlled cleanroom, but care in operating and maintaining these devices is required. Radiation can be a product contaminant primarily in electronic device and photographic film manufacturing areas, but it can be a personnel hazard in any location.

Only recently has vibration been considered a serious problem, except for work in some optical measurement or in precise metrology facilities. Techniques had been developed for isolation of working metrology components, using flexible supports for heavy, rigid structures. The same approach was found to aid in producing and using precise optical systems. The proliferation of high-density, small-component electronic devices has resulted in the need for extreme stability in component positioning when either visible light or electron beam lithographic processes are being used. Feature dimensions in the micrometer and smaller range make it necessary to keep the wafer position absolutely constant relative to the mask. Uncontrolled vibration, as a "contaminant," is now a concern primarily in microelectronic and optical system manufacture. Methods to control vibration range from use of special building foundation structures to minimize earth movement effects on the building to use of specially designed substrate structures in the cleanroom or within manufacturing tools. In some flow control systems, specially designed solenoid valves that operate quite slowly are used to reduce emission of particles by the shock and vibration caused by rapid solenoid and valve element movement. In addition, carts and other component and equipment transport devices are designed and used so they do not generate or transmit vibration to a building structure in which sensitive systems are being manufactured.

As noted, a variety of contaminants other than particles are a potential problem in the cleanroom. Their effects vary with the type of manufacturing operations. Organic chemical films can interfere with optical systems by modifying light transmission or can cause photographic degradation and obscure color brilliance. Formation of metal oxide or polymer reaction product formation on precise mating surfaces can cause mechanical interferences. These films can change a surface so that particulate contaminants can be attracted and/or retained to cause later damage. They can produce undesirable chemical reactions. Chemical film contamination on semiconductor surfaces can produce immediate catastrophic failure or poor or failing performance.

An additional point should be kept in mind. Even though some materials are of immediate importance for specific manufacturing operations, the possibility always exists that combinations of otherwise innocuous materials can produce contaminants that affect a manufacturing process. As an example, halocarbon solvents can react with warm, moist air to produce acid vapors. Synergistic combinations include such systems as electrostatic discharge in combination with low relative humidity, and secondary emissions of plasticizer or filler from plastic structures or filters. Particle growth from vapor adsorption or out-of-tolerance operating temperatures can cause physical interference between closely sized components with different thermal expansion coefficients. Consider the situation in which a small contaminant chemical particle in the air is exposed to high humidity. A salt particle absorbs water and becomes a larger salt solution droplet that may deposit on a surface where it could either react slowly, resulting in chemical damage, or produce a high conductivity path for possible future electrostatic discharge damage. Other possible combination effects from more than one contaminant can also occur. A common problem is that in which particulate contaminants abrade a component, resulting in additional erosion and abrasion.

References

Blitshteyn, M., & Martinez, A. M., 1986. Electrostatic-Charge Generation on Wafer Surfaces and Its Effect on Particulate Deposition. *Microcontamination* 4(11):55–61.

Burkman, D. C., et al., 1988. Understanding and Specifying the Sources and Effects of Surface Contamination in Semiconductor Processing. *Microcontamination* 6(11):57–62, 107, 110.

Cooper, D. W., et al., 1988. Deposition of Submicron Aerosol Particles during Integrated Circuit Manufacturing: Theory. Proceedings of the 9th International Committee of Contamination Control Societies Conference, pp. 19–26, September 26, 1988, Los Angeles.

Frey, A. H., 1986. Reduction of Formaldehyde, Ammonia, SO_2, and CO_2 Concentrations in Room Air. *Journal of Environmental Science* 29(4):57–59.

Harper, J., & Bailey, L., 1984. Flexible Material Handling Automation in Wafer Fabrication. *Solid State Technology* 27(7):89–98.

Librizzi, J., and Manna, R. R., 1983. Controlling Air Pollution from Semiconductor Fabrication Operations. *Microelectronics Manufacturing and Testing* 6(12):46–48.

Liu, B. Y. H., et al., 1987. Aerosol Charging and Neutralization and Electrostatic Discharge in Clean Rooms. *Journal of Environmental Science* 30(2):42–46.

McMahon, R., 1989. Monitoring Hazardous High-Purity Gases Used in Semiconductor Fabrication. *Microelectronics Manufacturing and Testing* 12(4):8–10.

Ohmi, T., Inaba, H., & Takenama, T., 1989. Preventing Electromagnetic Interference Essential for ULSI E-Beam Performance. *Microcontamination* 7(11):29–35.

Turner, T., 1983. Static in Wafer Fabrication Facility: Causes and Solutions. *Semiconductor International* 6(8):122–126.

Whyte, W., 1968. Bacteriological Aspects of Air-Conditioning Plants. *Journal of Hygiene* 66:567.

Whyte, W., 1986. Sterility Assurance and Models for Assessing Airborne Bacterial Contamination. *Journal of Parenteral Science and Technology* 40(5):188–197.

5

Contaminant Sources

Contaminant sources include almost every component in the manufacturing process: people, materials, processing equipment, and manufacturing environments. People can generate contaminating particles, gases, condensible vapors, and viable materials. People can also generate electrical charges and transport charged surfaces to critical areas. In most cleanroom operations, personnel interaction with the product is universal. However, the problems of maintaining uniform processing operations have resulted in the introduction of many robots for repetitive parts-handling steps in critical areas. The materials brought into the manufacturing process area can contribute particles, gases and vapors, and/or electrostatic charges. Even careful cleaning procedures have not completely solved this problem. Process equipment and tools used in production can also produce contaminating particles, gases and vapors, electrostatic charge, or vibration. The manufacturing environment can be a source of all of the contaminants that may cause problems. As product requirements change, the contamination control needs for those products also change. The importance of the several sources of contamination changes in order of concern as control techniques change in response to product changes.

Figure 5-1 shows a recent estimate of relative levels of contamination hazards from various contaminant sources in a 1990-period semiconductor production cleanroom operating at a high cleanliness level. Note that the major particulate contaminant sources in this very clean area are primarily due to equipment and process operations. Only a few years earlier, the major contaminant sources in any cleanroom were personnel and the environment. Better cleanroom procedures for isolation and protection of products, including control of air handling and filtration systems as well as

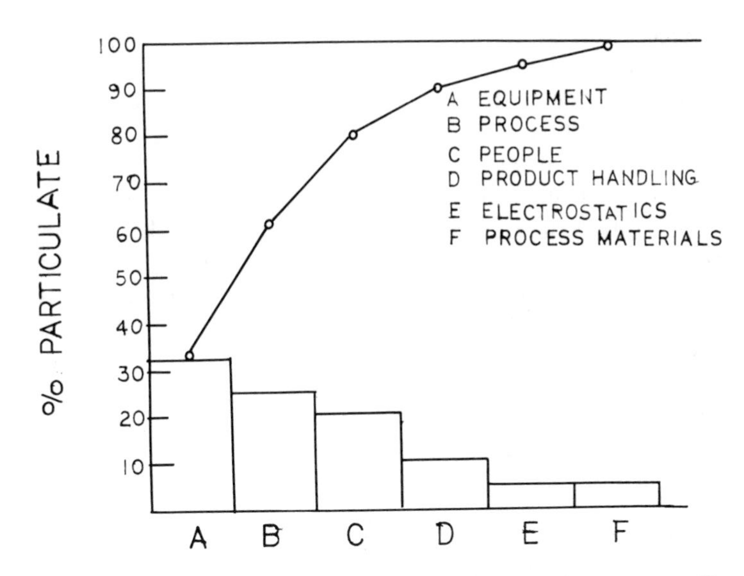

FIGURE 5-1. Semiconductor fabrication area particle contaminant sources. The percentages are typical of what one might expect in a good-quality electronic manufacturing area in the 1990 period. The relative levels change as designs and production skills are modified. (Courtesy A. Gutacker, ARGOT, Inc.)

requirements for improved personnel protective clothing, have reduced this problem. A very important means of reducing contamination has also been improvements in work procedures. Introduction of robot systems designed to minimize contaminant release during operation for in-process parts handling, better gowning, and personnel enclosure devices have reduced the severity of this problem. Even so, personnel problems are still a major concern in many areas. The contaminant source distribution illustrated in Figure 5-1 represents most of the cleanroom problem areas in 1988. It shows the importance of continual control of all contamination sources. In different product areas, the problem area importance varies. For example, as of 1989, it has been reported that particles from process and material sources cause 75–90% of the defects reducing yield in semiconductor production areas (Miller, Baker, and Ahn 1989). Similar work has shown that sources of contamination in the cleanroom contribute to product failures more than almost any other sources. Product-handling and processing equipment is a primary problem area and will continue to be a source of problems; even while present systems are improved, product requirements are becoming more severe (Hattori 1990). As time goes on, the overall contamination level should decrease. We can only hope that tolerance levels of new products will not decrease as fast as we can improve cleanliness levels.

Procedures for controlling contamination levels in cleanrooms today are based on identification of the contaminant material and empirical knowledge of the possible sources of specific contaminants. A similar problem exists in the air pollution control area on a larger scale. Consider that a city covering many square miles may contain a large number of potential sources of air pollution and that effective control of health hazard emissions may be accomplished by reducing specific source emissions. The problem is to identify and verify the location and effects of specific sources and to control their emissions. The analogy to many cleanroom contamination problems is obvious. Studies to identify specific source locations have been carried on for many years. A variation of chemical engineering practice for analyzing performance of inaccessible systems has been applied to cleanroom contamination source identification (Kern and Kunesh 1990). Accumulation of contaminant level is reduced to definition of inlet and outlet airflow, along with identification of inlet and outlet air cleanliness classes. The method aids in locating problem sources and showing the value of component isolation. In a similar manner, multivariate techniques, particularly principal component analysis, are applied to pollutant data obtained from measurements at many pollutant target locations to identify the sources to be controlled (Henry and Hidy 1982). Information about pollutant sources is obtained from analysis of pollution components, atmospheric conditions, aerosol composition, and spatial pollutant distributions. Application of the technique to defining particle source in cleanrooms with very little change has been shown to be feasible (Tian et al. 1983). Principal component analysis allowed positive definition of major sources in the room, thus allowing rapid specification of remedial measures.

People produce contamination in cleanrooms just by existing. In addition to their normal bodily excretions, their activity in any cleanroom is a continuous source of particulate and vapor contamination. They generate skin flakes that can deposit on products; they perspire, resulting in both water vapor and residual salt particle generation; they exhale quantities of organic vapors; both continuous and cyclic body functions result in various emissions. They transport contamination on their skin and within their body cavities. Their clothing may generate fiber fragments or emit debris that was previously collected. In normal activity, they move air into areas where it should not be moved. Data obtained for personnel in normal street clothing (Austin and Timmerman 1965) show high emission rates of inert particles $> 0.3\ \mu m$. Activity effects are shown in Tables 5-1 and 5-2.

TABLE 5-1 Inert Particle Emissions from Various Activities

EMISSION RATE	ACTIVITY
100,000/min	Seated, standing, motionless
500,000/min	Hand, forearm & head motion
1,000,000/min	Minor body motion (turn, etc.)
2,500,000/min	Major body motion (stand up)
10,000,000/min	Walking normally

In addition to contaminant emissions from activity in the clean area, people's external activities can also carry over contamination into the cleanroom environment. Tobacco smoke has been detected in exhalations hours after smoking; vapors from some foods persist in the breath long after ingestion; medication traces have been found in skin and hair fragments days after the medication has been used; cosmetics may not be completely removed from skin and hair during washing and gowning. As can be seen, the need for personnel to be well gowned and well trained in cleanroom procedures is vital. The need for good training and motivation is essential and will become more and more important as time goes on. The requirement for isolation of products from personnel emissions may require that all personnel are completely enclosed within protective devices that may be unpleasant for many susceptible people just to wear, let alone to work in.

A particular problem with personnel is the possible emission of droplets when they are coughing, yawning, sneezing, speaking, or shouting (Guichard and Magne 1984). These droplets can leave a residue after evaporation that can contain viable bacteria. A cough can produce more than 600,000 droplets $> 0.5\ \mu m$; a sneeze can produce more than twice that number. Normal speech produces relatively few droplets; a shout can produce up to 20,000 droplets. Although the mean size of such droplets is usually in the range of 2–5 μm, some droplets as large as 50–60 μm are found. Another problem with personnel is emission of skin and hair fragments. The skin continuously sheds particles in the form of dead cells and cell fragments (Moore 1985). The cells are rounded flakes, 20–40 μm in diameter and 2–4 μm thick. Tests have shown the extent of viable bacteria dispersion by overall body emissions (Tu and Cheng 1988); normal activi-

TABLE 5-2 Personnel Activity Emissions at Various Production Activities

ACTIVITY	PARTICLE SIZE MICROMETERS
RUBBING ORDINARY PAINTED SURFACE	90
SLIDING METAL SURFACES (UNLUBRICATED)	50 to 150
CRUMPLING OR FOLDING PAPER	60
RUBBING AN EPOXY PAINTED SURFACE	30 to 75
SEATING AND UNSEATING SCREWS	25 to 120
BELT DRIVE	5 to 35
WRITING WITH BALLPOINT PEN ON ORDINARY PAPER	15 to 30
HANDLING PASSIVATED METALS, SUCH AS FASTENING MATERIALS	10 to 20
VINYL FITTING ABRADED BY A WRENCH	5 to 60
RUBBING SKIN	5 to 300
CHIPS FROM WRENCH SEAT OF BROACHED ALLEN HEAD SCREWS	400

ties could release several hundred colony-forming units per minute per person, even when clean clothing is worn. The emission rate increased with activity, indicating that a combination of higher breathing rates and bodily movements generated greater bacteria emission rates.

Another major contaminant from personnel is residual cosmetic particles. These materials are a special problem in semiconductor manufacturing areas. Cosmetics are produced by using a variety of metal oxides and other compounds for color and texture. Commonly used cosmetics have been shown (Phillips et al. 1983; Lowry et al. 1987) to contain such elements as aluminum, barium, calcium, magnesium, iron, and sodium. Each application of cosmetics was shown to result in the transfer of up to 10^9 particles in sizes ranging from 0.5 μm upwards, both by air passage over the skin and by direct contact. Uncontrolled deposition of these materials on in-process semiconductor wafers can be disastrous to the quality of the final product. Emissions from clothing worn by personnel under cleanroom garb is still a problem in many areas where there is inadequate control over gowning materials and change procedures. This subject is discussed later when cleanroom fabrics are considered.

Many of these problems mentioned can be controlled by proper personnel gowning to protect the cleanroom environment. This means selecting the correct garments for proper personnel enclosure, as well as making sure that the garments are worn and handled correctly. Proper personnel training is even more important so that personnel emissions are kept under some control at all times.

The worst problems with personnel emissions result from improper work procedures for the cleanroom, including poor gowning procedures and improper personal activity such as scratching, moving the hands over the body, and walking too fast or with exaggerated arm movements near critical assembly areas. Incorrect tool and work item handling is also a contaminant-producing activity. These problems are mainly caused by carelessness and/or inadequate training and motivation for cleanroom work.

Figure 5-2 shows some measurements of operating cleanroom particle counts at various times of the day. Notice how the level increases at times when personnel activity is at a maximum. Even though many of the particles during work periods come from normal equipment operation, the effect of personnel activity is seen in the large counts while people are entering or leaving the room. Particle count in the quiescent room is quite low, as expected. The surge of contamination when the shift starts and

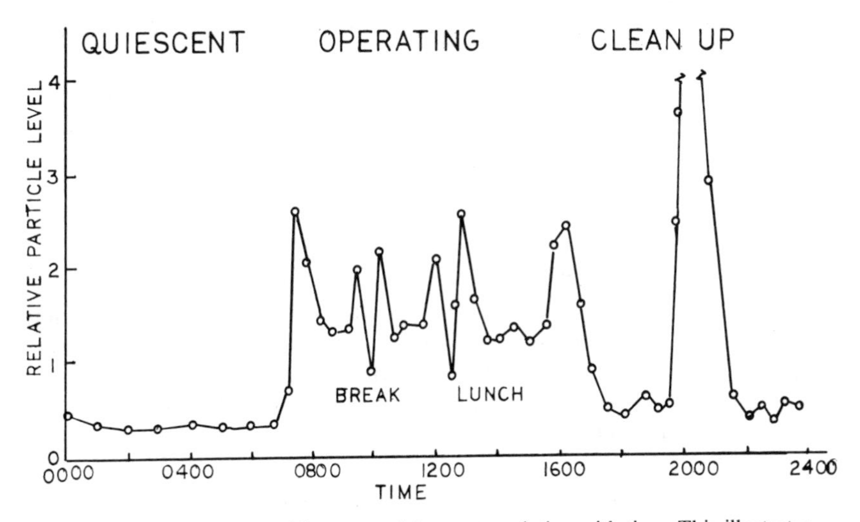

FIGURE 5-2. Cleanroom airborne particle count variation with time. This illustrates the relationships between personnel activity level and contamination generation rate. There is usually a decrease in particle generation just prior to leaving an activity. The enormous increase in particle level during cleaning shows how many particles deposited during work activity can become reentrained by cleaning operations.

ends results from the concentrated personnel movement into and out of the room. It is of interest to note that the particle count decreases just before the break and lunch periods. This change shows that personnel tend to prepare for departure from the work area by relaxing their activity level. This effect is not seen at the shift end. The reason for the increase in contaminant level at that time is ascribed to an increase in activity while personnel are covering work in progress and storing tools and components before departing the clean area for the day. Note how the room cleans itself immediately after the end of shift and remains clean until the cleaning personnel enter the room. Because their work patterns require vigorous activity, the particle count in the air reflects that activity.

Another important contaminant source is the materials used in all aspects of manufacture. These materials include all of the chemical and mechanical components used in the manufacturing process and the packaging for the final product. Process chemicals are an important contaminant source. These materials are in intimate contact with the product during most phases of manufacture, and any contaminant material can degrade the product quality. Many large-area semiconductor devices are being manufactured by "dry" processing methods, in which reactive gases are used to deposit materials on the wafer instead of using reactive liquid deposition sources. If the gases are not free of contaminants, then the deposited compounds are intimately contaminated by the unwanted materials that may be in the process gases themselves. Even though semiconductor-grade specialty gases are available with very low impurity content, storage and delivery systems for these materials may affect gas purity. Gaseous contamination levels well below 0.1 ppm are common, whereas particulate contamination limits for 0.02–0.1 μm particles of no more than 10 per standard cubic foot may be required. Handling and control systems for these process gases must be designed to prevent addition of contaminants by system components (Hardy, Christman, and Shay 1988). Pressure regulators and valves have been shown to be a major source of reported particle contamination from compressed gas cylinders. In addition, the cylinder internal surfaces must be treated to minimize particle emission. Even with these procedures, final filtration before use is still recommended. It has been shown (Kasper and Wen 1988) that particles can be entrained from contaminated surfaces in gas-handling components at ambient temperatures. If some surfaces in the system are heated, then contaminants can evaporate from the surface and recondense to form submicrometer particles.

A detailed survey of contamination sources in semiconductor processing (Burkman et al. 1988) shows how essentially every material used for processing and even for cleaning surfaces in semiconductor manufacture can deposit contaminants on the wafer surface. Both the materials used in the processes and the equipment used in applying these materials can cause problems. Specific chemical contaminants as well as miscellaneous particles and vapors can be deposited on the wafers by almost all of the materials used in their manufacture.

In pharmaceutical manufacturing, gases are seldom part of the process as such, but container cleaning and drying operations before filling often require the use of clean gases in sterilizing and drying tunnels to ensure package sterility. Integrity of container cleanliness, filling, sterilization, and sealing has always been a concern in pharmaceutical processing (Khan 1989). After the final rinse of containers for large-volume parenteral liquids, containers are dried with hot, sterile air. Even though the drying tunnels used for this purpose employ air at temperatures above 125° C, the presence of any particulate material in the drying air can result in deposits of materials that may be nutrient to any airborne bacteria in the cooling or storing areas before fill lines.

Acid and solvent liquids used in essentially all semiconductor manufacturing operations in the 1980–1990 time period show extremely high particulate levels. These particles can deposit and be retained on critical surfaces and can interfere with the manufacturing process for which they are used. Data collected on particle levels in a variety of process liquids show concentration levels as high as 10,000/ml for particles $\geq 0.5\ \mu m$ (Dillenbeck 1984). Significant variations in count data were seen between products from different vendors. Major contamination levels were seen in sulfuric acids and in photoresist strippers. As bad as these data appear, improvements in process chemical cleanliness are noted. Recent data (Grant and Min 1991) show that optimized bulk chemical delivery systems can deliver process chemicals with concentrations significantly less than 10,000/ml for particles $\geq 0.2\ \mu m$.

Many liquids used for processing or for cleaning have significant vapor pressures. Evaporation during use of these materials can add organic vapors to the processing environment. These vapors are not often a problem at the point of use, but if they are transferred to other manufacturing areas they may contaminate components that should not be exposed to process material that has vaporized elsewhere. In addition to controlling contamination within the manufacturing operation, present-day Environmental Protection Agency (EPA) regulations do not allow release of many processing material vapors to the atmosphere. Process liquid vapor release must be controlled. This problem is as serious in manufacture of pharmaceutical materials as it is in electronics manufacturing. Adsorption of unwanted organic vapors on powders to be used in tableting can degrade tablet physical properties as well as appearance. Emissions from processing equipment in the cleanroom must be carefully controlled and contained in all process areas.

Large quantities of high-purity water are used for cleaning semiconductor devices during the manufacturing process. It has been estimated that each wafer requires exposure to as much as 1,000 gallons of water during semiconductor processing. The manufacturers' concern with this material is the content of total oxidizable carbon (TOC), colloidal silica, some dissolved ions, bacteria, and particles. TOC level is related to the dissolved organic material in the water. Colloidal silica and dissolved ion content depend on the raw water source. Bacteria and particle content depend on the efficacy of the internal plant water-treatment system. Deionized water and process water used in pharmaceutical manufacturing may contain excessive amounts of both dissolved and ionic contaminants. Modern water supply systems with ion exchange, membrane processing systems, and high-efficiency filters are very effective in removing essentially all types of contamination from process water systems. Care is required to make sure that no bacterial contaminants grow in the clean water systems. Control methods are discussed later.

The manufacturing process can often result in contamination of the product. It involves structural modification by changing product or subassembly shape, composition, location, or energy content. Product elements are exposed to handling, cutting, grinding, coating, removal, or deposition of material. This is common in making a mechanical or electronic device, such as a precision ball bearing assembly or a semiconductor chip. The mechanical or pneumatic components used to handle delicate subassemblies may deposit contaminants on the clean surface or mar the surface. Contaminants can be produced by sliding or rolling surfaces in mechanical handling systems or by lubricant release from hydraulic or pneumatic systems. Whenever mechanical components are machined or cut in any way, particles are generated by the cutting tool action as well as by operation of the system used to move the tool or the part that is being machined. Large particles can be removed easily, but removal of particles smaller than 5–10 μm requires major cleaning efforts.

Semiconductor manufacturing is a major area where control of contamination has been shown to be a necessity for satisfactory yield in production. Many repetitive operations may be required in wafer processing. Some typical procedures and times for completion are

1. Lithography, with up to 11 repetitions at an hour per exposure
2. Etching, with up to 17 repetitions at an hour per etch
3. Ion implantation, with up to six repetitions at 45 minutes per operation
4. Heating steps, with up to 36 repetitions at 10 minutes each
5. Cleanup, wet and dry, with up to 40 repetitions at 10 minutes each

Thus, a wafer may be exposed to the air in the processing environment of the cleanroom for 40 hours or more during processing. In addition, days of storage can occur between process steps. If a 125-mm wafer is exposed to class 100 air for 40 hours, deposition of some 60 0.5-μm particles can be expected during that time. The need for effective controls is obvious.

Many semiconductor manufacturing processes involve mechanical handling and chemical processing of components for precise material addition and removal. Without going into great detail on the processes, consider some of the steps involved in wafer processing. New clean silicon wafers are delivered in clean packaging material. If the wafers are stored before use for very long periods, hydrocarbon contamination from outgassing of some packaging material or as a result of inadequate air cleaning may generate particles > 0.3 μm in diameter as a result of gas phase reaction or condensation effects (Fishkin and Baker 1988). The silicon wafers are introduced into equipment that should be clean. Some contaminants are transferred with and by the wafers. Wafer and wafer cassette handling in the equipment can result in abrasion of various surfaces to produce additional particles.

Surface processing of wafers is frequently carried out in low-pressure systems. The pumping operation to reduce pressure and the repressurization process after the operation is complete can also result in deposition of contaminant particles on the in-process wafer surface. Specially designed pumping systems may be required, with mechanical components isolated from the vacuum chamber and with pumping fluids, where used, selected for extremely low vapor pressure characteristics. In this way, neither wear particles nor potential vapor contaminants are allowed into critical areas. In addition, specially designed valves prevent rapid pressure changes in the chamber that might entrain previously deposited contamination materials.

If mechanical handling systems are used, then the equipment manufacturer must design all components, especially motors and bearings used in the mechanical components, for minimum particle generation. It has been shown that brushless stepping motors and spur gears can generate as many as 40,000 particles ≥ 0.1 μm per minute (Matsuo and Fukao 1988). The generation of large numbers of very small particles suggests that wear is a major mechanism in generation of particles during operation of mechanical handling systems (Nagaraj 1988). The processing materials and operations can produce additional contaminants both into the equipment and within critical process areas. Measurements have been made showing particle concentrations of several thousand per cubic foot of air existing within clean process equipment before needed remedial action takes place (Dillenbeck 1988).

Photographic film manufacture involves chemical processing, surface coating, drying, cutting, and packaging. Pharmaceutical manufacture may involve chemical processing, plastic molding, powder pressing, or liquid transfer. In all cases, the possibility of contaminant generation from any element of the manufacturing process exists and must be controlled. The manufacturing equipment is being examined as a possible source of contaminants that must be controlled. A few examples will show the importance of this concern. Some older clean benches were made of plastic with silica fill that emits both plastic and filler particles when it is abraded or stressed. This material is no longer being used for clean operations. Parts transport on belts is being examined carefully to make sure that emissions of particles and vapors from pulleys, bearings, and motors does not occur. Electronic test equipment used in cleanrooms is carefully cleaned before being brought into the room because any particles entrapped within the cabinet can be emitted into the cleanroom air when the electronic device cooling fan is operating. As an alternate, low-power devices that run cooler are being used. In areas where devices are being manufactured that may be very sensitive to contamination, robot handling devices are replacing personnel. Testing (Hardegan and Lane 1985) has shown that robotic devices generate less contamination than personnel carrying out the same repetitive parts-handling operations. Contaminant emission from these robotic systems is still examined very carefully before any new robot is used. Operation with robotic systems can be cleaner than personnel operations, but some emissions still occur and must be controlled.

The manufacturing environment was the first area where effective and meaningful contamination control efforts were applied. The knowledge obtained from many years of cleanroom operating experience can result in that environment being the smallest problem area for contamination. Both the materials of construction and manufacturing environment design used in modern cleanrooms have been refined to minimize contamination from this source. Good design, construction, and maintenance of needed control procedures will minimize contamination problems from the manufacturing environment.

Some cleanrooms have been installed in older buildings. Contamination problems are seldom considered in selecting the basic structure of such buildings. Many older building structures contain materials that will emit contaminants indefinitely. Cement, cinder block, or plaster wall material that has not been sealed well emits particulate material. Some painted surfaces can emit both particulate and vapor contaminants. Many older buildings contain vast quantities of construction debris within the walls, ceiling, and subfloor elements. This material can penetrate unsealed utility access panels and enter work spaces.

Care is also required when a new building is constructed for a new production facility. Tests have shown that concentrations of some volatile organic vapors, such as aromatic and aliphatic hydrocarbons, can be as much as 5–6 times higher in new buildings than in older buildings (Ember 1988). Sources of these chemicals are most common building materials, including latex caulking, telephone cable materials, and particleboard partition materials. Although the EPA study that was reported is directed toward health effects, the part-per-million levels can affect many sensitive products even before health problems occur.

Older HVAC systems may contain duct materials that emit particles, especially when heaters are used. Fiber insulation must be isolated from the clean areas. If the cleanroom is located in an area where temperature and/or humidity extremes may occur, then these materials may corrode when the HVAC system operates so that moisture condenses during a cooling operation. Even though final filtration can remove the corrosion-generated particles, the load on the HVAC system prefilters is not desirable. Most metal ducts are acceptable if final or point-of-use filtration is used. Otherwise, maximum cleanliness may dictate the use of more costly duct materials than galvanized iron or aluminum.

These problems are considered in design and construction of new cleanroom facilities. The building is designed so that the clean areas can be isolated and protected from any potential contamination source within the building structure; and the HVAC system is designed to ensure contaminant-free air supply as required. Particles in the environment are kept under control by the HVAC system, the HEPA and ULPA filter systems, and by product enclosure systems. Air and surface cleanliness monitoring verifies integrity of these systems. Otherwise, atmospheric particulate and gaseous contaminants would interfere with product quality. Figure 5-3 shows a typical particle size distribution seen in urban areas. Normal atmospheric air can have several million particles per cubic foot, whereas the cleanroom air concentrations may be 4 or 5 orders of magnitude less. The filters remove essentially all particles larger than 2–3 μm and smaller than 0.1 μm. Table 5-3 shows some particle concentration before and after HEPA filtration (Rivers 1984). Figure 5-4 shows particle size distributions seen in some unidirectional-flow cleanrooms with little or no activity to generate new particles in the room (Ensor, Donovan, and Locke 1987). Note that there are essentially no particles smaller than 0.1 μm present in the air in such a cleanroom. As the air is recirculated through the filter system, both smaller and larger particles are removed preferentially by the filters. The particles that are seen represent material that is generated within the room by local activity. The HVAC system can control air temperature and relative humidity, but local source generation of contaminants in the cleanroom must be controlled.

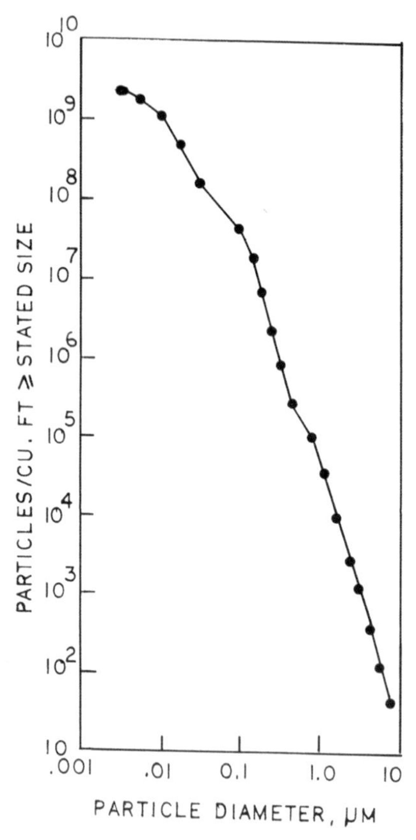

FIGURE 5-3. Typical urban aerosol particle size distribution. A variety of instrumental measurement methods were used to procure data over the wide particle size range shown.

The composition of particles in the cleanroom depends primarily on the nature of the activity in the particular room. The composition of ambient air particles varies somewhat with location, but the typical pollutant particles in most areas are generated by similar sources, and the composition remains fairly constant unless great changes in atmospheric conditions occur. Those materials shown in Table 2-1 are found in essentially all urban industrialized areas. In addition, metal and metal oxide particles are seen in heavily industrialized areas. Rural area particles are quite similar in content, with varying quantities of sea salt particles depending on the closeness to a maritime area of the point of measurement. Although the HVAC and filter systems designed for use in a cleanroom are quite effective in keeping the air cleanliness under control, the external atmospheric air cleanliness level affects the cleanroom air cleanliness (Lieberman 1971;

TABLE 5-3 Ambient Air Particle Counts Before and After Filtration with High-Efficiency Filter

Diam, μm	Upstream Air Particles per cu. ft.	After Filter Particles per cu. ft.
.010 - .015	130556714	3.518
.015 - .020	128698927	3.509
.020 - .030	127579693	3.498
.030 - .040	126162918	3.479
.040 - .050	125054886	3.469
.050 - .070	124106715	3.464
.070 - .10	122443811	3.458
.10 - .15	118645259	3.341
.15 - .20	96640822	2.317
.20 - .30	51903176	0.812
.30 - .40	12588190	0.068
.40 - .50	4096184	0.009
.50 - .70	1608079	0.002
.70 - 1.0	443698	0.001
1.5 - 2.0	33897	0.000
2.0 - 3.0	19666	0.000
3.0 - 4.0	8699	0.000
4.0 - 5.0	3976	0.000
5.0 - 7.0	1802	0.000
7.0 - 10	342	0.000

* Even though no particles larger than 1.5 μm are shown in this table after filtration, the actual number is more than zero but less than 0.001 per cubic foot. The concentration is simply too low to be measured in a reasonable time.

Source: (Courtesy R. Rivers, American Air Filter Corporation.)

Sinclair, Psota-Kelty, and Wechsler 1985; Viner and Donovan 1988). Even so, the major contributor to contamination in the cleanroom is the activity within that cleanroom. Contaminating vapors and gases in the cleanroom are usually produced by manufacturing processes. In addition to preventing penetration of external contaminant gases and vapors into the cleanroom, any high-vapor-pressure materials usedin the cleanroom facility must be stored in secure areas. While these materials are used, any vapor emissions must be controlled by containment, scrubbing, or adsorption.

In some cases, a nearby industrial operation can generate excessive quantities of these materials that enter the cleanroom air through the makeup air supply ducts. If the operator is not aware that contaminant gas

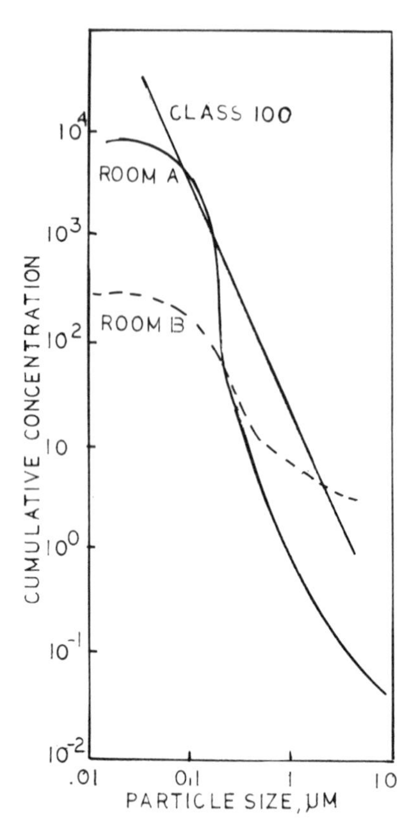

FIGURE 5-4. Quiescent cleanroom particle size distribution. The straight line is the extrapolated particle concentration and size distribution that would be expected from the tabulation of Federal Standard 209D. (From Ensor, Donovan, and Locke, Particle Size Distributions in Clean Rooms, *Journal of Environmental Sciences* 29[6]:44–49.)

may be present in the makeup air, then product quality can be degraded. A few examples are given next; note that even though gas emissions from such industrial facilities can cause problems in a cleanroom manufacturing operation, the emission levels may be well within any environmental protection agency limits. An upwind fossil fuel power plant burning high-sulfur coal with an inadequate gas scrubber can cause excessive sulfur oxides to enter the cleanroom HVAC system. A nearby petrochemical plant can emit a variety of organic vapors that can enter the cleanroom. If external contaminant vapor sources exist and cause product problems in the cleanroom, then the HVAC system must be modified to include means of removing the problem gases. Depending on weather conditions, even remote pollutant sources can result in challenges to the HVAC system. Windborne transport of particulate materials in size ranges around 0.1 μm over distances up to 120 km have been recorded (Reible, Ouimette, and Shair 1982). Observations of both prevalent and intermittent wind patterns can help in selecting remedial measures for externally generated pollutants (Rheingrover and Gordon 1988).

In most electronic, mechanical, and optical system manufacturing, viable airborne particles are no more or no less a problem than any other particles in the air. Particles can always deposit on products being handled, and unless the product substrate upon which the viable particles deposit is a nutrient material, viable particles are seldom a problem worse than any other particles. However, if the deposition location is in an area where the bacterial materials can find nutrients and are not disturbed for some period of time, then the viable particles multiply, and clumps of bacteria may be emitted for long periods of time. Some thought has been given to the possibility that HEPA filters may be a breeding area for bacteria, especially if airflow may be intermittent.

Contamination from viable particles in the cleanroom air is a special problem mainly in food and pharmaceutical manufacturing. In these areas, the deposited particles have the capability of growing on and in the product; the bacterial particles can spread throughout liquids and powder products very quickly. If any viable particles deposit on cleanroom floors or walls or on cleanroom manufacturing components, they must be sterilized before significant growth occurs.

Although airborne viable particles are not as serious a problem in mechanical and electronic manufacturing as they are in foods and pharmaceutical manufacturing, the presence of viable particles in liquid-handling components is a problem of equal severity in any cleanroom manufacturing situation. Bacteria can multiply easily in any environment where the proper temperature, adequate moisture, and nutrient material are present. Such an environment is present in any deionized water system in a location where little or no flow exists for extended times. When flow begins, bacteria colonies can break loose and deposit on any convenient surfaces in the flow lines, such as fittings, valves, filter media, and—eventually—on the product. Good deionized water system designs avoid "dead leg" lines where zero flow can exist.

Low-level radioactivity exists throughout the environment. Radon is a common material in all air. It emits alpha particles and has a short half-life, producing solid daughter products. Radon gas is a decay product of radium, present throughout soil and releasing radon to the atmosphere. It is estimated that radon is present in all air at a level of one part radon in sextillion parts of air; this quantity of alpha radiation is not a problem.

Radioactivity can be a problem in some cleanroom operations. The problem is exacerbated for products such as photographic films. Some testing instruments use high-energy radiation. Electron beam devices and x-ray analyzers are the types of instruments of concern. Unless damage has occurred to the instrument, radiation leakage is seldom a problem.

Vibration is a contaminant in areas where precise component alignment is needed. This situation can exist in optical fabrication facilities, in metrology laboratories, and in modern semiconductor photolithography areas. Vibration can cause misalignment of components even in rigid massive structures. Vibration occurs as a result of stress on building structures, roadways, and the like from the movement of heavy bodies. In addition, tectonic earth movements occur in many parts of the world. High winds can cause movement of some buildings as well. Heavy traffic on nearby roads can produce vibrations, transmitted through the ground to a building structure. Within the building itself, unless the cleanroom is built so that vibration is isolated or absorbed, even movement of carts carrying parts in a nearby hall can cause vibration problems in some situations.

After a product has been manufactured and has been tested and found to operate satisfactorily, immediate use seldom occurs. The product is packaged, stored, shipped, and installed before it can be used. No matter how well a product is manufactured, these operations are capable of destroying a badly packaged product through a number of mechanisms. The package used to contain the manufactured product must protect that product during all subsequent operations. Packaging material must be chosen that will not allow any contaminants to penetrate to the product. If the product is sensitive to moisture, the packaging material should not allow water vapor diffusion; in some cases, a suitable desiccant may be required. The package should be designed so that it protects the product against mechanical damage, especially during shipping; protective padding should be selected and built so that particulate contaminants are not released to the interior of the package from the padding when it is mechanically stressed. The package should be designed so that access to the product can be accomplished without contaminating the product during the process of opening the package. Low-strength adhesive coatings and bubble packs are very useful for this purpose.

References

Austin, P. R., & Timmerman, S. W., 1965. *Design and Operation of Clean Rooms*, p. 77, Detroit: Business News Publishing Co.

Burkman, D. C., et al., 1988. Understanding and Specifying the Sources and Effects of Surface Contamination in Semiconductor Processing. *Microcontamination* 6(11):57–64.

Dillenbeck, K., 1984. Characterization of Particle Levels in Incoming Chemicals. *Microcontamination* 2(6):57–62.
Dillenbeck, K., 1988. Application of Contamination Engineering Techniques for Solving Particle Problems in Process Equipment. Proceedings of the 34th Institute of Environmental Science Annual Technical Meeting, May 1988, King of Prussia, PA.
Ember, L. R., 1988. Survey Finds High Indoor Levels of Volatile Organic Chemicals. *Chemical and Engineering News* 66(49):23–25.
Ensor, D. S., Donovan, R. P., & Locke, B. R., 1987. Particle Size Distributions in Clean Rooms. *Journal of Environmental Science* 29(6):44–49.
Fishkin, B., & Baker, E. J., 1988. Particle Performance of CVD and Epitaxial Processes and Equipment. Proceedings of the 34th Institute of Environmental Science Annual Technical Meeting, May 1988, King of Prussia, PA.
Grant, D. C., & Min, D. P., 1991. The Effect of Operational Mode on Particle Concentrations in Bulk Chemical Delivery Systems. Proceedings of the 37th Institute of Environmental Science Annual Technical Meeting, May 1991, San Diego.
Guichard, J. C., & Magne, J. L., 1984. Contamination by Flugge's Droplets. Proceedings of the 7th International Committee of Contamination Control Societies Conference, September 18–21, Paris.
Hardegan, B., & Lane, A. P., 1985. Testing Particle Generation by a Wafer Handling Robot. *Solid State Technology* 28(3):189–195.
Hardy, T. K., Christman, D. B., & Shay, R. H., 1988. Measurement and Control of Particle Contamination in High Purity Cylinder Gases. Proceedings of the 9th International Committee of Contamination Control Societies Conference, pp. 294–300, September 26, 1988, Los Angeles.
Hattori, T., 1990. Contamination Control: Problems and Prospects. *Solid State Technology* 33(7):S1–S8.
Henry, R. C., & Hidy, G. M., 1982. Multivariate Analysis of Particulate Sulfate and Other Air Quality Variables by Principal Components. *Atmospheric Environment* 16:929–943.
Kasper, G., & Wen, H. Y., 1988. On-Line Identification of Particle Sources in Process Gases. Proceedings of the 34th Institute of Environmental Sciences Annual Technical Meeting, May 1988, King of Prussia, PA.
Kern, F. W., & Kunesh, R. F., 1990. A Methodology for Determining Particle Origins in Semiconductor Manufacturing Cleanrooms. Proceedings of the 36th Institute of Environmental Science Annual Technical Meeting, May 1990, New Orleans.
Khan, S., 1989. Automatic Flexible Aseptic Filling and Freeze-Drying of Parenteral Drugs. *Pharmaceutical Technology* 13(10):24–34.
Lieberman, A., 1971. Comparison of Continuous Measurement of Interior and Exterior Aerosol Levels. Proceedings American Industrial Hygiene Association Annual Meeting, May 24, 1971, Toronto.
Lowry, R. K., et al., 1987. Analysis of Human Contaminants Pinpoints Sources of IC Defects. *Semiconductor International* 10(7):73–77.
Matsuo, T., & Fukao, H., 1988. Experiment and Analysis on Particle Generation

of Moving Mechanisms. Proceedings of the 9th International Committee of Contamination Control Societies Conference, pp. 494–498, September 26, 1988, Los Angeles.

Miller, S. J., Baker, E. J., & Ahn, K. H., 1989. Integrating Techniques for Root Cause Analysis of Particulate Sources. *Microcontamination* 7(10):35–39.

Moore, E. W., 1985. Contamination of Technological Components by Human Dust. *Microcontamination* 3(9):65–71.

Nagaraj, H. S., 1988. Wear as a Significant Mechanism of Particle Generation in Devices. Proceedings of the 9th International Committee of Contamination Control Societies Conference, pp. 87–91, September 26, 1988, Los Angeles.

Phillips, Q. T., et al., 1983. Cosmetics in Clean Rooms. *Journal of Environmental Science* 26(6):27–31.

Reible, D. D., Ouimette, J. R., & Shair, F. H., 1982. Atmospheric Transport of Visibility Degrading Pollutants into the California Mohave Desert. *Atmospheric Environment* 16(3):599–613.

Rheingrover, S. W., & Gordon, G. E., 1988. Wind-Trajectory Method for Determining Compositions of Particles from Major Air Pollution Sources. *Aerosol Science and Technology* 8(1):29–41.

Rivers, R. D., 1984. Private communication.

Sinclair, J. D., Psota-Kelty, L. A., & Wechsler, C. J., 1985. Indoor/Outdoor Concentrations and Indoor Surface Accumulations of Ionic Substances. *Atmospheric Environment* 19(2):315–323.

Tian, Y., et al., 1983. Principal Component Analysis for Particulate Source Resolution in Cleanrooms. *Journal of Environmental Science* 32(6):22–27.

Tu, G., & Cheng, Q., 1988. An Experimental Study of Bacteria Dispersal from Human Body. Proceedings of the 9th International Committee of Contamination Control Conference Societies Conference, pp. 479–483, September 26, 1988, Los Angeles.

Viner, A. S., & Donovan, R. P., 1988. A Comparison of Clean Room Air Quality with Outdoor Ambient Air Quality. Proceedings of the 35th Institute of Environmental Sciences Annual Technical Meeting, May 1988, King of Prussia, PA.

6

Contaminant-Generation Mechanisms

In the last chapter, the areas where contaminants are generated were discussed. Knowing the location of contaminant generation is helpful in controlling that contamination, but understanding the mechanisms is equally important in that control process. Process materials, tools, equipment, and personnel control of operations are necessary in any production operation. Many of the contaminant sources cannot be avoided, but some control and remedial activities are possible and should be used. Knowing the contamination-generation processes and the mechanisms involved may aid in providing information to minimize the type, rate, and time of contaminant emission from specific sources. Understanding the contaminant-generation mechanisms can also help to show the nature of the specific contaminant material that is produced by a particular source. In this way, knowledge is available that may allow advantageous modification of a process or component to reduce contamination-generation rates.

Several generation mechanisms exist for particulate contaminants. These include gas phase reactions, condensation from gas, spray droplet production, comminution, surface erosion, and abrasion. Some examples may help in understanding these mechanisms: Atmospheric smog is frequently produced by oxidation of organic vapor to produce liquid droplets; gas phase chemical reaction between ammonium hydroxide and hydrochloric acid vapor produces finely divided ammonium chloride crystals. Droplets are produced by spray from a liquid or by condensation of a vapor when temperature or vapor concentration changes. Oil droplets can be emitted from rotating surfaces when excess lubricant is present; bubbles bursting at the surface of a pool of liquid can emit droplets. Comminution can occur whenever sufficient stress concentration exists in a solid body; fracture results in particle generation over a wide range of sizes. Erosion and abrasion result from frictional contact between solids. High stresses occur at the point of contact to produce wear particles that can be emitted into the nearby atmosphere or onto sensitive product surfaces that may be close to the generation point.

These mechanisms cover most of the normal particle-generation processes. The physical properties of the generated particles are related to the source properties and composition, as well as to the mechanisms producing the particles. The same production mechanisms result in droplet generation from a water vapor source or from a soldering procedure, but the liquid water droplets and the metal spheres are obviously quite different. Knowing these properties can assist in selecting control procedures. To illustrate procedures in use in other problem areas, some generation processes that produce industrial air pollutants are described here. Although they are seldom seen in a cleanroom, knowledge of the presence and source of these materials is important because the HVAC system must remove them from the air that enters the cleanroom; if the cleanroom operation is close to other industrial emission locations, then the HVAC system may have to be selected so that it can operate effectively and remove a wide variety of chemical pollutant species in the air.

Industrial smog arises from combustion aerosols, which are generated by gas phase reaction and/or vapor condensation. Most industrial air pollutant particles are the result of gas phase reactions and subsequent condensation and growth. These pollutants include metal oxide fumes from ore processing, vehicular smog particles, and many chemical manufacturing process fumes. Gas phase reaction product particles are initially very small, on the order of a few nanometers in size. They tend to agglomerate rapidly and grow to sizes of several micrometers quite quickly because their initial source concentration is usually very high. Because they are so small, they diffuse rapidly to contact other particles. Industrial air pollutants from ferrous and nonferrous refining operations are composed of metal oxide, metal salt, or organic materials; they are crystalline and can be abrasive; they are frequently electrically conductive; they are frequently found as chainlike agglomerates of single crystals; they usually have a surface film of water molecules; and some materials are sufficiently hygroscopic to absorb sufficient water from a normal atmosphere to form a solution droplet.

The gas phase reaction can also be polymerization of an organic monomer, as occurs in escape of industrial process fumes or organic vapor emissions from natural vegetation in forested rural areas. Ultraviolet radiation from the sun can catalyze polymerization of many monomers. These particles are normally not conductive and may be health hazards, depending on the source gas. Another process producing particles can be reaction of a surface with gas. Oxidation of ferrous materials or reactions with sulfur oxides are common on unprotected surfaces to form oxides (rust) or sulfides (tarnish). These materials can flake off and become airborne quite easily. This process occurs continuously on many metal surfaces exposed to the air and occurs more rapidly in moist air containing oxidizing gases. This situation exists in most urban areas and may be duplicated in the interior of HVAC systems in areas where relative humidity is normally high and in some HVAC systems using water spray for cooling. Selection of duct materials for HVAC systems must take into consideration the nature of the ambient air pollutants as well as the requirements for a specific cleanroom operation.

Condensation from a gas or vapor is a common generation process for many soot and smoke particles. Essentially all combustion product aerosols are formed by vapor condensation when the hot gas cools. These particles can be liquid or solid, depending on the condensation temperature and the type of gas or vapor. They can come from any high-temperature reaction or process that produces gas or vapor, such as fuel combustion, electrical arcs, metal ore processing, and soldering processes. In a cleanroom, vigorous personnel activity can result in generation of excessive perspiration and water vapor emission. As the vapor cools, condensation occurs upon nucleating particles, which are normally present in all atmospheres. The nuclei are ions or particles in the nanometer and larger size range. Vapor condensation in an environment where the vapor concentration is very high results in formation of droplets that can grow rapidly to 10–20 μm in diameter. Depending on the vapor source, the original droplets can be either liquid or solid when they cool to ambient temperatures. The particles produced are usually spherical in shape and very small. The particles are liquid originally and may solidify if the liquid freezing point is above ambient temperature.

Larger droplet formation results when a moving sheet or stream of liquid is generated and perturbed so as to produce finite-length filaments that condense into spheres because of liquid surface tension. This atomization process produces spherical particles that can range from submicrometer to several millimeters in diameter. The process can result from pneumatic (gas flow interaction with a liquid stream) or hydraulic (only liquid pressure) atomization. Droplet generation can also be produced by bubble formation and disruption, as occurs in sea spray generation. A similar process occurs in ultrasonic atomizer operation. It can also result from liquids thrown from surfaces: a wet tool moved rapidly, a wet spinning object of any kind, a sneeze, a stream of liquid striking a solid surface, or excess lubricant sprayed from bearing surfaces. Some droplets may contain dissolved solids that can be left behind as a small, solid crystalline or amorphous body when the droplet evaporates. If the solid residual particle is composed of a material that may be harmful to a cleanroom operation, it should be removed as soon as possible. Because the original droplet is normally much larger than the residue, removal of the original droplet is usually more easily accomplished than removal of the much smaller residue solid particle.

Webster's New Twentieth-Century Dictionary defines *comminution* as "the act of reducing to a fine powder or to small particles; . . . a gradual wearing away by the gradual breaking off of small pieces." Indeed, this process, along with erosion and abrasion, produces most solid particles found in the environment. This process operates in the normal world to produce road dusts, sand, tire rubber particles, animal and plant fibers, and other such solid particles. Comminution processes in the clean environment include such mechanisms as emission of skin fragments from people in their normal activities, generation of fiber fragments from fabrics rubbing against any solid object, emission of cosmetic particles from skin and hair during body movements, metal wear particle generation by friction during normal moving part operation or in threaded fastener operation, O-ring and seal fragment production from operation of control devices handling process chemicals, and secondary erosion from particles moving at high speed in fluid lines or from magnetic disks within unsatisfactory cleaned enclosures. These processes usually produce particles of irregular shapes. They may be flakes, cubic or rhombic crystals, or long shards, as are produced in machining operations. It has been shown that different wear mechanisms can be identified by the morphology of the wear particles they produce (Prater et al. 1989). Adhesive wear, or galling, occurs when smooth surfaces under high-contact stress slide over each other. Abrasive wear particles are frequently rounded, small particles that can agglomerate to larger clusters. Contact stress may draw some particles from one surface, which then adhere onto the other. These particles are usually flat. Agglomeration rate depends upon the size of the particles, but most stress-generated particles adhere to other flat surfaces easily.

In many cleanrooms, bipolar ion generators have been used to reduce electrostatic charge levels. Sharpened tungsten needle electrodes maintained at high voltage (but still below the corona-generation level) have been used for ion sources in many cleanrooms. In some situations, these remedies have been shown to cause particle problems (Liu, Rubow, and Pui 1985). Erosion of the needlepoints by sputtering has been reported to produce large quantities of particles in sizes from 0.02 to 5 μm in diameter; there may also be some electrochemically activated particle formation from contaminant gases in the cleanroom. This subject has been discussed in some detail since the point was first raised. Later work indicates that selection of different ionizer materials and better voltage control can almost completely eliminate this problem. It has also been suggested (Sebald, Stiehl, and Sigusch 1986) that controlled voltage pulsing and electric field structure can eliminate this particle source.

Gas and vapors are generated as exhaust products by people, by vegetation in the environment, and by high-temperature processes and are emitted from essentially all liquid surfaces and some solids. The mechanism occurs because all materials have a significant vapor pressure and therefore evaporate in a normal environment. Most materials used in cleanrooms have a very low vapor pressure; therefore, the quantity that is present in the air is small. For example, the dioctyl phthalate (DOP) liquid used for generating filter test aerosol has very low vapor pressure, boiling at 256° C. For that reason, it has long been accepted in many cleanroom environments as a challenge for filter leak testing. However, recent requirements for extremely low level vapor concentrations in some critical operations indicate that even the small quantity of DOP vapor emitted from tested HEPA filters may be excessive. For some time, there have been comments that DOP may also be a carcinogen, but the hazardous level has not been established conclusively. Emissions from tested filters are probably not a health hazard, but many operating personnel find the DOP odor distasteful. Replacement materials have been used for filter testing that are very similar to DOP but are reported as having no carcinogenic properties. However, the vapor pressure of these agents is still similar to that of DOP.

Water vapor is present in all environments, but its level must be controlled in cleanrooms because of its reactivity with many compounds. Gas and chemical material contamination is mainly a problem in optical and electronic system manufacturing. Gas can be released from many microelectronic manufacturing operations, where it is used as a process material or diluent carrier for materials used in an operation; residual waste gas may be vented after a processing step is finished. Organic cleaning solvent vapors can be released from cleaned surfaces or from open vessels used during the cleaning process. Vapor contamination can result in film deposition that can degrade quality of optical surfaces.

The generation and removal processes for gases and vapors are very difficult to control. A variety of gases and vapors are normally present in all environments, and many gases and vapors are used in manufacturing operations, particularly for electronic materials. A major difficulty arises from the fact that the gases and vapors diffuse very rapidly into the atmosphere and are seldom localized or concentrated for relatively easy removal. Where a gas or vapor must be removed in a cleanroom operation, frequently the removal process involves handling the entire airflow. If at all possible, gaseous contaminants should be removed as close as possible to the point of generation.

Electrostatic effects are a contaminant problem mainly for microelectronic device manufacturing and handling areas. The problem arises from the harmful effects of high electric fields and discharges on semiconductor properties. Electrostatic discharge can damage or destroy sensitive electronic components, particularly on devices with a very fine structure. One problem caused by electrostatic charge in any area is electrostatically augmented deposition of particulate contaminants on surfaces where they should not be found. This problem is annoying and can be controlled by bipolar ion insertion, but electrostatic discharge in semiconductor elements can be catastrophic unless it is similarly controlled.

Electrostatic charge arises from contact charging and triboelectric effects, by ion transfer mechanisms, by radioactive material decay, or by photoelectric charging. The charge may be positive or negative and can occur in either the presence or the absence of an electric field. Small particles carrying an electric charge are accelerated toward oppositely charged surfaces (or other particles) when an electric field is present. The usual fair weather electric field is about 1 V per cm, but stormy weather electric fields are much higher. These values are not effective in enclosed areas such as a cleanroom, but indicate external conditions in the open air.

Contact, triboelectric, or frictional charge is generated whenever two surfaces are separated. The charge level and polarity depend on the material composition and the friction levels during separation. If materials are separated with great force, the charge level increases with the rate of separation. The charge level is also affected by the location of the materials in the triboelectric table. In typical cleanrooms where relative humidity is 40–50%, charge dissipation from surfaces usually occurs much more slowly than in "normal" environments. Moisture film on surfaces is usually low in cleanrooms because of the low relative humidity. Charge generation and retention on surfaces can be quite high in these areas. Sliding surfaces across one another in separation can generate quite high charge levels on dielectric materials. Table 6-1 shows one version of the familiar triboelectric series, and Table 6-2 shows some voltage levels that can be produced by triboelectric charges. A familiar example of contact charge effect is the spark produced from the fingertips after one walks across a carpet on a dry day and touches a grounded surface.

Photoelectric charging occurs when short wavelength light absorption (ca. 250 nm) results in surface electron excitation with subsequent escape of the electron from the surface. Although this mechanism is minor, it can

TABLE 6-1 Triboelectric Series

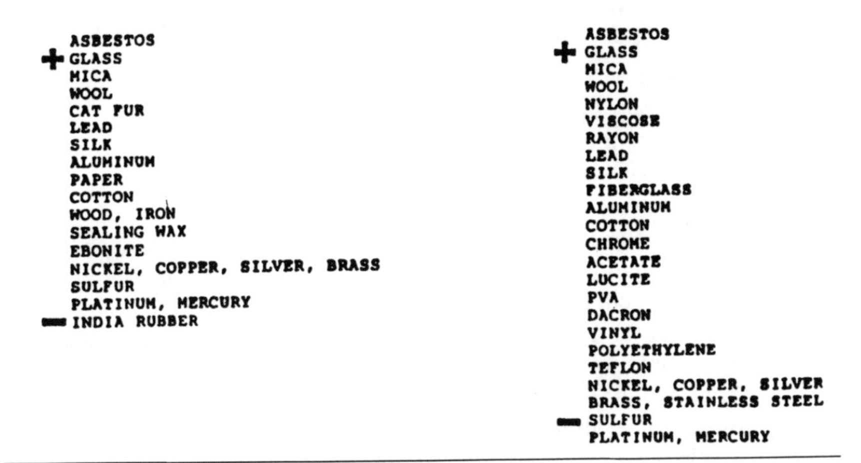

TWO OF THE MANY VERSIONS OF THIS TRIBOELECTRIC SERIES. THE MATERIAL UPPERMOST IN THE SERIES BECOMES POSITIVELY CHARGED WHEN RUBBED BY A MATERIAL LOWER IN THE SERIES: THE LOWER MATERIAL BECOMES NEGATIVELY CHARGED.

ASBESTOS
\+ GLASS
MICA
WOOL
CAT FUR
LEAD
SILK
ALUMINUM
PAPER
COTTON
WOOD, IRON
SEALING WAX
EBONITE
NICKEL, COPPER, SILVER, BRASS
SULFUR
PLATINUM, MERCURY
− INDIA RUBBER

ASBESTOS
\+ GLASS
MICA
WOOL
NYLON
VISCOSE
RAYON
LEAD
SILK
FIBERGLASS
ALUMINUM
COTTON
CHROME
ACETATE
LUCITE
PVA
DACRON
VINYL
POLYETHYLENE
TEFLON
NICKEL, COPPER, SILVER
BRASS, STAINLESS STEEL
− SULFUR
PLATINUM, MERCURY

TABLE 6-2 Some Triboelectrically Generated Charge Levels Obtained in an Area with no Bipolar Ion Generation Used to Reduce Surface Charges

Means of Static Generation	Relative Humidity	
	Low - 10-20%	High - 65-90%
Walking across carpet	35,000 V.	1,500 V.
Walking over vinyl floor	12,000	250
Worker at bench	6,000	100
Vinyl envelopes for work instructions	7,000	600
Common poly bag picked up from bench	20,000	1,200
Work chair padded with urethane foam	18,000	1,500

occur in areas where short wavelength light and certain sensitive materials are present. Ion charging occurs when mobile ions deposit on an object. Deposition can occur by diffusion or can be modified if a significant electric field is present. The normal ion distribution in the atmosphere is close to neutral, with slightly more negative than positive ions. If any radioactive material is present, it emits ions at a rate and polarity (or polarities) dependent on the particular material. These ions then transfer charge to any object that they contact. Ion generation is the most common mechanism used to control charge levels in cleanrooms. Bipolar emission from either high-voltage or radioactive sources is commonly used to reduce charge levels in these areas.

In the contamination-generation mechanisms considered here, it is usually assumed that the mechanisms are primarily based on steady-state phenomena. The generation mechanisms for specific sources are assumed to follow the general rules that each source emits at a uniform rate that may increase with time as tool parts wear or deteriorate to the point where cleaning, maintenance, or replacement is needed. However, many process tool operations are intermittent; temporary retention of contamination upon or within working parts, followed by sudden release of material, can

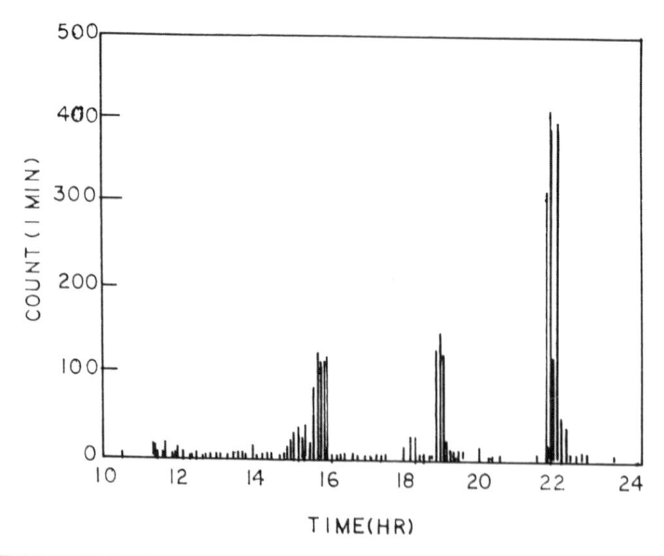

FIGURE 6-1. Episodic particle release during metal sputtering. This illustration shows the need to monitor continuously in critical areas when intermittent operations take place. Long-term mean levels may not be significant, but some instantaneous levels can be disastrous. (From Borden, P., 1990. Monitoring Particles in Production Vacuum Process Equipment: The Nature of Particle Generation. *Microcontamination* 8(1):21–26.)

occur; device operation procedures are carried out at varying rates. The end result is that episodic bursts of contamination are emitted with durations ranging from seconds to hours. An example is shown in Figure 6-1 (Borden 1990). Bursts of particle emission are shown with concentrations well over the accepted baseline level for the system being monitored, even though the long-term average concentration level is acceptably low.

References

Borden, P., 1990. Monitoring Particles in Production Vacuum Process Equipment: The Nature of Particle Generation. *Microcontamination* 8(1):21–26.

Liu, B. Y. H., et al., 1985. Characterization of Electronic Ionizers for Clean Rooms. Proceedings of the 31st Institute of Environmental Sciences Annual Technical Meeting, May 1985, Las Vegas.

Prater, W., et al., 1989. Preventing Contamination in Magnetic Disk Drives Through the Use of Wear Resistant Coatings. *Microcontamination* 7(4):31–39.

Sebald, T., Stiehl, H., & Sigusch, R., 1986. How to Avoid Particle Generation from Needle Electrodes of Ionizer Systems for Clean Rooms. Proceedings of the Fine Particle Society Technical Meeting, July 1986, San Francisco.

7

Contaminant Transport and Deposition Mechanisms

Gases, vapors, and ions are more or less uniformly distributed in the air and diffuse rapidly throughout the cleanroom. The molecules of these materials have nearly the same dimensions and mass as the air molecules in the cleanroom and respond to changes in the environment essentially the same as the air molecules. Particles, however, have significant mass and inertia and do not necessarily move at the same rate or as soon as the air parcel in which the particles are located. Further, they may be retained on a surface or within a liquid for some time with small changes in their format. During this time, the particles may also react with the substrate material or collect other particles on their surface and grow to the point where they are released to the atmosphere or transported to other locations on or near that substrate. The particles can change significantly with time in both size and composition as a result of reactions to environmental conditions. An understanding of the mechanisms controlling particle deposition and transport is quite important in contamination control. By knowing more about how particles are transported to and deposited upon critical products, the procedures and the basis for designs used in contamination control systems are better understood in terms of how they minimize particle deposition and maximize particle removal after deposition. For these reasons, this chapter emphasizes particle transport and deposition mechanisms, with lesser emphasis on gas and vapor movement mechanisms. Concerning particle motion in liquids, the differences in fluid properties, such as viscosity and density, result in the liquids transferring much larger shear forces to the particles than can gas to affect particle motion. For this reason, liquidborne particle motion is discussed separately.

For the discussion of airborne particle transport, several useful descriptive parameters have been developed. These parameters, which are used throughout any discussion of particle transport, are described here briefly.

The *drag coefficient* is the ratio of the gravity force to the inertial force on a particle in a fluid. Smaller particles have smaller drag coefficients because the mass of a small particle is also small. The drag coefficient can be considered as a quantitative indication of how a particle will resist any force that could cause a change in the particle velocity from that present before the force is applied. This applies to particles moving within a fluid when an obstacle is present within the fluid or to quiescent particles suddenly exposed to a moving fluid. Stokes law governs the motion of airborne particles. It states that the drag force on a particle is affected by the particle radius, the air viscosity, and the particle velocity in the air. For very small particles, a correction to the drag force values is required to account for the effect of gas slip. A particle in an electrically neutral environment has a dynamic mobility equal to the ratio of its speed to the drag force. In an electric field, the electric mobility is equal to the ratio of particle speed to the electric field.

The *Knudsen number* is defined as the ratio of the gas mean free path length to the particle radius. As the Knudsen number increases, gas velocity changes have less effect on the particle velocity. The smaller particles "slip" between the moving air molecules. For Knudsen numbers in the range below 0.01, the particle motion is defined as being in the continuum regime, where particle and air movements eventually coincide. In the Knudsen number range 0.01–0.02, particle motion is defined as being in the slip flow regime, where some slip occurs. In the range 0.2–10, the particle motion is in the transition regime to the free molecular flow regime; there, the Knudsen number is greater than 10.

The *relaxation time* is the time for a particle initially in equilibrium with a moving fluid to match a new fluid velocity. Small particles have a short relaxation time because their drag coefficients are low. Relaxation time is equal to the product of particle mass and dynamic mobility. The *stopping distance* is defined as the distance required for a particle initially moving within a gas stream to come to a stop when the gas flow is halted, as by an obstacle.

The *Stokes number* is the ratio of the particle radius to the dimension of an obstacle. This parameter is important in defining whether a particle in motion will be collected by an obstacle or will pass around that obstacle. Obstacles can be anything from a critical component to a sample tube inlet.

Deposition velocity is frequently used in summarizing deposition rates that are due to any effective mechanism. The *deposition velocity* is the ratio of particle flux (particles per unit area per unit time) to the ambient

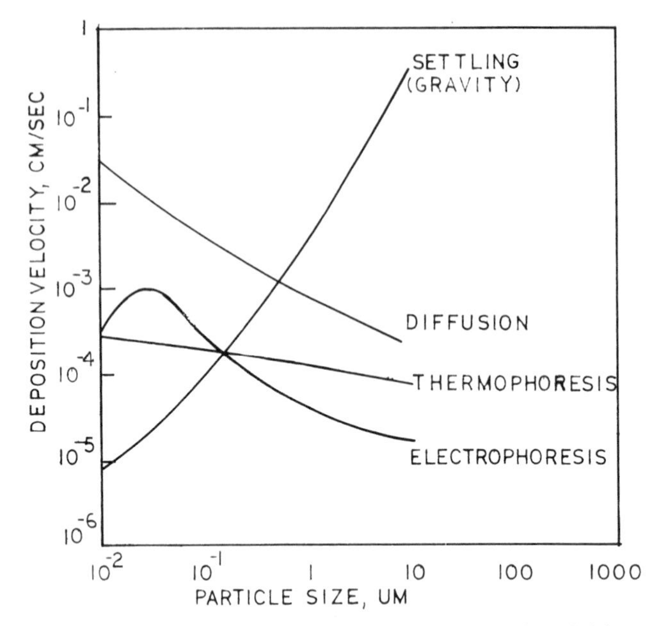

FIGURE 7-1. Deposition velocities versus particle size. The size of airborne particles affects the relative importance of specific forces controlling movement of the particles and requires consideration of optimum control methodologies. (From Cooper, D. W., 1986. Particle Contamination and Microelectronics Manufacturing: An Introduction. *Aerosol Science and Technology* 5(3):288–299.)

particle concentration. Figure 7-1 shows deposition velocities as a function of particle size for several deposition mechanisms (Cooper 1986). Note that Brownian diffusion becomes a major mechanism for particle motion for particles smaller than approximately 0.1 μm. For such small particles, air molecules moving in normal random manner, when impacting a particle, have sufficient energy to move the particle from its original location. There are other useful particle parameter definitions, but even some familiarity with these just described will aid in understanding the mechanisms of particle transport and deposition.

Aerosol particle transport is controlled by the forces on the particle and its response to those forces. The forces that affect particle motion in air are: the aerodynamic force from a moving stream of air, random impacts from individual air molecules in Brownian motion, gravitational force, electrostatic forces, and thermophoretic or photophoretic forces. The particle response to these forces is controlled by particle size, shape, charge, and mass. Although the same forces can affect particle transport on surfaces, the strong retention forces keeping the particles on those surfaces may modify that motion drastically.

The air in which particles are suspended is composed of discrete gas molecules moving randomly with mean free paths of about 0.066 μm (Jennings 1988) before collisions with other gas molecules. This value for the mean free path is for air at standard temperature and pressure. The mean free path varies inversely with atmospheric pressure. Knowledge of the Knudsen number allows characterization of the type of movement of a particle in the air. The aerodynamic or drag force on a particle with a small Knudsen number is a direct function of the particle size and the difference between the particle velocity and the fluid velocity as shown in Equation 7-1.

$$F_d = 6\pi a(U - V)/(1 + 0.86\,Kn) \tag{7-1}$$

where F_d is the drag force, a is the particle radius, U is the particle velocity, V is the air velocity, and Kn is the Knudsen number. For small particles with a large Knudsen number, the drag force is very small.

The gravitational force is a function of the particle mass, of the gravitational field, and of the difference between the particle and fluid densities, as shown in Equation 7-2.

$$F_g = g(\sigma_1 - \sigma_2) \cdot \mathrm{v} \tag{7-2}$$

where F_g is the gravitational force, g is the gravitational constant, σ_1 and σ_2 are the densities of the particle and the fluid, respectively, and v is the particle volume. For particles settling only under the influence of gravity, a terminal velocity is reached where the gravitational force and the air buoyant force are equal. As particle size decreases, the force on the particle becomes very small. For a particle of 0.1 μm radius, the settling velocity imposed by gravity is 0.000225 cm per second, whereas a 1-μm particle settles at 0.0128 cm per second. For some very small particles, atmospheric turbulence with upward airflows may overcome gravity forces and result in long-term suspension of small particles in the air.

In pharmaceutical processing cleanrooms, microbiological contamination is of most concern. Aseptic filling lines may be exposed to larger particles derived from personnel activities in the cleanroom. These particles may settle into product containers. It has been suggested that control in these areas may be aided by simply reducing the time of exposure for container openings and reducing the available target area of the container opening to minimize both gravitational and diffusional deposition of hazardous particles (Frieben 1989).

Electrical force on a particle is a function of its charge level and the strength of the electric field in which it is located. The size of the particle determines the maximum number of elementary charges it can hold. Previous exposure to ion sources with differing polarity affects the quantity of charge on a particle. If there is a predominance of ions of either polarity, then the electrical charge, F_e, usually has the same polarity as those ions. The force on the particle is shown in Equation 7-3.

$$F_e = n_p e E \tag{7-3}$$

where F_e is the electrical force, n_p is the number of elementary charges on the particle, e is the elementary charge unit, and E is the electric field strength.

Diffusion of airborne particles is the result of random molecular impacts from air molecule motions. The displacement of a particle is defined in terms of quadratic mean value of velocity, shown in Equation 7-4.

$$V_B = k \cdot \mathrm{T} \cdot (1 + 0.86Kn)/3\pi\mu\mathrm{a} \tag{7-4}$$

where V_B is the amplitude of diffusional motion, k is the Boltzmann constant, T is the temperature, Kn is the Knudsen number, μ is the gas viscosity, and a is particle radius. A particle of radius 1 μm settles in air at a rate 30 times faster than it diffuses, whereas a particle of 0.1-μm radius diffuses 8 times faster than it settles. For a particle whose size is similar to that of the molecular mean free path for air molecules at normal cleanroom temperature, the movement caused by sedimentation is about the same as that caused by Brownian motion, which means that a particle with at least unit density and smaller than about 0.5 μm almost never settles out of the air because of gravity alone. Other forces, such as electrostatic, thermal, or fluid dynamic forces acting singly or in concert, may still cause deposition onto vertical or horizontal surfaces.

Thermophoretic force is a relatively weak force compared with some of the other forces affecting particle motion. However, it is mentioned here because microelectronic manufacturing has recently become concerned with very small particles for which thermal forces can become significant. Particles move toward the lower-temperature area in a temperature gradient. Reduction of the effects of this phenomenon can be accomplished by heating critical components to a temperature somewhat above that of the ambient air. Equation 7-5 (Brock 1962) illustrates the factors affecting thermophoretic force.

$$F_t \approx K \cdot R \cdot a^2 \cdot Kn \cdot T'/T \tag{7-5}$$

where F_t is the thermophoretic force, K is a proportionality constant, R is the gas constant, a is the particle radius, Kn is the Knudsen number, T' is the temperature gradient, and T is the gas temperature; K contains constants referring to the thermal accommodation coefficients. Note that the thermal force varies with thermal gradient, the coefficients of thermal conductivity of gas and particle, and the Knudsen number. In summary, the smaller the particle, the less the thermal gradient required to produce significant motion of that particle. Similar work has been carried out to examine the effects of photophoresis on particle motion. Photophoresis occurs when a particle in gas is nonuniformly heated by an incident light beam. Because of anisotropic heating, gas molecules rebound from the surface with velocities affected by surface temperatures. This phenomenon is effective over a wide range of particle sizes.

In discussing forces on particles, one may imply that motion due to any or all of these forces occurs instantly. The inertia of the particle may cause some delay. The particle relaxation time, τ, given in Equation 7-6, tells the time required for the particle to achieve equilibrium velocity. The stopping distance l_d, given in Equation 7-7, shows how far a particle with an initial velocity V_0 travels when it is injected into a fluid at rest. This distance is of importance when particles are injected into a boundary layer directly above a surface in an airstream such as a wafer or a liquid surface where air may be directed at such a surface. Both of these equations are very important in defining particle motion, both in free airstreams and as the particles approach a surface.

$$\tau = (2/g) \cdot (a^2\sigma/\mu) \tag{7-6}$$

$$l_{\mathrm{d}} = V_0\tau \tag{7-7}$$

Because particles in the air may be exposed to any of the forces just mentioned, they may change in structure, location, and configuration. If these changes result in movement that causes particle-to-particle contact, then the particles may combine to form loose, temporary flocs or permanent agglomerates. The particles may also contact component or tool surfaces and adhere to them as well.

Agglomeration can be considered as a combination of several particles to form larger assemblies. *Cohesion* is the term used for like bodies in intimate contact; *adhesion* is usually used for unlike bodies. A suspension of polydisperse particles present in the air is exposed to any of the forces just discussed. Gravity causes the particles to fall at their terminal velocities, which vary roughly with the particle radius squared. The larger particles tend to overtake and impact upon the smaller ones. The smaller particles diffuse within the air due to Brownian motion, resulting in impact of smaller particles upon almost any surface in their path. In addition, relative motion is caused by airflow, with both axial and radial velocities changing with air turbulence levels.

The particles normally have a Boltzmann charge distribution because of interaction with bipolar ions in the atmosphere. Oppositely charged particles attract each other, whereas like charged particles repel each other. Thus, a variety of forces affect the particles suspended in the air. Once a particle has contacted another surface or another particle, particles may attach rigidly. The normal coagulation rate equation, assuming only Brownian motion, states that the particle number loss rate is proportional to the square of the number concentration of particles. The agglomeration rate constant value in a typical urban atmosphere results in the number concentration being halved in 1 second if the concentration is approximately 1 billion particles per liter of air. However, the effects of additional forces can increase the coagulation rate constant. For example, agglomeration of chainlike combustion agglomerates can occur because of Brownian motion, and the rate of agglomeration increases when the agglomerate structure is very fragmented rather than densely packed (Kaplan and Gentry 1988). Agglomeration in a cleanroom aerosol occurs very slowly unless charged particles and strong electrical fields are present.

Intermolecular or van der Waals forces are of great importance in both cohesion and adhesion. Where the separation distances between the particle and its contact are small, these forces become significant. Electrical forces are effective in moving particles to another surface. Once a particle has contacted either another particle or a surface such as a wall or a product, the electric charge that attracted the particle is largely dissipated or neutralized by the oppositely charged material. Other forces maintain the contact. Even after contact and charge transfer, residual electric charge can still affect cohesion. For example, it is shown (Yost and Steinman 1986) that the electrostatic contact force for a silicon wafer charged to 1000 V potential is about 40,000 times greater than that for a neutral wafer. If a liquid film is present between two particles or between a particle and a surface, then the cohesive force increases directly with the surface tension of the liquid. An adsorbed surface film on one or both surfaces can modify the contact force in either direction, depending on the materials. As extremes, a surface film can act as a lubricant to permit easier particle release or can act as an adhesive resulting in greater particle retention. Mechanical interference effects are also important. For larger particles, mechanical interlocking of surface irregularities can increase the cohesive forces. Figure 7-2 shows how small particles can interact with solid substrate surface irregularities.

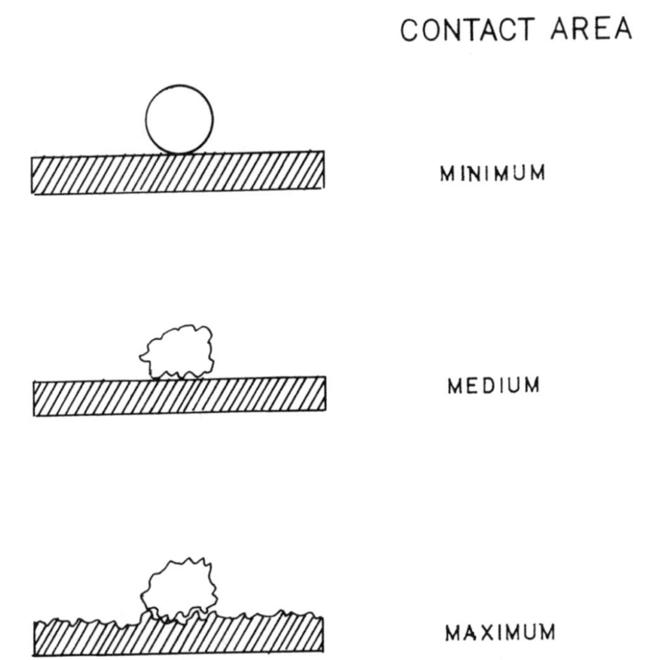

FIGURE 7-2. Particle interaction with surface irregularities. The degree of retention will be strongly affected by the configuration of both the particle and the substrate surface.

Particles suspended in air with relative humidity over 40% normally have adsorbed water molecules on their surface. If relative humidity increases much over 80%, then the particle can grow, eventually forming a solution droplet with properties and size depending on the physical properties of the particle material and on atmospheric conditions. If particles consisting of materials that react in solution agglomerate and are exposed to high humidity, then a solution droplet is formed wherein reactions can occur. If humidity is then reduced and the reactants crystallize, a new material can be released to the atmosphere.

In summary, particles in the air can change size because of collision and agglomeration with other particles or by water vapor transfer to or from their surface. In this situation, particles change not only in size but also in surface composition. They move because of a variety of forces, including air drag, molecular impact from Brownian motion, gravity, electrical forces, and thermal forces. The effective force and the amount of motion depend strongly on particle size. Some physical particle properties are also important, but size is the primary controlling parameter.

Once a particle is deposited on a surface, the degree of attachment depends on particle size, surface configuration, the forces that have caused the particle to reach the surface, and the opposing forces that are working to remove the particle from the surface. Until a few years ago, adhesion forces were considered to be operating on a static basis. Some recent studies have been directed toward a better understanding of the kinetics of the particle-removal processes from surfaces (Wen and Kasper 1989). A model is suggested wherein a particle is elastically bound; it is exposed to removal energy levels that oscillate around a mean value supplied by varying fluid lift forces. The model permits the particle to accumulate energy from successive fluid "impacts" until it has accumulated sufficient energy to cause release from the surface. Thus, the particle removal from a surface does not occur instantaneously but over some period of time that may vary with the energy level within the fluid. Data show reasonable support for this model.

Table 7-1 shows adhesive force levels for small particles. Particles are removed from surfaces mainly by shearing force, as from a fluid stream moving over the surface. The particle must extend out of the boundary layer before an airstream can be effective. At normal cleanroom air velocities, the boundary layer over any surface is several millimeters deep. Thus, few particles can be removed by airflow. Particles may be moved from one position to another on a surface by electrostatic forces, especially if the surface is not a good electrical conductor. If a particle is mechanically entrapped in a surface irregularity, then it will probably remain indefinitely. Table 7-2 shows some measured efficiency data for removing 5-μm particles from smooth optical surfaces. The data in both tables were obtained without complete control and measurement of electrostatic forces. Adhesion of particles is affected strongly by electrostatic force, and further work is necessary before complete understanding can be acquired (Ohmi, Inaba, and Takenami 1989).

Computer disk drive memories contain rotating disks with magnetic heads in near contact with the disk surface. The disks rotate at high

TABLE 7-1 Adhesion Forces for Small Particles on Surfaces*

	1 μm.		10 μm.		100 μm.	
	FOIICE	%	FORCE	%	FORCE	%
van der Waals	0.4	99	4	97	40	29
ELECTROSTATIC	0.005	1	.05	1	0.5	—
GRAVITY	0.0001	—	0.1	2	100	71
TOTAL	0.4	100	4.1	100	140	100

* This tabulation does not include the effects of the forces that occur when liquids are present; capillary effects, for example, are not considered here.

TABLE 7-2 Particle Removal Efficiencies for Some Cleaning Methods*

Method		Removal Efficiency for Particles >5 μm	Lowest Achieved Particle Concentration/cm² >5 μm
1. Scrubbing and dragging with lens tissue using ethanol and acetone		99.6 — 99.98%	2 — 40
2. Spraying with liquid solvent, 5-30 s duration Trichlorotrifluoroethane (Freon TF*)	345 kPa (50 psi)	97% 3% for >1 μm particles	1500
Trichlorotrifluoroethane (Freon TF*)	345 kPa (50 psi)	3% for >1 μm particles	
	6.9 MPa (1000 psi)	99.7% — 99.9% 81% for >1 μm particles	10 — 35
Water	17 MPA (2500 psi)	98 — 99.5%	2 — 60
3. Adhesive strippable coating Scotch 2253†		95 — 98%	500
4. Ultrasonic agitation of Freon TF, 1—2 min duration		24 — 92% 1% for >1 μm particles	9000 — 70000
5. Sequential cleaning operation Ultrasonic agitated TWD-602*/followed by Freon TF low pressure spray/vapor degrease/immersion in boiling solvent/ultrasonic agitation in rinse tank		92%	4000
6. Compressed gas jet for 10 s duration Micro-duster** 690 kPa (100 psi)		50 — 61%	5000 — 5800
7 Vapor degreasing in Freon TF		11 — 28%	65000 — 80000

* Particles are removed most effectively when sufficient force can be applied to the particle so as to overcome the adhesion force. This becomes more difficult as the particle becomes smaller.

Source: (Courtesy A. Gutacker, ARGOT, Inc.)

speeds, resulting in a very complex airflow field. These devices are sensitive to particles that may be present in the turbulent airflow. Studies indicate that particles deposit on disks by turbulent deposition with efficiency more than 90% for particles larger than 2 μm (Ananth and Liu 1988). For particles smaller than 0.1 μm, the dominant deposition mechanism appears to be Brownian diffusion (Fardi and Liu 1990). Deposition velocity for submicrometer particles increases as disk rotational speed increases because the boundary layer thickness decreases with rotational speed.

Theoretical and experimental studies have been carried out on combined effects of the several forces that affect particle movement on deposition velocity to semiconductor wafers (Donovan et al. 1988; Stratmann, Peterson, and Fissan 1988). It was found that thermal shielding effects can help to offset electrical capture of particles on wafer surfaces. Even so, electrostatic forces have a greater effect than inertia, gravity, and diffusion on deposition rates of submicrometer particles to wafer surfaces (Sakata et al. 1988; Fissan and Turner 1988). This is especially true when surface charges exist in the production area. To minimize particle deposition, the wafer surface should be free of electrical charge *and* held at as high a temperature as possible. Similar studies were carried out for particle deposition on components in compressed gas supply systems. An intuitive assumption might be that conditions in compressed gas systems result in greater molecular density; any high flow rates might result in convective and inertial forces from the "packed" gas molecules becoming an overwhelming factor in terms of particle motion. Some redistribution of the ranking of the effects of importance occurs, but this effect is minor because of the normally extremely low concentration of particles in compressed gas systems. In addition, the relative population of larger particles is very minor in these areas. Diffusion and electrostatic effects are still of great importance in particle deposition in compressed gas supply systems (Schwartz and McDermott 1988; Pui, Ye, and Liu 1988). These phenomena are of importance with compressed gases because most compressed gas supplies are filtered to the point where essentially the only particulate contaminants present are in size ranges less than 0.1 μm and are very sparse.

Many pharmaceutical processing operations are carried out in aseptic manufacturing areas, where bacterial contamination is limited by minimizing contamination in that area. Filling line air is carefully filtered, and the production area is regularly cleaned with microbiological control agents. Personnel exposure to operating production lines is carefully controlled. In this way, potential deposition of particles that may contain viable biological material from personnel is kept to a minimum. Some particulate material is still present, mainly brought in from containers and their seals, on moving machine parts, and from normal personnel sources. Significant transport routes to products in most pharmaceutical manufacturing areas are mainly airborne transport of bacteria, no matter what the source, along with some direct contact by liquid filling line contamination. It is assumed that aseptic conditions can be maintained for pharmaceutical product handling systems and for containers that will be used for those products. These conclusions are based on measurements made primarily by culturing of settling plate collection or swab wiping in the area; thus the transport mechanisms for most bacteria are mainly due to convective airflow and electrostatic attraction forces (Whyte 1986). In areas where a stable aerosol may be present, then some bacteria can be collected upon and can multiply within the inert dust particles. This is expected especially if any nutrient organic particles are present in that aerosol.

Gas and vapor transport mechanisms are relatively simple, compared with those for particles. The gas molecules are small enough so that effective transport mechanisms are essentially limited to diffusion in quiescent airflows. The electrical forces that affect particulate materials are relatively ineffective with the very small gas molecules. In moving airstreams, the contaminant gas molecules move in response to both static and dynamic pressure differences, just as the air molecules move. Even for gas or vapor molecules with densities much greater than that of air, the individual molecules tend to match air movement quickly. Other forces affect gas molecules, but the gas molecules are so small compared with most particles that their response is negligible. The deposition mechanisms are also controlled by the gas molecule size, so that adsorption and retention on surfaces are mainly affected by van der Waals forces, along with chemical reactivity. Once a gas molecule has been sorbed onto a surface or within a liquid, then removal is essentially impossible. If sufficient vapor has been deposited on a surface, then film formation may result. These comments implying that gaseous contaminants may be relatively innocuous apply only if the gas or vapor does not condense or react to form liquid droplets or solid particles after the gas has been released and if the molecular film layer is of no concern. As plasma or "dry" etching with chlorinated etching gases becomes more widely used for semiconductor processing, contaminant particles can result from generation and transport of volatile metal chlorides followed by chemical reaction with a variety of substrates. These can form and deposit particulate material throughout a dry etching system (Ruuskanen et al. 1990). In this way, the gaseous contamination can produce secondary particulate contaminants as a result of gas phase reactions. In a similar manner, some chlorinated fluorocarbons have been reported to react with water vapor to form hydrochloric or hydrofluoric acid vapors when the solvents have been heated. Subsequent corrosion can be expected in this situation.

Once an electrical charge has been produced, especially upon individual particles, then the major forces involved in its transport and change arise from interaction between the charge on the particle and those of the electric fields in the environment. Charges are transported upon fluidborne ions or particles. If the particles are within a totally nonconducting gas or liquid, the charge remains with the particle until the particle reaches a conductive surface or fluid. The particles carrying charge move as a result of any or all of the forces mentioned that can transport molecules and particles through a fluid. In addition, electrical fields also interact with the charge on a particle to affect its motion. Charged particles are then very much subject to forces as a result of electric fields in the environment. The electrical forces can often exceed the effects of many normal fluid or aerodynamic effects seen in most cleanrooms. Strong local electrical fields can exist as a result of the normal atmospheric electric field or as a result of polarized ion charging of surfaces in the cleanroom. Fair weather fields are normally much lower in charge level than those fields seen in bad weather conditions, particularly when electrical storms are also present.

The entries in Table 7-3 summarize the particle characteristics that can be and are important in transport and deposition of a variety of particle sizes. Note that this summary refers to spherical particles, as do all theoretical discussions of particle phenomena. Theoretical and experimental studies have been carried out to determine characteristics of particle deposition in cleanroom atmospheres and in clean area processing tools. This work has considered the fluid flow characteristics peculiar to these environments, the particle size distributions in these areas, and the extreme sensitivity to damage of the potential deposition targets in cleanrooms. Consideration of the forces that may result in deposition from the air of submicron particles on wafer surfaces was carried out on a theoretical and experimental basis (Cooper et al. 1988, 1989). Experimental data agreed reasonably well with theory, although the sparse particle levels normally seen in cleanroom air resulted in large variability about anticipated means. Deposition rates of most particles originally suspended in liquids upon wafers being cleaned in the liquid depended primarily on the concentration of particles in the liquid (Riley and Carbonell 1990). Time of exposure did not materially affect the final particle concentration on the surface. Further, it was noted that particle deposition on the wafer surface occurs mainly as the wafer is withdrawn from the liquid. Apparently, the detailed effects of electrical forces in liquid systems upon particle deposition on submerged surfaces are not adequately understood at this time (Donovan et al. 1990). Reentrainment of deposited particles from various surfaces has been shown to be important in transferring contaminant particles into gas streams, especially at high pressures (Wang et al. 1989). Significant particle transfer by reentrainment and generation by nucleation during gas cooling by pressure reduction was shown. Even at low pressure, gas flow–induced particle motion can transport particles to critical surfaces (Hablanian 1989). Vacuum processing operations have been shown to result in dislodgment of particles from surfaces in the low-pressure tool with subsequent deposition on in-process work. Control of gas molecule velocity appears to be very important in controlling contamination during pressure change operations. Calculations were made by Liu and Ahn (1987) for the deposition of uncharged aerosol particles on semiconductor wafers, considering convective diffusion and sedimentation as the major deposition mechanisms. They showed that the deposition velocity increases with particle size in the sedimentation regime and decreases with particle size in the diffusion regime. A minimum deposition velocity was found for particles near 0.2 μm. Wafer orientation and configuration were found to modify the deposition velocity. These calculations were verified by Otani and associates (1989). Calculation of particle deposition from a restricted stream that might be emitted from a filter leak (Stratman, Peterson, and Fissan 1988) indicated that thermophoretic effects could be significant for appreciable temperature differences between that of a wafer surface and the airstream.

The effect of particle charge on wafer deposition was calculated by Welker (1988). He concluded that surface contamination rates from air in an area with bipolar charge neutralization systems could be decreased by a factor of 50 to 100 in comparison to the rates in air where no bipolar ionizer system was used. This agrees with data obtained by Inoue and colleagues (1988). These observations lead to the conclusion that deposition velocities caused by electrostatic force can be greater than those from diffusion, gravity, or inertia for a wide range of particle sizes. It is believed that normal charge on even the few particles in a cleanroom can accelerate their deposition onto surfaces with different charges. Reduction of particle charge and/or electric field reduces particle deposition. Particle charge can be reduced by bipolar ion generation in the airstream; the same material can also reduce surface charges on potential target areas. A radioactive alpha source can be used for this purpose, as well as electrically charged point sources. The effectiveness of polonium-210 was described by Liu, Pui, and Lin (1986). Effectiveness was seen to decrease rapidly with the age of the source. Liu recommends that a krypton-85 source would be a better means for particle charge neutralization by radioactive emitters.

TABLE 7-3 Characteristics of Particles and Particle Dispersoids*

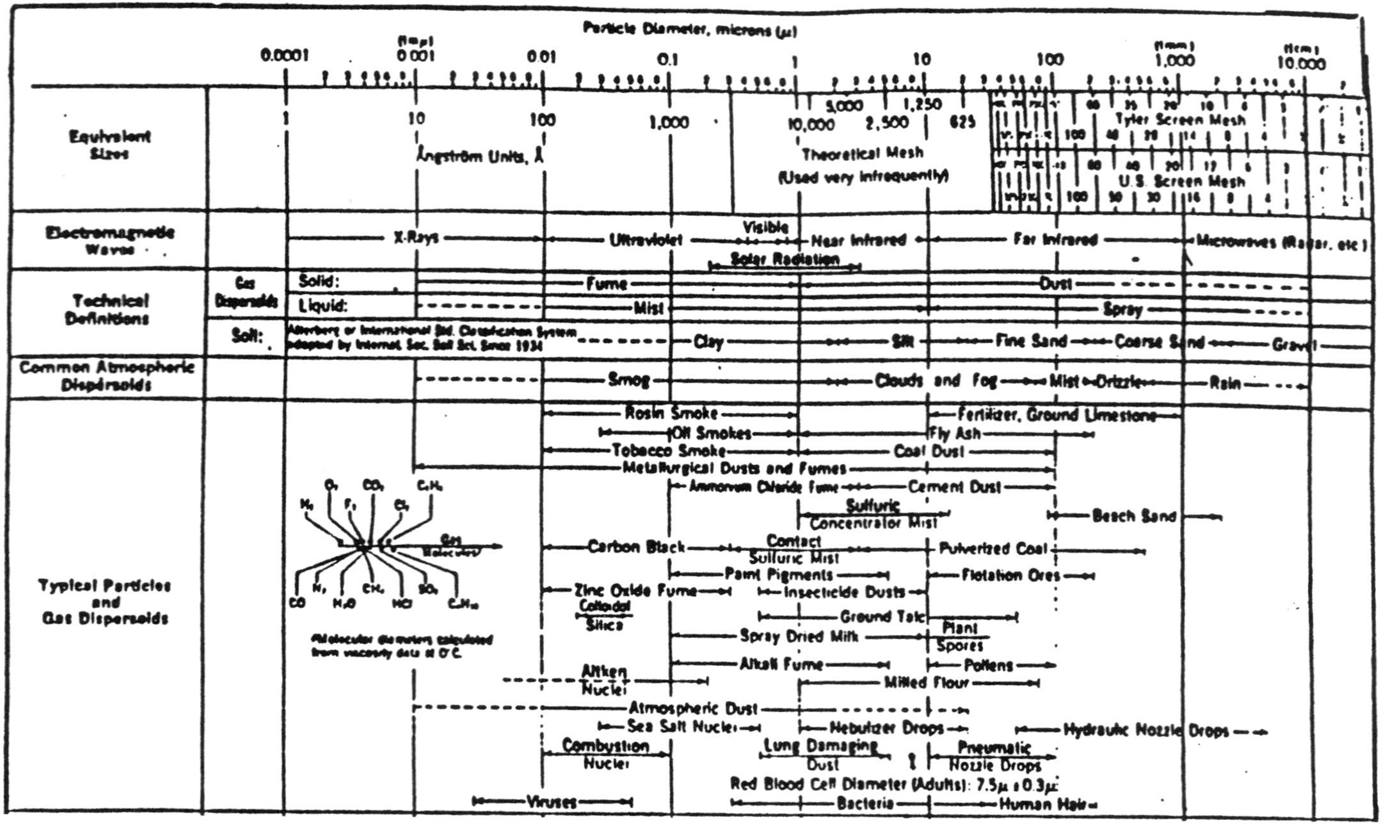

Particle Diameter, microns (μ)

0.0001 | 0.001 (1mμ) | 0.01 | 0.1 | 1 | 10 | 100 | 1,000 (1mm) | 10,000 (1cm)

Equivalent Sizes

Ångström Units, Å: 1 | 10 | 100 | 1,000

Theoretical Mesh (Used very infrequently): 10,000 | 5,000 | 2,500 | 1,250 | 625

Tyler Screen Mesh: 100 | 60 | 35 | 20 | 10 | 6 | 3

U.S. Screen Mesh: 100 | 80 | 60 | 40 | 30 | 20 | 16 | 8 | 4

Electromagnetic Waves

X-Rays — Ultraviolet — Visible — Near Infrared — Far Infrared — Microwaves (Radar, etc.)

Solar Radiation

Technical Definitions

Gas Dispersoids: Solid: Fume — Dust

Liquid: Mist — Spray

Soil: Atterberg or International Std. Classification System adopted by Internat. Soc. Soil Sci. Since 1934 — Clay — Silt — Fine Sand — Coarse Sand — Gravel

Common Atmospheric Dispersoids

Smog — Clouds and Fog — Mist — Drizzle — Rain

Typical Particles and Gas Dispersoids

Rosin Smoke; Oil Smokes; Tobacco Smoke; Metallurgical Dusts and Fumes; Ammonium Chloride Fume; Cement Dust; Sulfuric Concentrator Mist; Fertilizer, Ground Limestone; Fly Ash; Coal Dust; Beach Sand; Carbon Black; Contact Sulfuric Mist; Pulverized Coal; Paint Pigments; Flotation Ores; Zinc Oxide Fume; Colloidal Silica; Insecticide Dusts; Ground Talc; Plant Spores; Spray Dried Milk; Alkali Fume; Pollens; Aitken Nuclei; Milled Flour; Atmospheric Dust; Sea Salt Nuclei; Nebulizer Drops; Hydraulic Nozzle Drops; Combustion Nuclei; Lung Damaging Dust; Pneumatic Nozzle Drops; Red Blood Cell Diameter (Adults): 7.5μ ± 0.3μ; Viruses; Bacteria; Human Hair

Gas Molecules*: H_2, O_2, F_2, CO_2, Cl_2, C_3H_8, CO, N_2, H_2O, CH_4, HCl, SO_2, C_4H_{10}

*Molecular diameters calculated from viscosity data at 0°C.

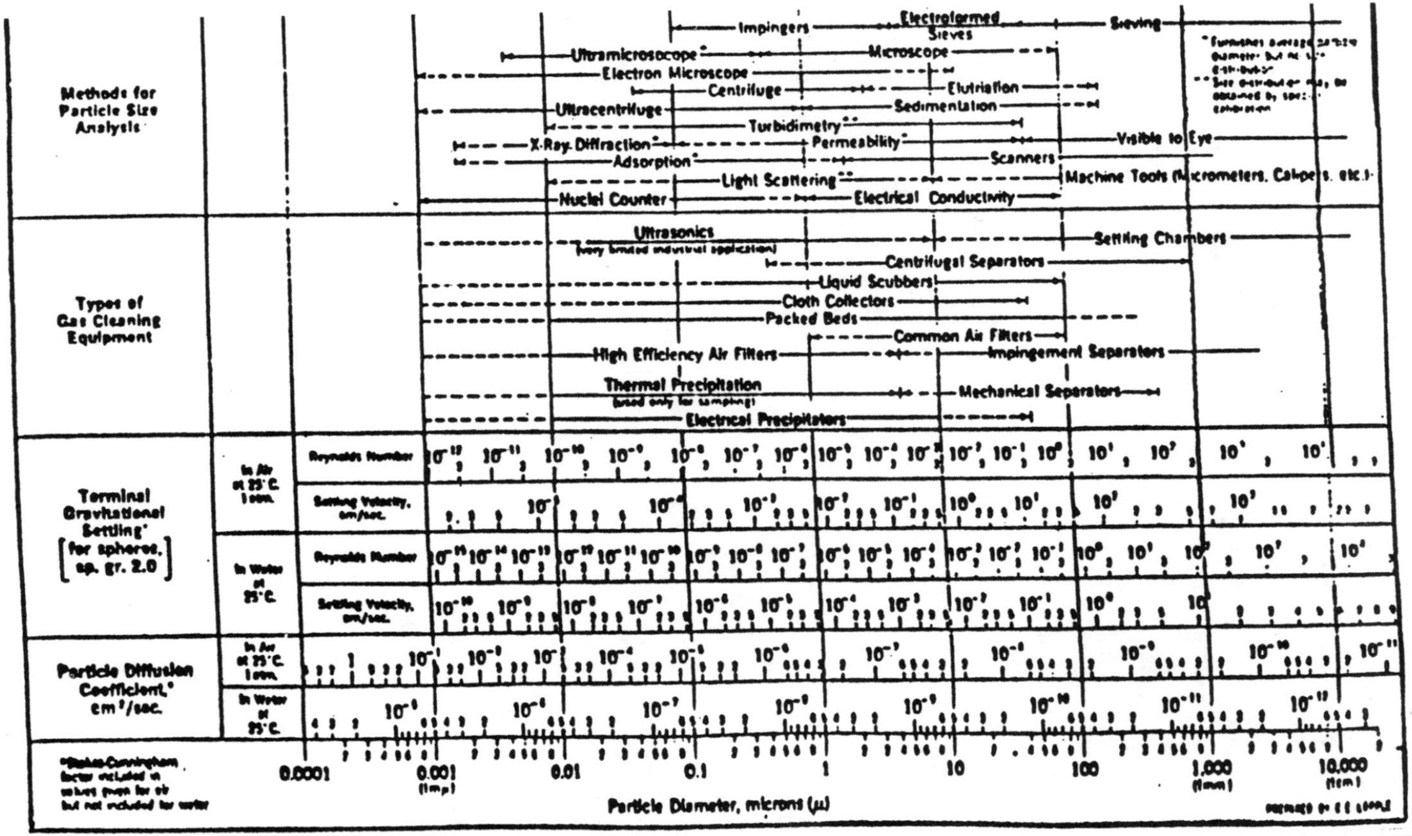

Methods for Particle Size Analysis

Impingers — Electroformed Sieves — Sieving
Ultramicroscope* — Microscope
Electron Microscope
Centrifuge — Elutriation
Ultracentrifuge — Sedimentation
Turbidimetry**
X-Ray Diffraction* — Permeability* — Visible to Eye
Adsorption* — Scanners
Light Scattering** — Machine Tools (Micrometers, Calipers, etc.)
Nuclei Counter — Electrical Conductivity

Types of Gas Cleaning Equipment

Ultrasonics (very limited industrial application) — Settling Chambers
Centrifugal Separators
Liquid Scubbers
Cloth Collectors
Packed Beds
Common Air Filters
High Efficiency Air Filters — Impingement Separators
Thermal Precipitation (used only for sampling) — Mechanical Separators
Electrical Precipitators

Terminal Gravitational Settling* [for spheres, sp. gr. 2.0]

In Air at 25°C 1 atm. — Reynolds Number: 10^{-12}, 10^{-11}, 10^{-10}, 10^{-9}, 10^{-8}, 10^{-7}, 10^{-6}, 10^{-5}, 10^{-4}, 10^{-3}, 10^{-2}, 10^{-1}, 10^{0}, 10^{1}, 10^{2}, 10^{3}, 10^{4}
In Air at 25°C 1 atm. — Settling Velocity, cm/sec.: 10^{-5}, 10^{-4}, 10^{-3}, 10^{-2}, 10^{-1}, 10^{0}, 10^{1}, 10^{2}, 10^{3}
In Water at 25°C — Reynolds Number: 10^{-15}, 10^{-14}, 10^{-13}, 10^{-12}, 10^{-11}, 10^{-10}, 10^{-9}, 10^{-8}, 10^{-7}, 10^{-6}, 10^{-5}, 10^{-4}, 10^{-3}, 10^{-2}, 10^{-1}, 10^{0}, 10^{1}, 10^{2}, 10^{3}, 10^{4}
In Water at 25°C — Settling Velocity, cm/sec.: 10^{-10}, 10^{-9}, 10^{-8}, 10^{-7}, 10^{-6}, 10^{-5}, 10^{-4}, 10^{-3}, 10^{-2}, 10^{-1}, 10^{0}, 10^{1}

Particle Diffusion Coefficient,* cm²/sec.

In Air at 25°C 1 atm.: 10^{-1}, 10^{-2}, 10^{-3}, 10^{-4}, 10^{-5}, 10^{-6}, 10^{-7}, 10^{-8}, 10^{-9}, 10^{-10}, 10^{-11}
In Water at 25°C: 10^{-5}, 10^{-6}, 10^{-7}, 10^{-8}, 10^{-9}, 10^{-10}, 10^{-11}, 10^{-12}

*Stokes-Cunningham factor included in values given for air but not included for water

Particle Diameter, microns (μ): 0.0001, 0.001 (1mμ), 0.01, 0.1, 1, 10, 100, 1,000 (1mm), 10,000 (1cm)

* Furnishes average particle diameter but no size distribution.
** Size distribution may be obtained by special calibration.

Prepared by C. E. Lapple

* After 30 years, this chart is still a valuable tool for particle physics work; it will never become obsolete.

Source: (From Lapple, C. E., 1961. Particle Technology. *SRI Journal* 5(3):94–102. Courtesy SRI International.)

Since the period when this work was carried out, additional development of electric field generators has been carried out with good results in reducing charge with no significant particulate emission.

Measurements (Fujii et al. 1984) showed that most particles follow the air streamlines approximately even when quite close to a deposition target. In a later study (Fujii et al. 1988), it was shown that air turbulence near target wafer elements in a cleanroom caused increased deposition of submicrometer particles on wafer surfaces as a result of added particle motion, even while passing through the boundary layer at the wafer surface. High-speed streams of filtered gas are used with "blow-off guns" for cleaning surfaces in clean rooms. Some comments have indicated that these devices are effective in removing only large particles from surfaces and may actually deposit many small particles. The mechanisms affecting particle interaction with surfaces were considered by Blitshteyn (1986). He concludes that gun efficiency increases with gas stream velocity, mainly for relatively large particles, and that ion neutralization is essential for smaller particle removal. However, effective filter maintenance is required to avoid particle deposition from gas streams.

References

Ananth, G. P., & Liu, B. Y. H., 1988. Particle Transport and Deposition in Computer Disk Drives. Proceedings of the 9th International Committee of Contamination Control Societies Conference, pp. 235–239, September 26, 1988, Los Angeles.

Blitshteyn, M., 1986. The Use of Blow-off Guns in Semiconductor Manufacturing. *Microcontamination* 4(1):20–28.

Brock, J. R., 1962. On the Theory of Thermal Forces Acting on Aerosol Particles. *Journal of Colloid Science* 17:768.

Cooper, D. W., 1986. Particle Contamination and Microelectronics Manufacturing: An Introduction. *Aerosol Science and Technology* 5(12):287–299.

Cooper, D. W., et al., 1988. Deposition of Submicron Aerosol Particles during Integrated Circuit Manufacturing: Theory. Proceedings of the 9th International Committee of Contamination Control Societies Conference, pp. 19–26, September 26, 1988, Los Angeles.

Cooper, D. W., et al., 1989. Deposition of Submicron Aerosol Particles during Integrated Circuit Manufacturing: Experiments. *Journal of Environmental Science* 32(1):27–31.

Donovan, R. P., et al., 1988. Experimental Study of Particle Deposition on Silicon Wafers under the Combined Effects of Electric Fields and Thermal Gradients. Proceedings of the 9th International Committee of Contamination Control Societies Conference, pp. 37–42, September 26, 1988, Los Angeles.

Donovan, R. P., et al., 1990. Investigating Particle Deposition Mechanisms on Wafers Exposed to Aqueous Baths. *Microcontamination* 8(8):25–29.

Fardi, B., & Liu, B. Y. H., 1990. Deposition of Particles on Spinning Computer Disks. Proceedings of the 36th Institute of Environmental Sciences Annual Technical Meeting, pp. 169–172, April 1990, New Orleans.

Fissan, H. J., & Turner, J. R., 1988. Control of Particle Flux to Prevent Surface Contamination. Proceedings of the 9th International Committee of Contamination Control Societies Conference, pp. 33–36, September 26, 1988, Los Angeles.

Frieben, W. R., 1989. The Effect of Cleanroom Design and Manufacturing Systems on the Microbiological Contamination of Aseptically Filled Products. *Journal of Environmental Science* 28(3):25–27.

Fujii, S., et al., 1984. Studies on Design Theory of Laminar Flow Type Clean Room: Particle Deposition to the Surface. Proceedings of the 7th International Committee of Contamination Control Societies Conference, September 18–21, 1984, Paris.

Fujii, S., et al., 1988. Measurements of Airflow Turbulence in a Clean Room and Particulate Behavior in the Boundary Layer on a Wafer. Proceedings of the 35th Institute of Environmental Sciences Annual Technical Meeting, King of Prussia, PA, May 3–5, 1988.

Hablanian, M. H., 1989. If You Rough Slowly, Do You Get a Clean Vacuum Chamber? *Research and Development* 31(4):81–86.

Inoue, M., et al., 1988. Aerosol Deposition on Wafer. Proceedings of 35th Institute of Environmental Sciences Annual Technical Meeting, King of Prussia, PA, May 3–5, 1988.

Jennings, S. J., 1988. The Mean Free Path in Air. *Journal of Aerosol Science* 19(2):159–166.

Kaplan, C. R., & Gentry, J. W., 1988. Agglomeration of Chainlike Combustion Aerosols due to Brownian Motion. *Aerosol Science and Technology* 8(1): 11–28.

Liu, B. Y. H., & Ahn, K. H., 1987. Particle Deposition on Semiconductor Wafers. *Aerosol Science and Technology* 6(3):215–224.

Liu, B. Y. H., Pui, D. Y. H., & Lin, B. Y., 1986. Aerosol Charge Neutralization by a Radioactive Alpha Source. *Particle Characterization* 3(3):111–116.

Ohmi, T., Inaba, H., & Takenami, T., 1989. Research on Adhesion of Particles to Charged Wafers Critical in Contamination Control. *Microcontamination* 7(10):29–42.

Otani, Y., et al., 1989. Determination of Deposition Velocity onto a Wafer for Particles in the Size Range Between 0.03 and 0.8 μm. *Journal of Aerosol Science* 20(7):787–796.

Pui, D. Y. H., Ye, Y., & Liu, B. Y. H., 1988. Sampling Transport and Deposition of Particles in High Purity Gas Supply System. Proceedings of the 9th International Committee of Contamination Control Societies Conference, pp. 287–293, September 26, 1988, Los Angeles.

Riley, D. J., & Carbonell, R. G., 1990. The Deposition of Liquid-Based Contaminants onto Silicon Surfaces. Proceedings of the 36th Institute of Environmental Sciences Annual Technical Meeting, pp. 224–228, April 1990, New Orleans.

Ruuskanen, J., et al., 1990. Contamination in an Experimental Gallium Arsenide Etch System. *American Industrial Hygiene Association Journal* 51(1):8–13.

Sakata, S., et al., 1988. Aerosol Deposition on Wafer Surface. Proceedings of the 9th International Committee of Contamination Control Societies Conference, pp. 65–72, September 26, 1988, Los Angeles.

Schwartz, A., & McDermott, W. T., 1988. Numerical Modelling of Submicron Particle Deposition in Pressurized Systems. Proceedings of the 9th International Committee of Contamination Control Societies Conference, pp. 78–85, September 26, 1988, Los Angeles.

Stratmann, F., Peterson, T., & Fissan, H., 1988. Particle Deposition onto a Flat Surface from a Point Particle Source. *Journal of Environmental Science* 31(6):39–41.

Wang, H. C., et al., 1989. Factors Affecting Particle Content in High-Pressure Cylinder Gases. *Solid State Technology* 32(5):155–158.

Welker, R. W., 1988. Equivalence between Surface Contamination Rates and Class 100 Conditions. Proceedings of the 35th Institute of Environmental Sciences Annual Technical Meeting, King of Prussia, PA, May 3–5, 1988.

Wen, H. Y., & Kasper, G., 1989. On the Kinetics of Particle Reentrainment from Surfaces. *Journal of Aerosol Science* 20(4):483–498.

Whyte, W., 1986. Sterility Assurance and Models for Assessing Airborne Bacterial Contamination. *Journal of Parenteral Science and Technology* 40(5):188–197.

Yost, M., & Steinman, A., 1986. Electrostatic Attraction and Particle Control. *Microcontamination* 4(6):18–27.

8

Product Protection Methods Summary

All working environments can and do contain significant quantities of contaminants. Some contaminants are brought into that environment by exchange of air, movement of personnel, transport of process materials, and the like; others are generated by normal operations within the work area. It is essentially impossible to remove all contaminating materials from any area and to prevent ingression and generation of contamination into that area both while the area is in operation and even while it is quiescent. Previous chapters have shown how some contaminants are generated and brought into the work areas from the external environment. Some are generated within that work area by the materials, processes, equipment, and people in the work area. A variety of processes can transport contaminants from their source to critical locations in the cleanroom. Experience has shown that critical products must be protected from harmful contaminants. Several methods and procedures have been developed to protect products from contamination. Contamination control for product protection both before and during processing requires that the product be isolated from contaminants, protected from those that cannot be kept out of the process environment, cleaned when unavoidable contamination occurs and monitored to verify integrity of the protection processes.

Isolation from contaminants means that the product is enclosed in containers that exclude contaminants during storage and while the product is in the manufacturing process. The isolation containers may be designed strictly for protective storage or so that processing can be carried on within a special enclosure. *Protection* means that the product is shielded from contaminants in the environment by packaging, by contaminant containment or removal, and so on. Packaging may include special tool design to prevent contamination from reaching critical areas while products are stored between processing operations or are exposed within a tool during processing. *Cleaning* means that contaminants within process fluids that may contact a product are removed, that contaminants that may be present on elements close to products are removed, and that contaminants deposited on products from any source are removed from sensitive substrates before they can cause harm. Monitoring verifies the effectiveness of the protective measures.

We protect against contaminants transported in environmental air by cleaning that air to the best of our ability. This type of protection is almost never completely satisfactory because the environment is only part of the source of contamination; however, it is effective against gross contamination and is needed to minimize the effects of any contaminant. Without protective measures to clean the manufacturing environment to the best of our ability, production of high-technology products would be severely curtailed. For example, even with aseptic filling techniques, it has been nearly impossible to keep viable bacteria excluded totally from many pharmaceutical areas. That is the basis for inclusion of sterilization procedures with many routine cleaning operations in the areas where pharmaceutical products are manufactured. Essentially all of the procedures used in every cleanroom are designed to protect the product from contaminants during manufacture and storage.

Experience has shown that all sensitive products do not require the same degree or kind of protection. As a primary example, many pharmaceutical products must be manufactured in cleanrooms. The cleanrooms are used to control particulate contamination that may contain microbiological materials. Inert contaminant particles are of minor concern in the pharmaceutical area if they are small enough, but if one could remove *all* contaminant particles, then one can assume reasonably that viable particles are not present. Regulatory agencies in the government have specified a range of cleanroom environments for preparation and packaging of these products. Terminal sterilization by radiation or by heating can also be used. It must be carried out over a time and temperature regime adequate for specific containers or environments where microorganisms may be shielded and/or insulated from direct exposure. Terminal sterilization may not be possible with some heat- or radiation-sensitive products or packaging. Thus, very clean air in the pharmaceutical industry may not be as serious a requirement as it is for other product areas where inert particles are a hazard. It was shown by Whyte (1983) that a major source for both inert and viable particles in pharmaceutical products is the product container, that autoclaving was adequate for reducing contamination particles, and that changes in cleanroom air quality had only minor effects on product quality as defined in terms of microbiological content. A later study (Frieben 1985) points out that emphasis on cleanroom design and operation may not be the most cost-effective means of protecting pharmaceutical products. Consideration of other methods of protecting the product should be explored. This broad approach is one that should *always* be considered in planning any contamination control effort. The product requirement should be defined first; the available methods of achieving that level of contamination control should be examined in terms of cost and effectivity; next the best method *for the specific product and situation* should be chosen, tested, and used.

The major protection activities are not necessarily isolated, one from another. Quite often, complementary activities for product protection may act to reinforce each other. For example, isolation within a smaller area from a contaminated environment reduces the amount of cleaning required after exposure to an activity that cannot provide total protection from that contaminated environment for the product. Care in production system operation always improves quality of sensitive products. A variety of activities, each of which reduces particulate contamination in a production facility, have been pointed out recently by Ohmi and associates (1990). Selection of fluorine-based rather than conventional greases to reduce organic material emission from heated moving parts, use of sufficient airflow velocity to contain emissions from nearby contaminant sources, and vibration control by both damping and isolation devices are only a few of the methods that can be incorporated with relatively low costs after examining the problem and carried out by adequate planning.

The earliest means of protection by isolation was to establish "ultraclean" work spaces. The first cleanroom, described in the 1950s as a *white room* (a term that has almost completely disappeared since 1970 except for some aerospace applications), was an area where work equipment and tools were cleaned as much as possible, where ventilation air was filtered, and personnel procedures were established to improve processing cleanliness. For some of the more rigorous requirements of the day, clean benches were used to produce a secondary stage of isolation. Two present examples of protection for electronic and mechanical products by isolation are the filtered air enclosed areas used in class 1,000 to 10,000 cleanrooms for final product operations and the standard mechanical interface (SMIF) system (Parikh and Kaempf 1984) used in many semiconductor manufacturing areas.

Some product processing involves a series of operations that may require different tools. In this situation, the tools are "clustered" within a single clean-air enclosure, and products are transferred from one tool to another without exposure to ambient air. The cluster tool concept is similar to the localized process control system that has been in use for many years (Gutacker 1984). This system uses varying levels of cleanliness maintenance to maximize environmental cleanliness in areas where products might be exposed and allows lesser contamination control levels in surrounding areas where personnel or operating tool surfaces are present.

Pharmaceutical product manufacturing is usually carried out in clean areas with specified maximum contamination control levels ranging from class 100 to class 10,000. Most critical production areas maintain cleanliness levels much better than the required ones. Both liquids and powders are commonly produced, along with many medical devices. Sterile products are sought, and aseptic production procedures are often used to aid in reducing microbiological contamination of the products. Along with isolation of products from environmental contamination, it is frequently necessary to ensure isolation of production personnel from contact with materials that may be hazardous. Class 100 areas are specified for liquid product filling lines; areas up to class 10,000 can be used for container processing and some material handling operations. Parenteral products are manufactured using filtered water to ensure both cleanliness and sterility. The product is contained in sterilized vessels until it is packaged. Containers to be filled are cleaned and sterilized in a sterilization tunnel, where they are exposed to steam at 121° C for 30 minutes. The final step in that operation is to dry the containers with heated and filtered air. The containers are then placed on a moving belt that transports them to the filling line. Filtered air exhausts directly over the line. The filling line is isolated from the overall cleanroom by polymer curtains that prevent air entry from the room and contain the filtered air directed over the filling operation. The only items between the filtered airflow and the container openings are the product fill systems. The containers are then sealed while still on the clean filling line.

During the critical period between container cleaning and sealing, the product is isolated from the external environment. It has been shown that use of such a system for ampoule filling can reduce rejects to half of those before introduction of the system (Stemmer 1983). Similar systems have been used for food products. Milk products have limited shelf life even with refrigeration. When containers are filled, some airborne bacterial contamination can occur, resulting in eventual spoilage. Use of class 100 HEPA filtered unidirectional airflow over a filling line for yogurt into radiation-sterilized plastic containers was used in one area (Bruderer and Schicht 1989). Longer shelf life was obtained without the need for chemical preservatives for this product. A positive isolation procedure has been described for filling small-volume parenteral products (Khan 1989). A fully automated high-output system that has completely removed personnel from the sterile environment was developed. Particulate cleanliness was improved by a factor of 2–3 over conventional filling operations. Careful design is needed to ensure reliable and economic operation. The same considerations apply to other manufacturing operations. Semiconductor processing by automated equipment has also been shown to reduce particulate contamination to products (Baker 1989).

For some products, production and testing is carried out in microbiological safety cabinets. They are designed so that filtered air protects the product from contamination and protects the operator from exposure to hazardous product materials (Clark and Ljungqvist 1989). Careful front panel design provides acceptable protection to the operator while still allowing access for work within the cabinet. When working with very toxic materials, safety cabinets must be completely enclosed, using recirculated filtered air and allowing personnel access only through glove ports.

The SMIF system is frequently used for isolation in electronic semiconductor fabrication. This system keeps the wafers isolated from environmental air during all stages of manufacture, once the wafers are placed in the cassette. The system encloses the cassette in a clean, minimum-volume container for transport from one operation area to another. The container is opened only within a clean interface port installed on process equipment. Figure 8-1 shows the basic operation of this type of system. It is very effective in working with relatively small products and subassemblies. It can be used to isolate the product from essentially all contaminant types. Tests have shown (Gunawardena et al. 1984) that the system reduces particle deposition on wafers by a factor of 10 over that seen on wafers handled in open cassettes in a cleanroom. Even storage of SMIF containers in an office space did not result in more particle deposition on the wafers than that seen during storage for the same time in a cleanroom. In fact, at least one study (Hughes et al. 1990) has shown that the use of SMIF isolation technology allows acceptable semiconductor production within an open area if all components are clean and carefully handled. SMIF isolation technology is used for wafer handling and processing. It has been estimated that use of the SMIF system reduces wafer defect density by a factor of more than 2 (Titus and Kelly 1987; Sayre et al. 1989). Note also that use of this and similar systems also isolates the products from at least some personnel handling steps that can degrade products in many ways. In planning any manufacturing process, capital and operating costs must be considered. A cost study was made for a traditional class 10 semiconductor wafer fabrication area compared with costs for a class 100 or class 1,000 area fitted with SMIF material handling systems (Briner and

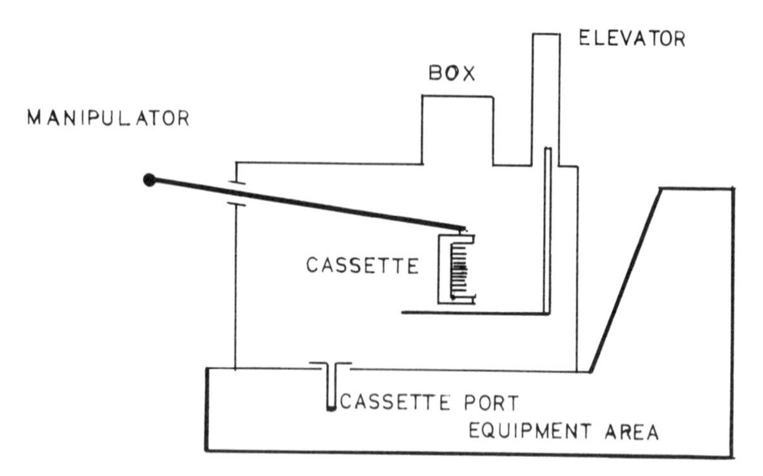

FIGURE 8-1. SMIF components schematic. This rough sketch illustrates how the SMIF system isolates the product from the environment during transfer from one operation to another. (From Gunarwardena, S., et al., 1984. The Challenge to Control Contamination: A Novel Technique for the IC Process. *Journal of Environmental Sciences* 27(3):23–30.)

Yeaman 1988). Capital costs were higher for the class 100 SMIF area and only slightly smaller for the class 1,000 SMIF area. Operating costs (mainly for fan power) for the class 100 SMIF area was 87% of those for the class 10 area, and those for the class 1,000 SMIF area were about 50% of those for the class 10 area. Therefore, a complete evaluation of costs and benefits must be made before making this type of change in operation.

In addition to isolation systems that can be used during manufacturing processes, such as those just mentioned, adequate packaging containers are required that isolate products between production steps and after manufacturing. Packaging containers have certain requirements for specific products. The containers cannot shed any particulate materials from interior surfaces. The containers should not retain electrostatic charge for long time periods. If the containers are plastic, then the material should not emit organic vapors. Container design should not allow movement of contents, which may result in abrasion of either the product or the container materials. The process of opening and closing the container should not result in surges of air caused by excessively rapid closure movement or interior volume shift when the cover is moved. The containers should be cleanable, and interior surfaces should be configured so that contaminant trapping does not occur.

Many medical devices must be supplied with flexible packaging materials. For example, many surgical and medical devices are supplied within easily opened, flexible packages to ensure easy access in emergency situations. In this situation, the major requirement is that the packaging material act as a barrier to penetration of microorganisms from the air. Basically, the packaging material is used as a barrier to airborne bacteria.

Hardware to be used in space also has special requirements for packaging. Cleaned rigid containers are used to protect a device from physical damage during transportation and in process handling. Polymer sheets are used to protect the devices from atmospheric contaminants that may penetrate the containers and before the device is placed in the rigid container. The polymer sheets must be free of particles, and molecular films must not be deposited on the device. The contaminant films may be produced by condensation of external gases or vapors, including water vapor, that penetrate the packaging material or are emitted from the organic plasticizer used with some polymers. Because many electronic devices are on board space-related hardware, the films must have sufficient conductivity so that electric charge buildup does not occur. The film should have minimum flammability. The film must have sufficient tensile strength so that gravitational stresses from load shifting within the package does not cause rupture. A more complete discussion of space-related packaging requirements, along with a list of applicable standards and specifications, has been given by Shon and Hamberg (1985).

In the semiconductor industry, the wafer and chip fabrication processes use the most advanced methods to isolate the products during fabrication. In some cases, the same care is not used to protect the completed system in its final package. For example, dual inline packages (DIP) are used commonly for semiconductor devices. These packages consist of a chip mounted on a lead frame and the assembly, along with contact leads, sealed within a polymer case. Quite often a DIP is used in a dirty, hot, and vaporous environment. If the case seal integrity is compromised, operation of the completed chip can be seriously degraded after use (Ellis 1988).

Product protection is a very general term that covers a variety of processes. For some products, complete isolation from the environment is impossible. For example, protection from contaminants is the objective when processing is carried out in a class 100 clean bench located in a class 10,000 room. Essentially all normal manufacturing is carried out in an ambient-temperature, ambient-pressure environment. Most manufacturing operations can result in contaminant generation in the process itself. Because the products cannot be isolated from contaminants that may be generated in the manufacturing process, protection from these contaminants is required.

Basically all areas where contamination may affect the product must be addressed in order to achieve acceptable product yield. The environment, the equipment, the process materials, the fabrics used in the cleanroom, the personnel, the storage devices, the test equipment operation, and the packaging materials must all be considered in product protection. Some of the requirements are discussed here. Cooper (1985) suggests a systems approach to contamination control in which standard systems analysis methodology is used to determine optimum protection procedures. He points out the advantages of protection procedure selection by first defining the contamination source, transport, deposition, and damage paths and then selecting the optimal cost-effective remedy.

Different products may require different protection methods for some of their problems. In wafer fabrication, effects of contamination can be reduced significantly by reducing wafer exposure to particles. Much contamination is transferred to products by process materials that are contaminated by inadequately cleaned containers. Use of specially designed and cleaned containers can reduce this problem significantly (Cole, Van Ausdel, and Waldman 1989). Not only is the container system cleaned by special methods but also the material removal and handling system is designed to minimize contamination release. Because many particles are deposited from process materials, particle control in working with photoresists and other liquids is one area where product protection can be highly effective (Long 1984). This protection operation not only must include maintenance of photoresist cleanliness but also must control all the handling procedures during all the process steps for photoresist application, etching, and resist removal. As "dry" processing for semiconductor manufacture becomes more widespread, cleanliness of compressed gas used for fabrication processes becomes more important. The materials of construction, methods of control, and measurement all become extremely important in protecting the product when these dry processing materials are used (Zuck 1984). Automation of handling systems is another method for protecting the wafers during fabrication (Harper and Bailey 1984). For these and similar products, the rapid advances in technology require that any material handling system be flexible enough to follow product modifications; otherwise, it will not be cost-effective. Modern electromechanical instrument maintenance requires extreme cleanliness (Stepien 1990). Automated cleaning and assembly systems have been found necessary to meet class 1 area requirements for assembly and maintenance of operable modern inertial control instruments.

In protecting products from particle deposition, the first approach is to ensure adequate filtration for the cleanroom or clean work space air. In almost all processing operations, the product may accumulate contaminants as part of the process. Examples include deposition of particles on semiconductor wafers during processing, depositon of particles in minimal clearance components in valves and balancing elements, and accumulation of particles in pharmaceutical liquids from improperly cleaned containers. In such cases, wafers or containers must be cleaned during the manufacturing operation.

Great care is taken in designing cleanrooms so that the overall airflow in the room sweeps out contaminants without passing them over a product where they may be deposited. Tool layout and equipment configurations that might lead to eddy formation in airflow are avoided as much as possible. Figure 8-2 shows how normal cleanroom components and activity can degrade the air quality itself. Overhead light fixtures and filter retainers can cause airflow variations that cause eddy formations where contaminants may accumulate. Work benches and/or tools can block uniform airflow and generate eddies, as well as be a source of contaminants themselves. When high-level tools or other tall structures are located near work areas, then turbulent eddies that may contain contaminants can persist below these structures. Sensitive materials should not be exposed to the space where such conditions can exist. Localized process control air flow systems can be used to direct a stream of clean air over a critical location.

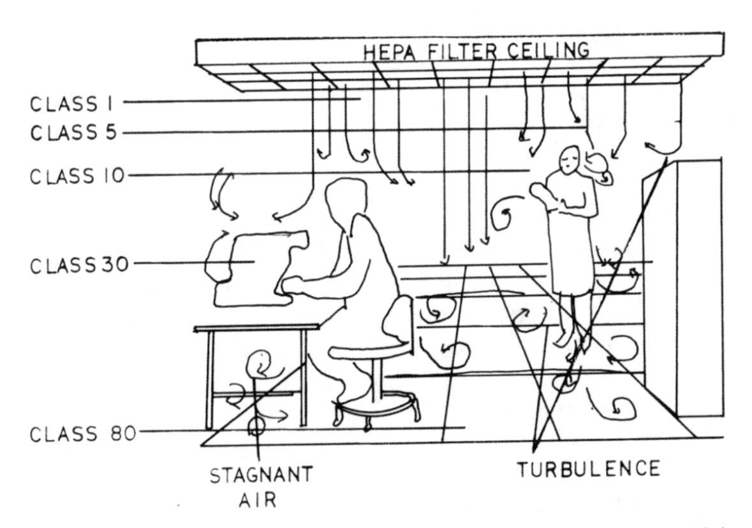

FIGURE 8-2. Cleanroom air quality degradation. Personnel and equipment activity produce contamination that degrades air quality. Both moving objects and airflow obstacles cause turbulence that can retain contamination or cause it to concentrate and deposit in critical areas. (From Larrabee, G., 1985. A Challenge to Chemical Engineers: Microelectronics. *Chemical Engineering* 92(12):57.)

As product sensitivity to smaller and smaller particle sizes and particle numbers increases, and particle generation from operations and equipment in the cleanroom becomes more of a problem, maintenance of adequate air cleanliness by filtration alone is less and less cost-effective. The HVAC system cost requirements for equipment and fan power keep increasing with little increase in product protection. For this reason, reducing particle deposition by air ionization to control electric charge on surfaces and electric field levels in the room air has been studied and found to be effective in many situations. Cleanroom air ionization sources can use pulsed DC, DC, or AC high-voltage sources. It has been found (Suzuki, Matsuhashi, and Izumoto 1988) that ions from pulsed DC ionization systems induce the most rapid charge decay rates on surfaces in the cleanroom. Data obtained at the same time indicate that particle emission from the pulsed DC system appears somewhat less than that from the other two types.

In addition to electrostatic charge control, thermal gradient establishment is being investigated as a protective measure. For very small particles, thermophoresis has been shown to be a promising protective mechanism. Recent work indicates that a dust-free space at least 100 μm thick can be established over a flat plate, such as a wafer surface (Stratmann et al. 1988). A temperature difference between the wafer surface and surrounding air of 10° K to 30° K with air movement at typical cleanroom velocities of 0.5 m per second is adequate for control of 0.1–0.5-μm particles.

Semiconductor manufacture is particularly sensitive to the effects of electrostatic discharge. Passage of an electrostatic discharge through an electronic device can cause instantaneous power levels that can cause either instantaneous catastrophic failures or degraded device performance. Some high-voltage gradients can cause holes in element layers in a device, resulting in insulation failures due to local deposition of metal layers. Assemblies containing metal oxide semiconductors, field effect devices, MOS capacitors, and hybrid microcircuits are among the devices susceptible to electrostatic discharge (ESD) damage. It is well known that triboelectric charging that can occur as a result of normal material handling can cause surface charges at the kilovolt level. A favored protection method against ESD effects is to prevent charge accumulation by dissipating charge as fast as it is generated. As part of these efforts, a standard for definition of garments specifically designed for control of electrostatic charges is being prepared (Cranston 1988). Conductive, grounded surfaces and containers are provided for handling devices during the fabrication process. These include work surfaces, protective containers for in-process storage, conductive clothing for employees, grounding of all moving elements whenever possible, and isolation of parts from ungrounded potentially charged materials, surfaces, and personnel. Obviously, an important element in this process is adequate personnel training for correct use of the systems (Woods 1985). It has been stated that the most damage from electrostatic discharge occurs as a result of personnel failure to comply with listed procedural requirements to minimize the problem.

Ion addition to the room air has been shown (Huffman and Nichols 1987) to accelerate charge decay rates on wafer surfaces by a factor of 500. At the same time, particle deposition rates are decreased by a factor of 200 or more. Some work has been reported (Frey 1988) that indicates that application of complex electric fields to ducted air flows can even reduce concentration of contaminant gases. Formaldehyde, ammonia, sulfur dioxide, and carbon dioxide measurements showed reductions from 10 to 40%. The mechanism of this operation is not clearly understood.

Sooner or later the production operation includes a high-efficiency cleaning stage in the process. No matter how much care is taken in the manufacturing process to protect the product from contaminants, some deposition of contamination occurs. In that situation, the product will have to be cleaned as well as practically possible. At higher levels of assembly, product cleaning may not be practical. Therefore, assembly elements must be kept as clean as possible. The cleaning operations include removal of particles from solid surfaces and from product gases and liquids. Many high-efficiency liquid filters are used in the electronics and pharmaceutical industries to clean liquids such as product liquids, process liquids delivered for subsequent manufacturing processes, and water used for flushing parts. At this time, the problem of cleaning 0.1-μm particles from semiconductor wafer surfaces has become important, and the usual washing processes are completely inadequate for particles of this size. The scrubbing energies needed for effective conventional cleaning may destroy the wafer. New methods are being investigated. Solid CO_2 particle streams appear to be very effective in removing some materials. Combinations of ultrasonic energy with ozonated water are being studied to remove ionic contaminants.

In any situation, the product and/or the environment should be monitored for cleanliness as one way of ensuring that the protection procedures are effective. We can define ambient air or a process fluid cleanliness in terms of concentration of particles ≈ 0.1 μm. We cannot define patterned surface cleanliness from particles in those size ranges in a short time period, and we have no practical way of defining particle compositon in real time. We can state the composition of gaseous contaminants in concentrations down to a few parts per billion, if we know what to look for. Again, the time necessary for sample collection, processing, and measurement may still be quite lengthy.

After the product has been manufactured, it is realized that all the efforts for isolation, protection, and cleaning may or may not have been adequate. Verifying freedom from contaminants and from failures in general is then required. We may use a brute force method to do so by operating the device and observing failure rates, but this is scarcely a cost-effective method. A better way is to monitor the product during and after manufacture for contamination level and to define an acceptable sample or surrogate method for performance testing. Semiconductor process fabrication defect-monitoring studies have been carried out with electrical test structures (Gill and Dillenbeck 1989). This monitoring process observes simple electrical system operation and defines contamination effects in terms of degradation of performance that can be ascribed to contamination collection on or in the structure. In addition to monitoring the product for freedom from contamination, the standard contaminant-monitoring methods used in cleanrooms are aimed at verifying the performance and integrity of the product isolation and protection systems. These monitoring methods are primarily used to determine the contaminant levels present in the processing environments. The methods are discussed in more detail later.

Pharmaceutical liquids are monitored for the presence of inert particles by either microscopic or instrumental observation to define whether or not the liquid cleanliness is adequate. In the United States, the legal requirement for control of particles in parenteral liquids and medical devices is only for those $\geq$ 10 μm and $\geq$ 25 μm, as of 1991. Bacterial content is determined by placing a sample of solids filtered from the liquid on a nutrient medium and determining the viable bacteria number in terms of colony generation rate. Particulate content of electronic manufacturing process fluids is normally determined by instrumental observation of particle concentration. Particles 0.05 μm and larger are of concern in this area. Dissolved contaminants and ionic impurities are monitored instrumentally, using devices best suited for a specific contaminant. Electronic product monitoring is usually carried out by instrumental measurement of particles and ionic and/or elemental contaminants deposited on test wafer surfaces during the manufacturing process. A scanning electron microscope has been shown to be extremely useful in ensuring deionized water quality (Hango 1989). It is possible to define composition and particle size by direct observation of collected materials. A variety of material characteristics can be defined by this device, including particle shape, elemental composition, and some compound identification.

To summarize, we protect our sensitive products from contaminants, clean them as well as we can, and verify that our procedures are effective by monitoring performance and cleanliness of products and processes. When poor yield or degraded operational conditions indicates problems, then we reexamine our efforts to find and solve the problems. Our approach has been to look for procedures that will eliminate contaminants from the work and assembly environments, to make sure that the fluids to which the products will be exposed are as clean as possible, and to verify our practices as well as possible.

References

Baker, E. J., 1989. Automation Reduces Process Equipment Particles. *Microelectronics Manufacturing and Testing* 12(1):18–19.

Briner, D., & Yeaman, M. D., 1988. Cost/Benefit Analysis of Two SMIF Alternatives Compared to a Conventional Class 10 Cleanroom. Proceedings of the 9th International Committee of Contamination Control Societies Conference, pp. 137–144, September 26, 1988, Los Angeles.

Bruderer, J., & Schicht, H. H., 1989. Laminar Flow Protection for the Sterile Filling of Yoghurt and Other Milk Products. *Swiss Contamination Control* 2(1):41–45.

Clark, R. P., & Ljungqvist, B., 1989. Containment Performance of Laboratory Fume Cupboards. *Journal of R^3–Nordic* 18(3):27–31.

Cole, M., Van Ausdel, R., and Waldman, J., 1989. Improved Container and Dispense System Leads to Reduced Defects. *Microcontamination* 7(11):37–41.

Cooper, D. W., 1985. Contamination Control Management: A Systems Approach. *Microcontamination* 3(8):49–55.

Cranston, J. A., 1988. Tentative Standard for Protection of Electrostatic Discharge Sensitive Items: Personnel Garments. Proceedings of the 9th International Committee of Contamination Control Societies Conference, pp. 435–440, September 26, 1988, Los Angeles.

Ellis, B. H., 1988. How Direct and Indirect Semiconductor Packaging Contamination Influences Assembly Reliability. *Microcontamination* 6(6):35–37.

Frey, A. H., 1988. Using In-Duct Electrical Fields to Reduce Particulate and Gaseous Contamination. *Microcontamination* 6(6):27–32.

Frieben, W. R., 1985. The Effect of Cleanroom Design and Manufacturing Systems on the Microbiological Contamination of Aseptically Filled Products. *Pharmaceutical Manufacturing* 2(11):13–17.

Gill, P., & Dillenbeck, K., 1989. Using Snake Patterns to Monitor Defects and Enhance VLSI Device Yields. *Microcontamination* 7(2):23–30.

Gunawardena, S., et al., 1984. The Challenge to Control Contamination: A Novel Technique for the IC Process. *Journal of Environmental Science* 27(3):23–30.

Gutacker, A. R., 1984. Localized Process and Product Contamination Control. In *Contamination Control Technologist Handbook,* chapter 10. Webster, NY: ARGOT, Inc.

Hango, R. A., 1989. DI Water Quality Monitoring for Very Dense Electronic Component Manufacturing. *Ultrapure Water* 6(4):14–21.

Harper, J. G., Bailey, L. G., 1984. Flexible Material Handling Automation in Wafer Fabrication. *Solid State Technology* 27(7):89–98.

Huffman, T. R., & Nichols, G. H., 1987. Reduction of Particle and ESD Damage by Room Ionization. *Solid State Technology* 30(11):127–130.

Hughes, R. A., et al., 1990. Eliminating the Cleanroom: More Experience with an Open-Area SMIF Isolation Site (OASIS). *Microcontamination* 8(4):35–38.

Khan, S., 1989. Automatic Flexible Aseptic Filling and Freeze-Drying of Parenteral Drugs. *Pharmaceutical Technology* 13(10):24–34.

Long, M. L., 1984. Photoresist Particle Control for VLSI Microlithography. *Solid State Technology* 27(3):159–161.

Ohmi, T., et al., 1990. Controlling Wafer Surface Contamination in Air Conditioning, Particle Removal Subsystems. *Microcontamination* 8(2):45–51.

Parikh, M., & Kaemph, U., 1984. SMIF: A Technology for Wafer Cassette Transfer in VLSI Manufacturing. *Solid State Technology* 27(7):111–115.

Sayre, S., et al., 1989. SMIF Reduces Defect Density in Class 100 Production Facility. *Solid State International* 12(10):104–107.

Shon, E. M., & Hamberg, O., 1985. Packaging Films for Electronic and Space-Related Hardware. *Journal of Environmental Science* 28(4):46–52.

Stemmer, K., 1983. Sterile Filling of Ampoules with Integrated Sterile Room Technology. *Swiss Pharma* 5(11a):33–39.

Stepien, T. M., 1990. An Automated System for Contamination-Controlled Processing of Precision Instruments. *Journal of the Institute of Environmental Sciences* 33(1):80–86.

Stratmann, F., et al., 1988. Suppression of Particle Deposition to Surfaces by the Thermophoretic Force. *Aerosol Science and Technology* 9(2):115–121.

Suzuki, M., Matsuhashi, H., & Izumoto, T., 1988. Effectiveness of Air Ionization Systems in Clean Rooms. Proceedings of the 35th Institute of Environmental Science Annual Technical Meeting, pp. 405–412, King of Prussia, PA, May 3–5, 1988.

Titus, S., & Kelly, P., 1987. Defect Density Reduction in a Class 100 Fab Utilizing the Standard Mechanical Interface. *Solid State Technology* 30(11):119–122.

Whyte, W., 1983. A Multicentred Investigation of Clean Air Requirements for Terminally Sterilized Pharmaceuticals. *Journal of Parenteral Science and Technology* 37(4):138–144.

Woods, W. R., 1985. Experience, Problems and Subtleties of Electrostatic Discharge at JPL. *Journal of Environmental Science* 31(6):42–46.

Zuck, D. S., 1984. Particle Control in the Construction of a 1 Mbit DRAM Gas Distribution System. *Solid State Technology* 32(11):131–135.

9

Liquid Cleaning Methods

Clean liquids are required in all cleanroom operations, whether the clean liquids are the product or are used in the fabrication of other cleanroom products. Processing and final cleaning of liquids to be used for semiconductor devices is frequently carried out in cleanrooms to provide clean process material for a later operation. Many pharmaceutical products are liquids used for injection or ingestion. Plastic films to be used for photography or for packaging elements are polymerized from organic liquids, which must be free of contaminating particles. These and other similar products must be free of both dissolved and suspended contaminants. Very clean reagent-grade water is required for container cleaning before products are placed in the containers. In the semiconductor industry, many manufacturing processes result in deposition of contaminants that must be removed from components by flushing with clean liquids. In addition, semiconductor manufacturing often requires the use of process chemicals that may be received with a high contamination level. Most aerospace device surfaces must be cleaned before use. This requirement applies both to operating devices and especially to the interior of fuel and oxidizer storage and flow systems. Liquids of all types are used in these areas. They may be inorganic or organic, with high or low vapor pressures, inert or reactive, and single component or mixtures.

Many large-volume process liquid cleaning operations are performed out of the product-processing cleanroom enclosure. Bulk liquids are transported in 2,000–10,000-gallon containers or in 55-gallon drums. Prepackaged chemicals are bottled in glass or polymer containers holding 1–20 L. High-purity semiconductor process chemicals are supplied in containers that have been cleaned and filled by automated systems operating in class 100 facilities (Gallagher 1989). Cleaned process liquids are delivered to the user for addition to a processing and/or storage system. Most users verify cleanliness for critical liquids, and some facilities include additional cleaning to reduce contaminant levels in liquids to be used in critical operations. Equipment is available for recycling and purifying liquids that may become contaminated in a processing operation. Some process tools that use liquids include a point-of-use cleaning system through which the liquid passes just before it is used. Although many process chemicals and solvents are delivered after some cleaning operations by the manufacturer, water for flushing or other use in most plants is cleaned on site.

Both the process liquids and cleaning solvents are much dirtier than process environment air. The air in a class 100 area has fewer than 100 0.5-μm particles per cubic foot (28.3 L). Cleaning most liquids to a level of no more than 100 0.5-μm particles per L even before use is difficult; many electronic process liquids used in 1990 contained particles in levels to 100,000 per L and greater. Standard clean and well-filtered deionized water contains less than 100 particles per L, and high-quality water is much cleaner; prepackaged "clean" sulfuric acid may contain 15,000–400,000 0.5-μm particles per L, whereas the particle levels in bulk acids may be greater by a factor of 10–200 (Jones 1986). Similar levels are seen in other process chemicals. These levels were not uncommon in 1990. In the period shortly afterwards, the use of controlled bulk chemical delivery systems reduced the particle levels in many process chemical liquids by several orders of magnitude.

Most of the particles in process liquids consist of sand or soil particles, metal fragments from flow systems, polymers from seals and valve elements, bacteria, and pyrogen particles. All liquids can contain large quantities of dissolved materials. Salts, metals, organic materials, or gases can dissolve in most liquids to percentage quantities. Ionic contaminants in liquids have been a serious concern in cleanroom industries only since the 1960s, with the development of semiconductors. The technology for large-scale liquid cleaning is relatively recent, and the technology for accurate measurement at the low levels required for some manufacturing operations is even newer. Some contaminants, such as organic carbon in deionized water, were not previously considered a serious problem.

A variety of methods are used for cleaning liquids. Much attention has been given to purification of process water for use in either the electronics or pharmaceutical industry. Water is used in very large quantities in semiconductor manufacture. The water must be clean in order to obtain reasonable product yield. Contaminants that *must* be removed include total oxidizable carbon (TOC), colloidal silica, dissolved ions, bacteria, and "inert" particles. TOC is related to the dissolved organic contaminants in the water; it has been found to correlate extremely well with wafer defect density (Craven, Ackerman, and Tremont 1986). Colloidal silica is basically residual silica that has not been completely removed from the original water supply system. It is composed of solid particles in the nanometer size range and larger, usually with a weak electrical charge. Excessive quantities of colloidal silica can degrade wafer deposition processes. Dissolved ions that are not adequately removed result in low-resistivity water. Bacteria in a water system can rapidly plug filters, coat ion exchange resins, and cause local defects and pinholes in oxide layers. Particles interfere with deposition on wafers and cause open circuits by blocking portions of a pattern; if they are conductive, they may cause short circuits by deposition across an area where there should be no material. Final rinse water quality requirements in England, as of 1987 (Stewart et al. 1987), are targeted at a resistivity level of 18 megohm-cm, TOC no more than 50 μg/per L, living organisms at no more than 1 per 5 ml, particle counts at no more than 50 per L $\geq$ 0.5 μm, silica at no more than 5 ppm, and ionic impurities (both cations and anions) less than 1 ppm. Each year, requirements for semiconductor process water quality have become more rigorous. This trend will certainly continue as semiconductor devices use smaller elements more closely spaced on the wafer surface and as the capability for detecting smaller quantities of contamination continues to improve.

Purification processes include filtration, ultrafiltration, distillation, reverse osmosis, ion exchange, and ultraviolet irradiation. Filtration, used for particle removal, involves passage of the liquid through a porous body with pores smaller than the particles to be removed from the liquid. Filter elements with pore sizes ranging upward from 0.02 μm are used. Ultrafiltration uses membranes with 0.002–0.1-μm pore sizes to remove small particles and some large contaminating molecules from the liquid. Distillation has been used to purify liquids for centuries. The major contaminants in the liquid are materials with boiling points different than the liquid of concern. When the liquid is vaporized and the vapor is condensed, only pure liquid is produced. The liquid may contain other materials that can be vaporized at temperatures close to the boiling point of the liquid of concern. The method is seldom used for water purification in the electronics industry because the energy and time requirements for processing are much greater than those needed for other purification methods. Distillation may also not completely remove some organic azeotropes that may be collected with the condensate. Membrane processing is the favored method in that area. Reverse osmosis uses membranes as molecular filters; liquid is pressurized to pass through the membrane with retention of dissolved minerals, organics, and colloids on the high-pressure side of the membrane. Ion exchange involves passage of liquid over resins; a contaminant ion is replaced by a nonhazardous ion from the resin. Ultraviolet (UV) irradiation of the water at 254 nm destroys bacteria and oxidizes some organic contaminants. The method is quite effective in relatively clean water for removal of bacteria, but the source must be kept clean because this radiation can be easily blocked by a relatively thin layer of debris on the source and does not penetrate an agglomerated dust particle to destroy encapsulated bacteria.

These methods are used in combinations for many water purification applications. A typical water-processing operation would operate by bringing water from the municipal supply through a process similar to the following:

1. Filtration through coal and sand bed filters
2. Removal of organics by activated carbon adsorption bed
3. Passage through degasifier
4. Reverse osmosis
5. Mixed-bed ion exchanger
6. Filtration through micrometer filter
7. Ion exchange, followed by resin trap
8. Sterilization by UV
9. Final filtration

Other sequences are in use now and, as the technology develops, still more sequences and processes will be adapted. Because most electronic fabrication process water is derived from municipal water supply systems, the residual chlorine used to reduce microorganism levels in the water must be reduced. In areas where chloride ion residue may be a problem, other methods have been used to control bacterial contamination (Pengra, Hogsett, and McCall 1990). Once the active hypochlorous acid or hypochlorite ion level is reduced to a point suitable for semiconductor processing, microorganism control methods are still needed. Design of deionized (DI) water return loops to eliminate dead legs in the line, daily particle monitoring to ensure that bacterial fragments are not appearing in the water, and routine flushing of use areas with dilute hydrogen peroxide as required are some of the means of improving bacterial cleanliness levels. Combining ultraviolet radiation with ozone is used both for oxidation of some organic materials and for destruction of bacteria (Zoccolante 1987). A synergistic action appears to occur that is more effective than would be expected from the two control methods alone.

Materials of construction for the water-handling system are chosen so that the water does not leach any materials from the components that can cause harm to the product to which the water is applied. Water flow systems are designed so that stagnation points, where bacterial growth can occur, are avoided. Almost any means to ensure recirculation can decrease bacterial levels at point-of-use locations by up to 2 orders of magnitude (Siegmann 1987). Once bacteria begin growing in a water system, terminal sterilization may destroy bacteria before the water is used, but organism fragments smaller than 0.2 μm may penetrate some filters rated at that size and cause problems downstream from the filters (Rechen 1985).

Some common liquid cleaning methods are shown in Tables 9-1 and 9-2. Particles are removed from liquids primarily by filtration. Both depth and surface types are used. Most cleanroom liquids are cleaned by replaceable cartridge filters. Depth filters are used for large flows up to several hundred gallons per minute to remove particles larger than a micron or so in size. Surface filters, usually membrane polymers, are used for sub-

TABLE 9-1 Water Purification Methods and Specific Contaminant Control Function*

Purification Method	Contaminants Removed or Function	Remarks
Distillation	Removes gross amounts of common contaminants.	Original purification method; slow; not practical for large quantities.
Activated charcoal	Removes odors, dissolved gases and residual organic material.	
Chlorination	Kills micro-organisms.	
Ultraviolet germicidal lamp	Kills exposed micro-organisms	Lamp should be checked periodically for effectiveness
Settling	Removes coagulated materials and suspended particulate matter.	
Softening	Removes calcium, magnesium, and iron ions.	Zeolite type beds; increases sodium content and leaves neutral salts.
Demineralization	Removes cations and anions to a high-purity level.	Dual-bed for heavy contaminants; mixed-beds for lighter contaminants or combined systems.
Filtration	Removes suspended particulate and coagulated matter, and bacteria above the filter pore size.	Filters must be replaced periodically to remove contaminants and assure required volume.

* Selection of the optimum method depends on the contamination control requirement for a specific product need and on the contaminant type in a particular area.

TABLE 9-2 Contaminant Control Processing for Liquids

PROCESS FOR SEPARATION

Reverse Osmosis (Hyperfiltration)

Microfiltration

Ultrafiltration

Particle Filtration

Scanning Electron Microscope

Optical Microscope

Visible to Naked Eye

Ionic Range

Molecular Range

Macro Molecular Range

Micro Particle Range

Macro Particle Range

MICROMETERS (LOG SCALE)

0.001 0.01 0.1 1.0 10 100 1000

APPROX. MOLECULAR WT.(SACCHARIDE TYPE-NO SCALE)

100 200 1000 10,000 20,000 100,000 500,000

RELATIVE SIZE OF COMMON MATERIALS

Aqueous Salts

Metal Ion

Pyrogen

Virus

Albumin Protein

Bacteria

Lung Damaging Dust

Red Blood Cells

Yeast Cells

Human Hair

Beach Sand

Source: (Courtesy A. Gutacker, ARGOT, Inc.)

micron cleaning. Removal efficiency up to 99.999% is commonplace with acceptable pressure drops to 5–10 pounds per square inch (psi). Liquids can be passed over activated carbon beds to remove odors, many dissolved gases, and organic materials by adsorption. Care is required to make sure that the available surface is not saturated with contaminants, or further removal will not occur. Distillation can remove many common contaminants, but ion exchange processes can be even more effective with less energy cost.

Bacteria can be removed by several procedures. Chlorination or ozonation kills most bacteria, but residual pyrogens and bacterial debris must still be removed by filtration or other procedures. Ultraviolet irradiation also destroys bacteria, but the penetration of effective UV radiation through most solids is very short; if a bacteria clump is irradiated, then penetration through more than a few micrometers of the bacteria may not be sufficient to destroy all of the cells. Care is required when ultraviolet irradiation is used in combination with membrane filtration for control of bacteria in water systems. Filtration by charge-modified membranes with positive charge is very effective in removing particles from deionized water supplies because most particles seen there are negatively charged. When ultraviolet irradiation or ozone is used to control bacterial contamination, it has been noted that the charge-modified filters may degrade quite rapidly if exposed to the strong oxidizing effects of these treatments (Shadman, Governal, and Bonner 1990).

For large-scale deionized water systems treatment, a first stage may be a packed coal-and-sand bed. Dissolved materials are usually removed by ion exchange columns, reverse osmosis, or adsorption beds. Particle removal by use of final filters with 0.02–0.2-μm pore size may be required. Experience has shown that bacterial growth can occur frequently in water-treating systems. Once this happens, then the system must be cleaned and disinfected. The disinfection process may require heroic measures. Sanitizing a process water system requires first determining the nature of the bacteria contaminant so as to select the best sanitizing agent and treatment. Such operations as flushing with up to 5% hydrogen peroxide or 10 ppm sodium hypochlorite for an hour or so may be required. These operations must not compromise incompatible equipment, such as some resin or carbon beds that might be affected by these strong oxidizers. All system areas, especially static dead legs, must be exposed to the sanitizing agent. Next the potentially contaminating sanitizing agent must be flushed from the system and its absence verified before the process water can be run through the system. The user of a large deionized water system learns very quickly to make sure that no dead leg lines are allowed to remain in the system. In addition, any filters in the system that do not have adequate flow can also act as a bacteria breeding point. Inadequate system design can cause particle levels and bacteria counts at a point of use to be 10–50 times greater than that in the incoming water (Hango and Pettengill 1987).

Installation and operation of any liquid cleaning method are usually expected to provide clean liquids with few problems. Sometimes the operators of cleaning systems do not eliminate all problems. "Point-of-use" filtration is highly effective if done correctly. Sometimes the filter is installed before a valve or short section of tube that is not properly maintained. In this case, in-line contaminant generation negates the filter performance. Cartridge filters are not always installed properly, and seals and gaskets may not be fitted correctly; operations requiring replacement of a large number of cartridge filters have produced occasions where some of the housings do not contain a filter! A process modification may result in changing liquid use to a material that may rapidly degrade an existing filter medium. Therefore, the same care is required in operating the liquid cleaning system as in any cleanroom installation. The methods commonly used for cleaning liquids are described briefly to aid in selecting the optimal method for clean liquids.

Superclean liquid filtration terminology describes *filtration* as a process in which particles larger than 0.02 μm or so are removed primarily by a sieving action. Ultrafiltration usually refers to removal of particles in sizes smaller than 0.1 μm. Ultrafiltration membranes can remove molecules with molecular weights in the 10,000 range. In ultrafiltration, a cross-flow technique is frequently used. The liquid feed stream flows tangentially along the surface of the membrane, whereas the liquid feed stream in normal filtration is into the filter medium surface. Normal filtration pressures can be as high as 10–20 bars, whereas ultrafiltration pressures are generally in the 3–4-bar range.

In the electronic and pharmaceutical industries, many of the filters used are cartridges with either nonwoven or membrane media. The media are replaceable in housings placed in the process lines. Cartridge filter housings are available in sizes small enough to accommodate a single cartridge 10 cm long or as many as 400 cartridges 1.8 m long. Flow rates up to 300 L per second at pressures to 200 bars can be handled. Care is always required to ensure that the housing and cartridge seals are correctly assembled so that leaks do not occur (Lawton 1987). Some nuclear power plants use cartridge filter systems to maintain cleanliness of coolant water streams at very high flow rates. These systems are always open to the possibility of some radioactive contamination. For this reason, handling problems always exist in terms of in-process leakage control and of after-use disposal. Remote handling systems for containing hazardous components are used as much as possible in this industry (Crinigan 1989).

Nonwoven or fibrous mat depth filters are used mainly as prefilters for final membrane filter systems. A membrane filter is a thin polymer film with pores that may be nearly cylindrical or may be convoluted. Pore sizes range from 40 nm to 10–20 μm. The filtration mechanisms are similar to those that are effective in gas cleaning. In gas cleaning, the mechanisms that are most important for small particles are interception and diffusion; impaction, electrostatic capture, gravitational settling, and sieving play minor roles in that process. The mechanisms that control particle removal in liquid filtration systems are more complex. The physical chemistry of the particle-liquid-filter system controls the operative mechanism. The effects of liquid physical properties, such as increased viscosity, shearing force, decreased boundary layer thickness, and electrostatic contact potentials alter the relations between particle size, fluid flow rate, and filtration mechanism (Grant et al. 1989). For example, the use of positively charged membrane filters was shown to remove essentially all active bacteria and prevent release of bacterial fragments when such a 0.2-μm pore size filter was tested (James et al. 1988). Charge modification of filters is accomplished by covalently bonding additional charge groups to the material surface. The result is a chemically stable surface with numerous cationic charge locations. Negatively charged materials, such as colloidal silica and many bacteria, are attracted to the positively charged surface and retained by multiple bonds at the cationic charge sites.

Reverse osmosis is used for water cleaning as an extremely energy-efficient method. If a pure liquid is placed on one side of a permeable membrane and liquid containing a soluble contaminant is placed on the other side, the pure liquid flows through the membrane to dilute the solution until the pressure difference across the membrane is equal to the osmotic pressure of the equilibrium diluted solution. If a pressure greater than the osmotic pressure is applied to the solution, the system forces pure liquid out of solution through the membrane until the system with applied pressure is again in equilibrium. Figure 9-1 illustrates the process. This is basically the same process seen in ultrafiltration, except that reverse osmosis is effective on smaller molecules, and higher pressures are used (Gooding 1986).

Ion exchange resins are polymers with fixed ionic charges that can exchange with free ions of opposite charge in the liquid to be cleaned. Ion exchange systems can consist of either packed beds of resin particles or of resin sheet membranes. Either cation or anion exchange resins can be used, depending on the ions to be removed. In mixed-bed systems, both types are in use.

A variety of methods are used to control bacterial growth in pure water supply systems. One of the most important procedures is adequate monitoring to ensure that the control methods are effective. The microbiological problem materials may range in size from less than 0.01 μm (pyrogens)

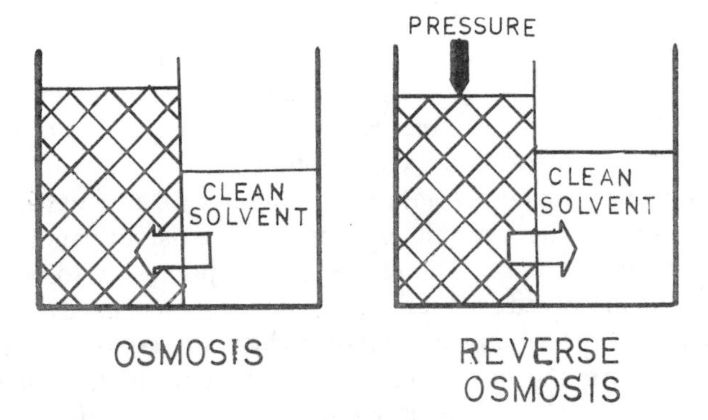

FIGURE 9-1. Reverse osmosis basis. A simplified illustration of the mechanism used to reverse the normal diffusion path through a membrane. Sufficient pressure causes the solvent to move from the concentrated solution to the more dilute one. (From Gooding, C. H., 1986. Reverse Osmosis and Ultrafiltration Solve Separation Problems. *Chemical Engineering* 92(1):56–62.)

to nearly 10 μm for some bacteria. Effective monitoring methods include bacterial colony growth observation, observation by scanning electron microscopy or epifluorescent microscopy, and limulus amebocyte lysate testing for pyrogen and endotoxin profiling (Carmody and Martyak 1989). Most pure water bacterial control methodology uses ultraviolet radiation at 254 nm in some portion of the flow line. This method is effective on bacteria that are irradiated, but most bacteria are in clumps on or in interstices of other contaminants or otherwise covered, so that the radiation cannot affect at least some of the bacteria. Common sterilization treatment for potable water involves addition of sufficient chlorine compounds so that a residual chlorine level exists in the water as it passes through water lines. In many cleanroom applications, the chloride ion content is unacceptable. For this reason, other methods have been investigated. In the pharmaceutical industry, ozone concentrations of 8–10 μg per L have been shown to be effective in preventing microbial growth (Werner 1983; Nebel 1988). It has been also recommended that continuous low-level ozonation be used to control biological contamination in semiconductor process water systems (Pittner and Bertier 1988). As a variation on this method, photolytic ozonation has been found to destroy many organic compounds that are refractory to oxidation by ozone alone (Peyton and Glaze 1988). The procedure irradiates ozonated water with ultraviolet radiation. It is found that photolysis of the ozone produces hydrogen peroxide, followed by secondary reactions to produce hydroxyl radicals that remove organic compounds. The combination of exposure to the energetic oxidant and to the radiation accelerates the oxidation process markedly.

Although water represents the largest use of liquids in cleanrooms, many other liquids must be cleaned, including organic and inorganic process liquids and solvents used to clean materials for which water is either ineffective or incompatible. Even though the quantity of water used in contamination control areas exceeds other liquids, many manufacturing operations require long-term intimate contact with process liquids that must also be clean. The same cleaning problems exist for handling process chemicals as for any liquid. The liquid must be processed to remove both particulate and dissolved contaminants. The containers used for shipment, storage, and use must be cleaned thoroughly; any contaminant addition that occurs before and/or during use must be controlled. For example, hydrofluoric acid is widely used in semiconductor processing. This aggressive material can be filtered with a fluorocarbon resin medium filter in a fluorocarbon container to levels below 10 0.5-μm particles per L (Kikuyama et al. 1989). This level was not exceeded when the material was stored in carefully cleaned perfluoroalkoxy (PFA) containers. A linear relationship has been reported for particle concentration in acid baths (Milner and Brown 1986) and the particles per unit area on semiconductor wafers processed therein.

Significant quantities of process chemicals are used, particularly in the semiconductor industry. A wafer fab may require use of several tens of thousands of gallons of acids such as sulfuric, hydrochloric, and hydrofluoric acid. Each wafer processing operation adds relatively small quantities of contaminant to the acid, but even those quantities may be unacceptable. Disposal of the used process liquid results in costs that may equal or exceed the original purchase cost and that may cause environmental damage both in and out of the plant. For these reasons, many semiconductor manufacturers reprocess chemicals for reuse.

Acid reprocessing requires removal of unwanted metallic ions and of contaminant particles from the acid. In some reprocessing operations, proprietary separation technology is used to remove metallic ions. A photoresist stripping material (commonly referred to as *piranha*) used to remove photoresist after lithographic processing consists of a combination of sulfuric acid and an oxidant such as hydrogen peroxide, ammonium persulfate, or peroxydisulfuric acid. After first use, the piranha can still be used for stripping, but it must be cleaned before reuse. Commercial reprocessing systems contain a stripper column for removing water and a distillation column for trace metal purification and, with filtration, for particle removal. Significant savings in acid costs and reduced waste chemical disposal requirements are reported (Cleavelin and Jones 1988). Particle removal from process chemicals can be accomplished by filtration through appropriate membranes, using pumps and flow control systems also resistant to chemical attack by the liquid being filtered. It is necessary to select the correct materials for the filter medium, the filter housing, the flow control system, and all components so that the process liquid is not degraded by reaction products formed with any of the filter system components. The filter manufacturers can usually define correct materials for specific operations, but care in verifying compatibility of all materials is always required. A variety of fluorocarbon membranes with pore sizes of 0.1 or 0.2 μm can be used to remove particles from either acidic or basic liquids and from organic solvents. The membranes are frequently used as stacked disks in cartridge assemblies to accomplish desired flow rates. A sulfuric acid reprocessing system consists of a stripping column to remove low-boiling components, such as water and carbon dioxide, followed by a high-temperature distillation column to purify the sulfuric acid for reuse (Jones and Hoffman 1988).

Most organic liquids can be processed for reuse by distillation at somewhat lower temperatures than required for sulfuric acid distillation. In either case, adequate safety procedures must be used throughout. In addition, care is required that the molecular structure of an organic process liquid is not changed by the cleaning process. For example, photolithographic processes are carried out with photoresist liquids that may be viscous, containing molecules that may polymerize in storage to form gelatinous precipitates, which can result in defects during semiconductor processing. Microfiltration is often used to remove these materials. Filtration rates should not be so high that flow shear stresses can cause changes in the structure of some high-molecular-weight compounds in viscous photoresist liquids (Blazka and Hegde 1989).

References

Blazka, S., & Hegde, R., 1989. Effect of Microfiltration on Photoresist Quality and Integrity. *Microelectronic Manufacturing and Testing* 12(7):33–35.

Carmody, J. C., & Martyak, J. E., 1989. Controlling Bacterial Growth in an Ultrapure Deionized Water System. *Microcontamination* 7(1):28–35.

Cleavelin, C. R., & Jones, A. H., 1988. The Manufacturing Impact of Continuous Point-of-Use Chemical Reprocessing. *Solid State Technology* 31(12):53–56.

Craven, R. A., Ackerman, A. J., & Tremont, P. L., 1986. High Purity Water Technology for Silicon Wafer Cleaning in VLSI Production. Proceedings of the 4th Millipore SEMI Symposium, May 19, 1986, San Mateo, CA.

Crinigan, P. T., 1989. Liquid Filtration in Nuclear Power Plants. *Ultrapure Water* 6(1):52–54.

Gallagher, D. J., 1989. High-Purity Semiconductor Process Chemicals. *Microelectronic Manufacturing and Testing* 12(4):21–22.

Gooding, C. H., 1986. Reverse Osmosis and Ultrafiltration Solve Separation Problems. *Chemical Engineering* 92(1):56–62.

Grant, D. C., et al., 1989. Particle Capture Mechanisms in Gases and Liquids: An Analysis of Operative Mechanisms in Membrane/Fibrous Filters. *Journal of Environmental Science* 32(4):43–51.

Hango, R. A., & Pettengill, N. E., 1987. A Cost Effective Approach to DI Distribution System Upgrade. Proceedings of the 6th Pure Water Conference, January 15, 1987, Santa Clara, CA.

James, J. B., et al., 1988. Bacterial and Endotoxin Retention by Charge-Modified Filters. *Semiconductor International* 11(4):80–82.

Jones, A. H., & Hoffman, J. G., 1988. Point-of-Use Acid Reprocessing for Semiconductor Applications. Proceedings of the 35th Institute of Environmental Science Annual Technical Meeting, pp. 474–477, May 19, 1988, King of Prussia, PA.

Jones, D., 1986. Submicron Particle Levels in Process Chemicals. Proceedings of the 4th Millipore SEMI Symposium, May 19, 1986, San Mateo, CA.

Kikuyama, H., et al., 1989. Cleanness Technology of Hydrofluoric Acid. Proceedings of the 35th Institute of Environmental Sciences Annual Technical Meeting, pp. 369–376, May 1989, Anaheim, CA.

Lawton, D. J., 1987. What Makes Cartridge Filters Perform Effectively. *Chemical Engineering Progress* 83(11):20–25.

Milner, T. A., & Brown, T. M., 1986. A Model for Predicting the Effect of a Processing Bath on Wafer Particle Contamination. Proceedings of the Microcontamination Conference, November 1986, Santa Clara, CA.

Nebel, C., 1988. Ozone, the Process Water Sterilant. *Ultrapure Water* 5(8):40–49.

Pengra, D., Hogsett, L., & McCall, D., 1990. Correcting Bacterial Contamination of DI Water in GaAs Wafer Prep Sinks. Proceedings of the 36th Institute of Environmental Science Annual Technical Meeting, pp. 217–220, April 23, 1990, New Orleans.

Peyton, G. R., & Glaze, W. R., 1988. Destruction of Pollutants in Water with Ozone in Combination with Ultraviolet Radiation: 3. Photolysis of Aqueous Ozone. *Environmental Science and Technology* 22(7):761–767.

Pittner, G. A., & Bertier, G., 1988. Point-of-Use Contamination Control of Pure Water through Continuous Ozonation. *Ultrapure Water* 5(4):16–22.

Rechen, H. C., 1985. Microorganism and Particulate Control in Microelectronics Process Water Systems: Pharmaceutical Manufacturing Technology. *Microcontamination* 3(7):22–29.

Shadman, F., Governal, R., & Bonner, A., 1990. Interactions between UV and Membrane Filters during Removal of Bacteria and TOC from DI Water. Proceedings of the 36th Institute of Environmental Sciences Annual Technical Meeting, pp. 221–228, April 23, 1990, New Orleans, LA.

Siegmann, R. P., 1987. Point-of-Use Contamination Control of Ultrapure Deionized Water for Semiconductor Processing. *Microcontamination* 5(5):22–26.

Stewart, D. A., et al., 1987. Contamination Control in Chemicals and Water. Proceedings of the 6th Pure Water Conference, January 15, 1987, Santa Clara, CA.

Werner, R., 1983. Anti-Microbial Processing of Deionized Water in the Context of Clean Area Technology. *Swiss Pharma* 5(11a):27–31.

Zoccolante, G., 1987. Innovations in Water Purification. *Semiconductor International* 10(2):86–89.

10

Gas Cleaning Methods for Ambient Air and Compressed Gases

Cleaning air or compressed gases in cleanroom installations requires removal of particulate and/or gaseous contaminants. The technology used for cleaning gases for the cleanroom is derived from processes long used in industrial applications for dust and smoke removal. Removal of particles from gases in industrial applications may involve methods such as centrifugal cleaning, wet scrubbing, electrostatic precipitation, or filtration as used for fossil fuel power plant emission control. For cleanroom operations, filtration is essentially the only process by which particles are removed from both ambient air and from compressed gases; contaminant gases are removed either by dry bed adsorption or by wet scrubbing. Electrostatic precipitation is practically never used in cleanroom operation for control of ambient air cleanliness. The method is considered suspect because operators are concerned about possible particle emission from precipitator elements in the case of power failure. However, there is increasing interest in electrostatically augmented filtration. This process operates by improving particle deposition to the filter fibers. Even if a power failure occurs, the material already collected within the filter remains there. More details are given later.

Modern high-efficiency ambient air filtration system component designs are derived from research that was carried out before World War II. At that time, development of better respirator filters for protection against chemical and biological warfare agents was carried out, and sensitive filter test methods were investigated. The use of small-diameter glass-fiber particle filters for protection against possible bacterial and chemical warfare agents, development of high-efficiency activated carbon adsorption beds, and development of high-sensitivity, light-scattering, small-particle detection systems used for defining penetration through filters was advanced at that time. This work was carried out first in Europe and in England and then in the United States. Essentially all membrane filters are based on developments begun in Germany before and during the war years.

During the latter stages of that war and afterward, extensive work was carried out by the U.S. Atomic Energy Commission (now Department of Energy) on application of better and safer high-efficiency particle filters and radioisotope gas adsorbers. These devices were required to protect the environment against emission of radioactive compounds from plants working on nuclear weapons and nuclear power sources. Possible emission of radioisotope-contaminated air was minimized by keeping the plant air pressure below that of the ambient environment and by venting work area air through high-efficiency gas cleaning devices. Much of the technology used to protect the cleanroom from the environment is derived from that developed in the 1950s to protect the environment from contamination produced in a manufacturing operation. The major difference is that the radioactive material processing and manufacturing operations were maintained at a lower static pressure than that of the environment that was being protected, whereas present-day cleanrooms operate at a higher pressure than that of the environment from which the cleanrooms are protected. In this way, air flows from the protected area to the dirtier area, rather than the reverse.

Cleanroom air is filtered by HEPA or ULPA filters after it has passed through prefilters to remove larger particles. A HEPA filter has a removal efficiency of $> 99.97\%$ for 0.3-μm particles, and an ULPA filter has a removal efficiency of $> 99.997\%$. These filters are made of small-diameter (ca. 0.1 μm) fibers randomly dispersed in the filter medium. Prefilters commonly used have efficiencies of 80–95% for airborne particles.

The filter medium has a very high porosity to permit maximum airflow at minimum pressure drop. Pore sizes are usually significantly larger than the particles that are removed. As is pointed out later, the pore size for high-efficiency air filters is not of primary concern in defining the minimum size of particles removed. Particle removal by a sieving action is effective only for particles much larger than those of concern in the cleanroom.

For work in areas that do not require extremely clean air, the air is filtered and supplied to diffusers in the cleanroom ceiling. The clean air moves through the cleanroom in random flow paths and is eventually exhausted through floor-level vents at various locations in the room. For very clean operations, the filtered air is directed toward the work area. The filters in operational areas are located directly upwind of the work area so that the flow from the filters does not transport contaminants from local particle sources to the work area. Filters may be placed in a wall upstream of the work area or in the ceiling space directly above the work area. Installation and operation of such systems are discussed later.

Many theoretical and experimental studies have been carried out on filtration. Models were first developed to understand the collection mechanisms for single fibers (Langmuir 1942). Next, several models were developed to study interactions among several fibers in a filter body. The models considered airflow through the fibrous matrix of the filter as well as the mechanisms of particle transport, deposition, retention, and release from the fibers over a range of particle and airflow parameters. Lee and Liu (1982) presented a model that has shown good correlation with experimental data. Needless to say, the two references just pointed out represent only a small part of the reported work on fibrous filter operation. The physical filtration mechanisms that result in particle removal are discussed briefly here. These mechanisms are gravitational sedimentation, impaction, interception, electrostatic effects, and diffusion. Because these filtration processes involve particle transport and deposition, the discussion in chapter 7 of forces affecting particle motion and the fluid and particle parameters affected must be kept in mind.

The operative filtration mechanisms for air filters are shown in Figure 10-1. The relative importance of the several mechanisms varies with particle size, and each filtration mechanism changes with fluid flow rate. Sedimentation is of importance mainly for particles larger than 5–10 μm in diameter. Particles in that size range are seldom seen in cleanroom airflow to and from the filter. They are generated locally and usually deposit on substrates close to their point of generation. For that reason, this mechanism is not discussed here. Impaction of a particle upon a fiber from an

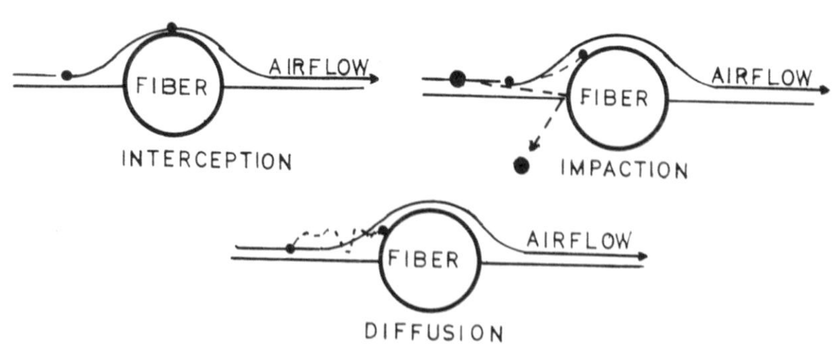

FIGURE 10-1. Particle capture mechanisms in fibrous filtration. The filter pore size that controls removal of particles by "sieving" is of minor importance. The major mechanisms involve particle collection *upon* the fiber surface rather than entrapment within a pore.

airstream moving through the filter medium occurs when a particle is large enough and has sufficient inertia to continue on in its original direction while the air streamline changes direction in passing around the fiber. The particle then impacts upon the fiber and is captured. Interception of a particle by a fiber occurs when the particle is embedded within an air streamline passing close to the fiber. The particle is large enough to collide with the fiber while the particle is still within the streamline. Again, once the particle contacts the fiber, it is assumed that the particle will be retained by that fiber. Electrostatic effects can occur with particles of any size, as long as the electric forces exceed the aerodynamic forces tending to keep the particle within the streamline flow path. The electrical force then moves the fiber out of the airflow path to result in deposition upon the fiber surface. Diffusion is of importance mainly with small particles, which can diffuse sufficiently out of the gas streamline as it is passing the fiber to result in contact with the fiber. In essentially all cases, once a particle strikes a fiber, it is retained by that fiber. Solid particles are retained by van der Waals forces, mechanical entrapment, or electrical effects. Liquid droplets are also retained by surface tension force.

The filter efficiency changes with particle size as the filtration mechanism varies, as shown in Figure 10-2. As particle size decreases below 1 μm in diameter, the filtration efficiency due to impaction and interception decreases. At the same time, the filtration efficiency due to Brownian diffusion begins increasing as the particle size decreases. For that reason, there is a particle size at which maximum penetration occurs through any filter. When a HEPA filter is operated at normal airflow of a few cm per second face velocity, the maximum penetration is about 0.03% for particles of 0.1–0.3-μm diameter. The particle size for maximum penetration through HEPA filters increases with increasing face velocity, as shown in Figure 10-3. This occurs because the time for diffusion to occur varies inversely with air velocity through the filter. A more detailed discussion of efficiency variation for fibrous filters was given by Lee and Liu (1980).

It is emphasized that filters are not absolute particle removal devices. Particle retention of more than 99.97% or so, even for the most penetrating particle size, is not total. However, reference to Figure 10-3 shows that particle penetration through a normal HEPA filter at the worst level is still less than 0.03% per pass. Consider that the particle challenge to the HEPA or ULPA filter has been reduced from ambient air concentrations by a prefilter and that recirculation of air through the HVAC and filter system means that most of the air in the cleanroom has been filtered many times. One can see how particle concentrations in cleanroom air can be reduced to barely measurable levels when the internal sources of particles in the cleanroom are not significant.

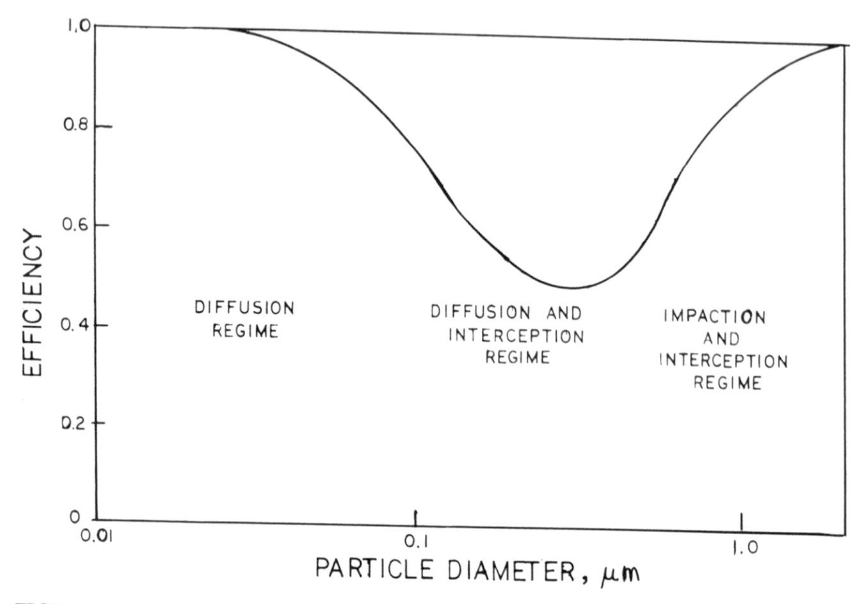

FIGURE 10-2. Fibrous filtration mechanisms and efficiency versus particle size. Diffusion becomes less effective as particle size increases, whereas impaction and interception become less effective as particle size decreases. This results in a minimum in filtration efficiency for various filter media. A HEPA filter media efficiency curve is illustrated here with a most penetrating particle size of approximately 0.3 μm.

The filters used in cleanrooms are usually 2 feet square or 2 feet by 4 feet across, with depths of 6–12 inches and rated airflows of 500–1,000 cubic feet per minute (cfm). Figure 10-4 shows the structure of a typical commercial HEPA or ULPA filter. The filter medium is pleated to provide for maximum filter medium area in a minimum filter element face area. The depth of the filter varies as filter medium area and filter airflow selection is changed. The filter medium is glued to the filter frame. Most modern cleanroom filter frames are metal. The filters are sealed at the frame edges to structural retainers by either a flexible gasket material or an edge sealed in a polymer gel. The filters are located in the ceiling for downflow rooms, in one wall for horizontal-flow rooms, or in the HVAC ducts for many mixed-flow rooms. Filters are placed in clean benches just ahead of the work surface. Air velocity at the filter face exhaust is normally 90–100 feet per minute, although some installations have been operated at lower flows in special situations in which economy or gas flow problems may exist. The 90–100 feet per minute velocity is used to maintain unidirectional airflow in a cleanroom area. If the flow decreases much below 75 feet per minute, the

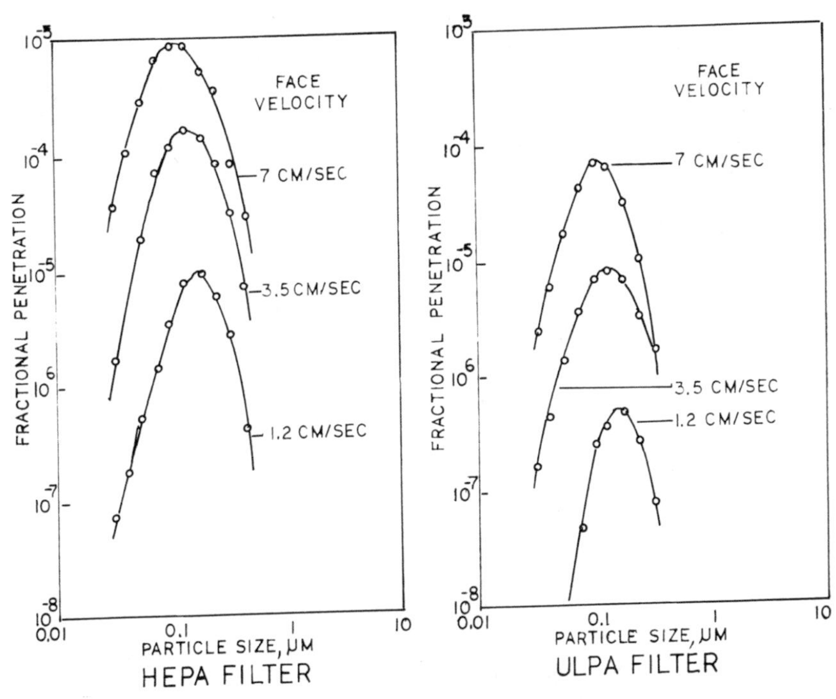

FIGURE 10-3. Penetration versus particle size for HEPA and ULPA filters. The two penetration plots show that the finer fiber ULPA filter results in decreased penetration for particles of equivalent size than does the HEPA filter. In addition, the most penetrating particle size is smaller for the same airflow velocity. (From Lee, K. W., & Liu, B. Y. H., 1980. Theoretical Study of Aerosol Filtration by Fibrous Filters. *Aerosol Science and Technology* 1(2):147–162.)

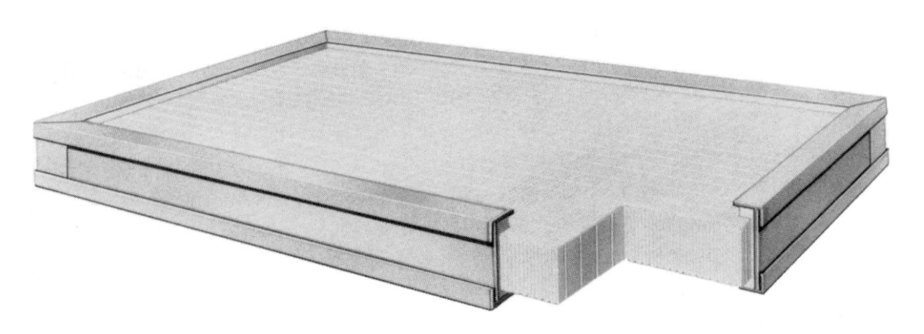

FIGURE 10-4. HEPA/ULPA filter structure. This structure has been used for many years. Some newer filter structures do not use corrugated separators between the filter pleats, as shown here. The newer filter media folding and pressing procedure adds spacers to the pleat structure that separates the folds. (Courtesy American Air Filter Corp.)

unidirectional flow structure of the air mass appears to weaken, with recirculating eddy flows appearing in the flow pattern. The importance of maintaining unidirectional flow in cleanrooms is discussed later. If the gas that is being filtered changes in composition or pressure during that process, the filtration efficiency for particle removal also changes, depending on the particular mechanism most important for the particle size of concern and the effects of the pertinent gas properties of density, viscosity, and gas molecule mean free path length.

Pressure drop for a clean filter is normally no more than 1 inch of water. For good performance and economy, a cleanroom filter system should include a 90–95% prefilter upstream of the HEPA or ULPA filter to remove gross contaminant particles before final filtration. The filter installation is normally carried out with filters whose integrity has been verified previous to delivery and installation at the cleanroom facility. The major problems in installation are to make sure that there are no leaks in the medium or at the edges of any filter and that a filter has not been damaged before installation. A leak may be present that permits gas flow of a few hundred cubic centimeters of unfiltered gas per minute. This leak rate represents one part per 100,000 of the total gas flow in a 1,000 cfm filter. The desired airflow condition downstream of the filter would probably be unchanged whether a leak is present or not. The particles present in this quantity of unfiltered gas may not be noticeable in defining the overall filter penetration, but if a critical operation is carried out just downstream of the leak area, serious problems can result. The leak can act as a source of concentrated contamination focused upon a critical operation. Filter-testing methodology for overall penetration and for leak detection are discussed in more detail later.

For the most part, filter testing is designed to verify the integrity of the filter medium and of its installation in the filter frame (freedom from leaks). HEPA and ULPA filters in cleanrooms also provide unidirectional airflow as well as cleaning the air. For this reason, overall cleanroom filter performance can be well described by use of a "figure of merit." This rating can consider both particle removal efficiency and minimum pressure drop so that airflow control and cleaning can be maximized with minimum cost. One expression has been suggested (Rubow 1981) as "a means of expressing the efficiency of a filter medium in relation to the pressure loss created." This figure of merit is directly proportional to penetration and inversely proportional to the ratio of pressure drop and air velocity. The direct term indicates the filter particle control efficiency, and the inverse terms indicate electrical energy cost for airflow. The concept of a figure of merit can be developed for other filter operating parameter terms.

As contamination from particles smaller than 0.1 μm becomes an increasing problem in semiconductor manufacturing, some of the "weak" forces that can affect only small-particle transport in the air become of interest. In particular, thermal effects are being examined more and more at this time. Consider that air molecules that are heated, as by contact with a warm surface, move more rapidly than cool molecules. The faster-moving air tends to move toward cooler surfaces. As the air molecules move in a preferred direction, they can transport small particles in that preferred direction. Thus, the particles can be removed from a particular area of the cleanroom into an area where they are of no concern. This consideration shows the potential importance of localized temperature control in the cleanroom as another contamination control operation for the near future.

Although some positive and negative ions are normally present in all gases, they are considered as potential contaminants in many cleanrooms and must be neutralized. Ions can be especially troublesome in areas where semiconductors are the product because of the electrical damage that can occur. However, airborne ions can attach rapidly to airborne particles and result in charge buildup on the particles. When this occurs, the charged particles are attracted by electrical forces to surfaces with the opposite charge. The surface can also be that of another particle. The particle-deposition problem can be aggravated by the presence of excessive quantities of monopolar ambient air ions. This is a special problem when surfaces that may be grounded or oppositely charged may be present. It has also been reported (Fray 1984) that application of complex electrical fields can modify particle concentrations in airstreams, although no specific mechanism was presented. A more recent discussion (Huffman, Nichols, and Bossard 1988) studied a model based on Brownian motion of particles and of electrostatic forces that could be applied through ionization. It was shown that an ionization system that reduces electrostatic energy levels on products can reduce surface particle contamination considerably.

The methods for eliminating an excess of high-energy, single-polarity ions from cleanroom air result in neutralization by supplying sufficient bipolar ions to reduce the net charge in the air to a level that approaches Boltzmann equilibrium, where the net charge is very low and slightly negative. Two procedures can be used. In one, an array of metal needles is charged by an alternating high-voltage (7–15 kV) source; the needlepoints emit bipolar ions as the charge polarity alternates. In the other, a radioactive source such as polonium-210 in the airstream emits alpha particles. These radioactive particles produce large numbers of both positive and negative ions when they collide with air molecules. Airflow carries the bipolar ions away from the source and reduces electrostatic discharge hazard in cleanrooms by bringing the charge level in the air close to Boltzmann equilibrium value.

Some high-voltage needle materials may produce small-particle emission with use. Recent work indicates that this problem can be controlled by suitable needle material selection and voltage application control. Tests (Jots and Liberia 1989) showed essentially zero emission of 0.02-μm particles from an operating needle type of ion source system for at least an 800-minute test period.

To clean contaminant gases or vapors from processing gas, either scrubbing or adsorption can be used. Gas molecules have sufficient mobility to diffuse through air quite rapidly. The processing operation is normally carried out at one specific location. However, if gas is released to the air during operation, it can spread throughout a large air volume quite rapidly. In this case, an entire cleanroom air system may have to be cleaned to remove a small amount of contaminant gas unless a localized vapor control system is used to remove the gas directly as it is generated. Obvious savings in HVAC and energy costs can be attained in this manner. Although scrubber designs vary, the overall operating principles are similar. The gas to be cleaned is passed through a spray of rather coarse (ca. 5–50 μm) droplets composed of chemically reactant material that absorbs and retains the contaminant vapor. The coarse droplets are easily collected by an inertial collector in the scrubber, and the liquid is recycled for further use or is discarded after the reactant is saturated. Gas collection efficiencies of 80–99% can be expected. A schematic scrubber design is shown in Figure 10-5. Even though the absorption liquid can be eventually recycled, safe *and legal* disposal of the saturated scrubbing liquid after use is necessary. This liquid is usually defined as hazardous by local safety and health authorities.

Adsorption is used for more complete removal than can be accomplished by scrubbing, usually of organic gases or vapors. The gas to be cleaned is passed through a bed of adsorbent granules, each of which has a high surface area. Activated carbon granules, for example, can have a surface area of many hundred square meters per gram. The gas to be removed is adsorbed on the interior or exterior surface of the granules until no further free surface is available. The sorbent granules can be discarded after use or, for more costly materials, the sorbent can be reactivated by desorbing the material that has been collected. Heat is often used to reactivate sorbents. The adsorbed contaminant gases are vaporized and released from the sorbent material. The released gases must be disposed of

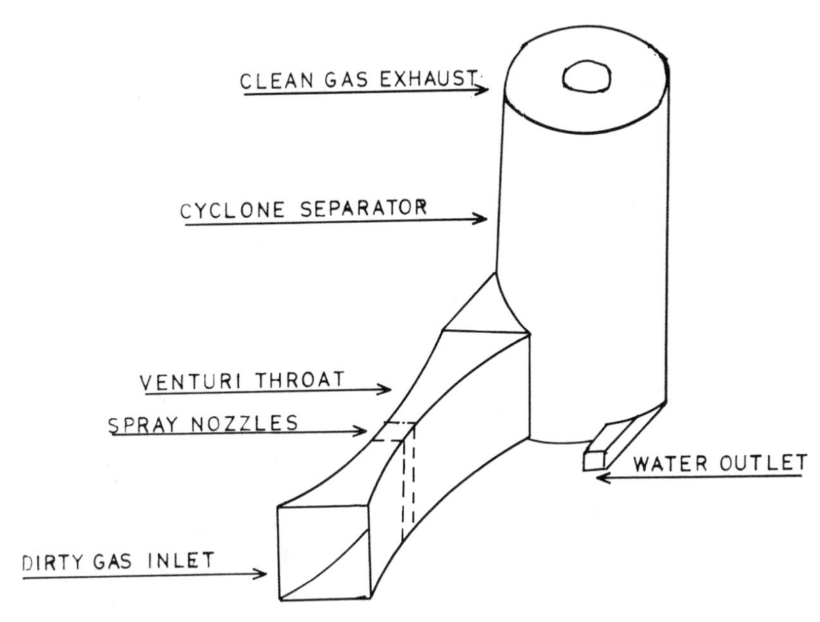

FIGURE 10-5. Wet scrubber schematic. The droplet generator produces large drops in a highly turbulent zone. Particles in the dirty gas impinge upon the droplets and are collected. The large droplets are collected by centrifugal force upon the walls of the cyclone separator, removing the collected particles as a slurry in the water outlet line.

in a safe manner that does not pollute the environment. Catalytic combustion or high-temperature incineration of toxic gases has been used to convert the toxic materials to relatively inert oxidized materials.

Another method of producing certain very clean gas, such as nitrogen, oxygen, or some noble gases, begins with cooling air to the point where the gas condenses to liquid. The various constituents are then separated by a modified fractional distillation process, taking advantage of the differences in the boiling points (Sugiyama and Ohmi 1988). Hydrocarbon impurities must also be removed before cooling. Impurities that must be removed vary with each process gas. Quite frequently, a particular material that is required for one process may become an impurity in another process. For this reason, many cleaning methods are required for process gases. For extremely critical operations, point-of-use gas purification methods may be required in addition to in-process cleaning operations (Hardwick, Lorenz, and Weber 1988).

Cleanroom process gases are usually provided in clean condition by the manufacturer or supplier. Any waste and effluent gases produced during processing must be cleaned before venting either in the ambient air or to the external environment. The use of acutely toxic materials in semiconductor manufacturing has made waste gas disposal a serious problem. Pharmaceutical manufacturing operations include use of many solvents and process materials with high vapor pressures and may use or produce viable products whose emissions must be controlled. Plant personnel health and safety needs make it necessary to ensure that the work environment is not contaminated with any of the very toxic materials. Municipal environmental control agencies monitor waste disposal processes to control emission of harmful materials of all types. For these materials, selection of the optimum disposal and control measures is required. Cost and effectivity must be considered. Simple dilution can be used when the waste gas streams do not contain very high concentrations of toxic gases and if an adequate safety factor is provided to ensure that the vented gas concentration of toxic material is well below the published threshold limit values. Wet scrubbers are effective for many materials, but the material to be removed must be soluble in the scrubber liquid, and the resulting solution must be handled safely. Thermal processing of harmful waste gases is effective for combustible materials, but it is also necessary that the combustion products not be harmful to the environment. In addition, effective safety measures are required for this method. Gas adsorption is very effective for relatively small quantities of waste gas purification. Maintenance and scheduled replacement of the sorbent material may be required. For more information, a detailed discussion of gas phase adsorption requirements is available (Institute of Environmental Sciences 1983) that can be helpful in disposal system design. In any disposal system consideration, it is helpful to consider all of the potential problems and fine details in designing a toxic gas abatement system. Scrubbers, adsorbers, thermal decomposition, dilution, and chemical dissociation can all be helpful and should be examined carefully in the design process.

Compressed gases are used in many semiconductor fabrication areas as process materials. The gas compositions vary over a wide range of materials. A complete description of all the process gases, their applications, and the contaminants of concern is beyond the scope of this discussion. The gases used include nitrogen, argon, helium, hydrogen, oxygen, hydrogen chloride, arsine, phosphine, silanes, boron compounds, carbon dioxide, and mixtures of gases. As can be seen, the process gases include both inert gases and gases that are extremely hazardous. Very clean inert process gases are often supplied in large quantities from either on-site production facilities or from large tank transporters. Hazardous gases are usually supplied in smaller cylinders at high pressure. Bulk gas supplies are usually operated at pressures of 150–200 pounds per square inch gauge (psig) (1,400 kPa), whereas cylinder gas pressures may be as much as 2,000 psig (13,900 kilo Pascals). In either case, it is necessary that the gases be clean. Particulate removal is usually achieved by filtration, and gas-handling components are designed and prepared to minimize particle emission as a result of gas flow and/or mechanical abrasion. It has been shown (Hardy, Christman, and Shay 1988) that changes in mechanical design and electropolishing of regulator, valve, and cylinder components can significantly reduce particle contamination of the gas stream from a high-pressure cylinder.

Process gas composition and cleanliness can affect a variety of semiconductor manufacturing steps. The gas may be used as an etchant or oxidizer for processing deposited substances on wafers; it may be a source of material for deposition on the wafer; it can be used as an inert blanket during thermal processing. Potential contaminants vary with each application. For example, moisture contaminates hydrogen chloride used for etching purposes, hydrocarbons must be removed from oxidant gas, and unwanted metal compounds must be removed from all gases used for chemical vapor deposition processes.

The requirements for cleaning compressed gases differ from those for cleanroom air in that the cleanroom air is recirculated through the contamination control devices while the compressed gas is passed through the devices only once. Following use in the process, the gases are seldom used again. Relatively inert gases, such as oxygen, nitrogen, and helium, as well as reactive, toxic, and corrosive gases such as silane, hydrogen chloride, and phosphine are used. Because the gases are in intimate contact with products as part of the manufacturing operation, they must be very clean. Some process gases are supplied by evaporation from a pure liquid or are delivered to the manufacturing facility by a gas producer. In either situation, the compressed gases are treated so that they are normally quite clean, as received. As an example of gas preparation requirements, the high purity nitrogen used in electronics manufacturing is produced by low-temperature distillation of atmospheric air, which contains many materials that must be removed before use. A variety of organic and inorganic compounds are present in concentrations ranging from 0.01 to 1,000 ppm. Metallic compound particulate contamination may be present in concentrations of several milligrams per cubic meter of air. Low-temperature distillation and filtration removes contaminants to less than ppb levels (Thorogood et al. 1986). For removal of gaseous impurities such as oxygen, moisture, or hydrocarbons, catalysts can be used. Palladium catalyzes a reaction between hydrogen and oxygen to form water so as to remove trace oxygen. Some gases can be removed by sorption on molecular sieve to collect water, carbon dioxide, and some hydrocarbons.

For removal of particles from compressed gases in use, membrane filters are widely used. Fluorocarbon membranes, such as polyvinylidene fluoride (PVDF), were shown capable of removing particles in the size range of 0.035–0.089 μm with efficiency greater than 99.9999997% (Accomazzo, Rubow, and Liu 1984). The actual efficiency could not be defined because measurement methodology was limited. Long-term testing has shown that compressed gas cleanliness levels for particles from 0.01 μm to 3 μm down to 10^{-5} particle per cm^3 can be attained (Kasper and Wen 1986). This type of filter can be characterized in a way similar to the way in which atmospheric air HEPA or ULPA filters are characterized (Rubow, Liu, and Grant 1988). A most penetrating particle size can be defined at 0.035–0.045 μm for face velocities in the 5–50 cm per second range. Penetration varies inversely with face velocity, whereas the most penetrating particle size increases as face velocity increases. Different gases affect filtration differently with various filter designs and media (Chahine et al. 1989). Some aggressive gases degraded performance of many filter media significantly, even that of normally inert fluorinated hydrocarbon media.

Normal procedures for process gas cleaning and use involve passage through piping, valve, and flow controls and operation within process tools. In all of these devices, the gas can become contaminated from operation of these items. For that reason, final point-of-use filters are recommended. However, care is required in filter selection and installation. Aside from the normal concerns with leakage and seals, some studies have shown that some compressed gas filters can shed particles that have been incorporated in the filter during manufacture or that are released from the filter element during use when line pressure or vibration affects the filter. Mechanical shock appears to produce particle shedding from most filters when new, but fiberglass elements are reported to shed particles after more than 20 hours of airflow (Accomazzo 1986; Jensen and Goldsmith 1987). After the compressed gas has been cleaned, then it is also necessary to be sure that no handling components can release particulate contamination into the clean gas. Component surface processing and treatment must assure freedom from particle release during gas flow (Bourscheid and Bertholdt 1990). Materials of construction must be especially selected for use with clean gases, surfaces must be treated so that particle deposition is minimal and component design should be carried out to minimize particle collection and subsequent reentrainment. With care, it is possible to remove essentially all particles from a compressed gas (Liu and Hsieh 1989). Figure 10-6 shows a schematic of the gas purification system used. A series of filters and an activated carbon adsorption bed were used to remove all particles from a compressed air supply. In addition, the flow system materials and design were very carefully chosen.

Finally, any operator of an electronics maufacturing facility that uses process gases must keep in mind the potential problems in disposing of gases after use. Many of these materials are extremely hazardous in terms of both flammability and toxicity. Concern for the environment and for safety within the plant can never be relaxed. Precautions must be implemented at all times. Gas containers should be stored outside the work area in a well-ventilated location. The containers should be inspected for visible damage that may result in leaks before being stored. Flammable gases should be stored in safe areas and should be protected from potential ignition sources. Backup cylinder placement in the work area should be

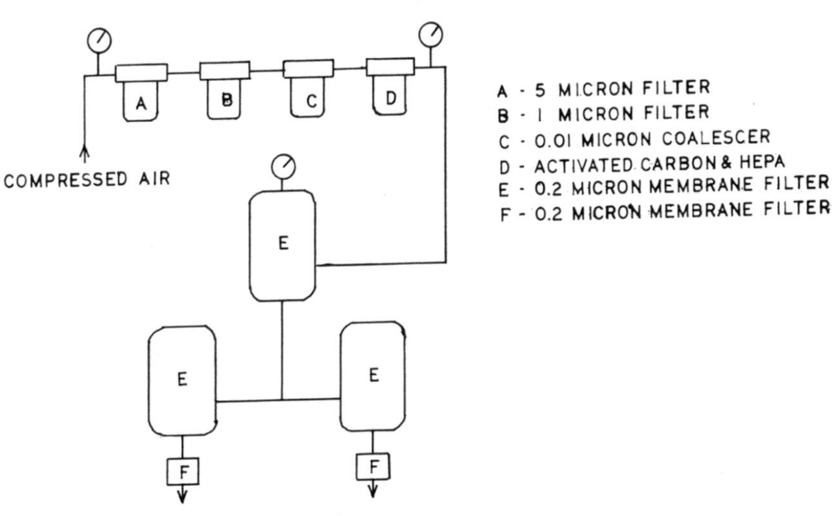

FIGURE 10-6. Zero-particle gas filtration system. This system removes both particles and a large portion of the condensible gases, which may produce droplets if adiabatic expansion during passage of the gas through the system reduces gas temperature to the point where nucleation occurs. (From Liu, B. Y. H., & Hsieh, K. C., 1989. Progress towards an Absolute Zero Particle Gas. Proceedings of the 35th Institute of Environmental Sciences Annual Technical Meeting, pp. 397–400. September 1989, Anaheim, CA.)

minimized. Hazardous gas containers in the work area should be installed in safe gas cabinets that may contain nitrogen purge systems, as well as detector devices. Work area monitoring should be carried out on a regular basis when using toxic gases. Finally, personnel who may handle these materials should be trained in potential problem areas and remedial measures (Kroll 1984).

References

Accomazzo, M. A., 1986. Particulate Retention and Shedding Characteristics from Point-of-Use Process Gas Filters. Proceedings of the 4th Millipore SEMI Symposium, May 19, 1986, San Mateo, CA.

Accomazzo, M. A., Rubow, K. L., & Liu, B. Y. H., 1984. Ultrahigh Efficiency Membrane Filters for Semiconductor Gases. *Solid State Technology* 27(3):141–146.

Bourscheid, G., & Bertholdt, H., 1990. How Production Technologies Influence Surface Quality of Ultraclean Gas-Supply Equipment: Requirements for Surface Quality. *Microcontamination* 8(2):41–43.

Chahine, J. J., et al., 1989. Evaluating the Effect of Various Process Gases on Filter Performance. *Microcontamination* 7(8):19–25.

Fray, A. H., 1984. Change in Room Aerosol Concentration by In-Duct Complex Electric Fields. *Journal of Environmental Science* 27(1):34–36.

Hardwick, S. J., Lorenz, R. G., & Weber, D. K., 1988. Ensuring Gas Purity at the Point-of-Use. *Solid State Technology* 31(10):93–96.

Hardy, T. K., Christman, O. D., & Shay, R. H., 1988. Measurement and Control of Particle Contamination in High Purity Cylinder Gases. *Solid State Technology* 31(10):83–87.

Huffman, T. R., Nichols, G., & Bossard, P. R., 1988. Room Ionization: Can It Significantly Reduce Particle Contamination? Proceedings of the 9th International Committee of Contamination Control Societies Conference, pp. 304–312, September 26, 1988, Los Angeles.

Institute of Environmental Sciences, 1983. Recommended Practice for Gas-Phase Adsorber Cells, IES-RP-CC-008-83T. Mt. Prospect, IL:Institute of Environmental Sciences.

Jensen, D., & Goldsmith, S., 1987. Evaluation of Critical Gasline Filters. *Journal of Environmental Sciences* 30(6):39–43.

Jots, M. G., & Liberia, A., 1989. Method for Measuring Particles from Air Ionization Equipment. Proceedings of the 35th Institute of Environmental Science Annual Technical Meeting, pp. 328–332, May 1989, Anaheim, CA.

Kasper, G., & Wen, H. Y., 1986. A Gas Filtration System for Concentrations of 10^{-5} Particles/cm^3. *Aerosol Science and Technology* 5(2):167–185.

Kroll, W., 1984. Contamination Prevention and Protection for Process Gases. *Solid State Technology* 27(5):220–227.

Langmuir, I., 1942. Report on Smokes and Fibers, Sec. I. U.S. Office of Science Research and Development. No. 865, Part IV, pp. 394–436.

Lee, K. W., & Liu, B. Y. H., 1980. On the Minimum Efficiency and the Most Penetrating Particle Size for Fibrous Filters. *Air Pollution Control Association Journal* 30(4):377–381.

Lee, K. W., & Liu, B. Y. H., 1982. Theoretical Study of Aerosol Filtration by Fibrous Filters. *Aerosol Science and Technology* 1(2):147–162.

Liu, B. Y. H., and Hsieh, K. C., 1989. Progress towards an Absolute Zero Particle Gas. Proceedings of the 35th Institute of Environmental Sciences Annual Technical Meeting, pp. 397–400, September 1989, Anaheim, CA.

Rubow, K. L., 1981. Submicron Aerosol Filtration Characteristics of Membrane Filters. Ph.D. Thesis, University of Minnesota Mechanical Engineering Department, 1981.

Rubow, K. L., Liu, B. Y. H., & Grant, D. C., 1988. Characteristics of Ultra-High Efficiency Membrane Filters in Gas Applications. *Journal of Environmental Science* 31(3):26–30.

Sugiyama, K., & Ohmi, T., 1988. ULSI Fab Must Begin with Ultra-Clean Nitrogen System. *Microcontamination* 6(1):49–54.

Thorogood, R. M., et al., 1986. Production of Ultrapure Nitrogen for the Electronics Industry. *Microcontamination* 4(8):28–35.

11

Surface Cleaning Methods

In any cleanroom application, the critical area most affected by contaminants is always either an exterior or an interior surface of a product. These surfaces are affected by inert and viable particles and by radiation and/or chemical film contamination. They are capable of collecting such materials continuously from any source where they may exist. For that reason, it is necessary that the product surfaces be kept clean and that any potential contamination be removed as soon as possible. Otherwise, the contaminating material can react and/or form extremely strong bonds with the critical surface material. In many present-day imaging systems, a change in surface topography as small as 0.2 μm can be detected and result in image degradation or data loss.

In semiconductor manufacturing, the wafer surface cleanliness is critical at all stages of the operation. The integrity and reliability of the deposited materials that allow for good operation of the semiconductor device depend on the quality of the wafer surface before and during the deposition and processing operations. For all electronic device production, the surface cleanliness is critical so that good bonding can occur for solder adhesion or so that no high-resistance paths are formed to interfere with electron transfer processes. In any precise mechanical systems such as valves or in many energy transfer devices, rolling, sliding, or contacting surfaces are always present. In the manufacture or handling of any liquid used in pharmaceuticals, foods, or fine chemicals, the exposed surface of the liquids is critical because it is the potential entry point for inert or viable contaminants. This applies whether the liquid is the final product or is used in cleaning containers to be used for storage or for delivery of product. In addition, as work is carried out on all of these devices, surface contamination of the tools used in manufacturing, of the containers for liquids and gases, or of the work surfaces where components and subassemblies are processed and handled can all contribute to potential contamination in the product. For this reason, it is necessary to keep both the product *and* processing system surfaces clean.

Before discussing cleaning processes, some of the mechanisms involved in contaminant retention and removal from surfaces are mentioned here. The deposition mechanisms were discussed in some detail previously in chapter 7. For very small particles—the ones that are most difficult to remove—deposition rates are affected more by electrostatic forces than by other forces, such as inertia, gravitation, diffusion, and thermophoresis. This situation exists in both turbulent and laminar airflow environments (Fissan and Turner 1988). For this reason, the forces required to remove small particles become quite large. The charge-to-mass ratio becomes very large for very small particles, even if only a single charge is present on a particle.

The primary considerations in high-efficiency surface cleaning are divesting the soil from the immediate location on the surface and removing the soil to a location where it will not be redeposited, both without damage to the surface. Adhesion between solids can occur as a result of several mechanisms. Small, smooth particles can become embedded within surface asperities or impurities, as shown in Figure 11-1. Particles with rougher surfaces can increase the number of surface contact points for more interaction locations. The primary forces affecting particle adhesion are van der Waals or intermolecular forces and electrostatic forces. Intermolecular forces are predominant for particles smaller than 10–20 μm in diameter; electrostatic forces are the main retentive forces for particles larger than ca. 50 μm. If any liquid is present, then strong capillary adhesion can also occur at the particle surface interfaces. Electrostatic forces become of lesser importance; the liquid molecules at the interface between particle and substrate tend to reduce van der Waals forces by shielding effects (Ranade 1987).

High-efficiency cleaning of particles from surfaces is usually carried out by a physical removal process. This process can be wet or dry, with or

FIGURE 11-1. Particle entrapment on surfaces.

without the aid of surface-active materials or physical devices. Dry process cleaning is usually applied for removal of large particles because the energy transfer is normally poor without the liquid film as a carrier. High-pressure, high-speed gas streams can be used to move particles larger than 30–50 μm from most surfaces.

A clean liquid jet under very high pressure (up to 1,000 psi) is effective in removing particles as small as 1–2 μm (Stowers 1978) from a surface. Mechanical scrubbing can move particles from one point to another on a surface, but complete removal is difficult. Most substrate materials cannot tolerate mechanical scrubbing actions well. Particle removal methods include such processes as plasma cleaning, ultrasonic agitation of solvents, detergent additions, solvent wiping, abrasive cleaning, vapor degreasing, and high-pressure solvent spray. Because particle removal from surfaces is related to the adhesion energy with which the particles are attached, the previous discussion on particle transport and deposition can be referred to in order to indicate the retention energy that must be overcome to remove materials from surfaces. Adhesive forces between particles and surfaces result from molecular and electrostatic interactions. The forces are also affected by the surrounding medium—gas or liquid—and the way in which this material affects the particle-to-surface interactions. Although detailed knowledge of these interactions is not well known, some generalizations can be made. The magnitude of the adhesive force relative to particle size increases for very small particles, making them most difficult to remove. As particle size decreases, retention force decreases linearly, and the area on which the removal force is effective decreases quadratically.

Table 7-2 in chapter 7 showed relative cleaning efficiency for some typical cleaning processes. Note that the first cleaning method described in this table is extremely effective under certain conditions. The method involves transmission of high shearing energy to the particles *and* penetration of the liquid boundary layer when the wet tissue is moved across the surface to be cleaned. This method of cleaning is very effective for small-scale systems with easily accessible surfaces. However, it is essentially a manual method requiring skilled personnel for effective cleaning; no satisfactory means of expanding the operation to more automated procedures has been found.

Studies on cleaning effectivity of gas flow over dry surfaces have shown that this method has relatively low removal efficiency for particles under a few micrometers in size. If the gas flow is accomplished by aspiration, using a vacuum cleaner, then particles smaller than 10–20 μm are barely affected; of course, any such particles that are removed from the surface must be contained within the vacuum cleaner filter system. Figure 11-2 (Gutacker 1991) shows particle removal efficiency for surface vacuuming.

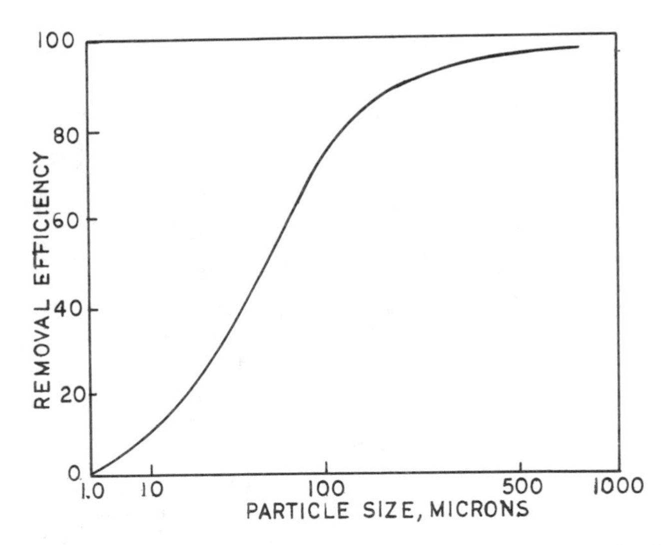

FIGURE 11-2. Particle removal efficiency by vacuuming. Only gas shearing forces are available with a maximum differential pressure no more than a few hundred millibars. Particles embedded within the immobile boundary layer are not affected by the vacuuming. (Courtesy A. Gutacker, ARGOT, Inc.)

The reason for the low efficiency of this type of cleaning is that the gas flow energy is limited by the available differential pressure of one bar. The available energy at this pressure is not sufficient to allow the moving gas to penetrate the gas boundary layer on the surface. If the vacuuming procedure incorporates dry scrubbing or rubbing the surface with the vacuum probe nozzle, then the sensitive surface probably becomes abraded. Use of a fluorocarbon material nozzle may aid in reducing this problem. In any case, touching the surface is definitely *not* recommended.

If the gas flow is derived by use of compressed gas from a nozzle, then the gas flow energy can be increased significantly. Again referring to Table 7-2, note that use of a 100-psi compressed gas jet is still not very effective in removing particles, even though the gas velocity at this pressure is sonic. Figure 11-3 shows how small particles are retained within a static gas boundary layer that cannot be penetrated, even with a high-pressure gas jet. Any larger particles removed from a surface by high-pressure air jets become suspended within the flow of ambient air and can be redeposited on other surfaces. Any large particles removed from a surface by the high-velocity airflow from the jet may be moving fast enough so that they can impact and erode other surfaces at that high speed. This process can generate additional particles that may be transported to other critical areas by airflow or, in the case of larger particles, ballistically. The gas pressure

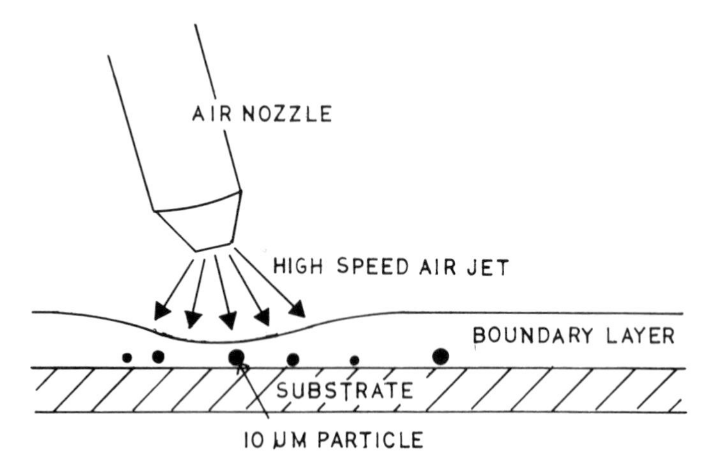

FIGURE 11-3. Gas jet effects on surface-retained particles within boundary layer. Somewhat smaller particles can be removed by a pressurized gas jet than by vacuuming because available compressed gas pressures can produce greater shearing forces than can be generated from a vacuum system. (Courtesy A. Gutacker, ARGOT, Inc.)

should be kept below the point where damage to fragile surfaces can occur from high-velocity gas impact. In addition to possible damage to the product surface, high-pressure gas (> 3 bars) jets can produce very high noise levels in an already excessively noisy cleanroom. This method is not recommended for use in clean operations. In addition to the noise, any hard particles removed from one surface can be impacted upon other surfaces with possible erosive damage to the target, along with generation of additional debris that can cause further damage. Use of an air jet for cleaning should be considered only if the operator is sure that debris removed from a surface will not deposit on other critical locations.

Solvent cleaning is a very effective method for many contaminant materials present on surfaces. Solvents can remove either film or particulate contaminant. The solvents can be either organic or inorganic fluids. For both solvent materials, addition of judiciously selected compatible surfactants usually accelerates the cleaning action. Surfactant concentration in the range of 0.01–0.001% is usually sufficient. The solvents affect bonds between the substrate and the contaminants (Jackson 1978). Good removal of process chemicals, residues, or films from surfaces requires clean solvent flushing of that surface. This procedure is effective but may require large quantities of solvent. The surfaces are usually cleaned by spraying with clean solvent; either batch or continuous solvent addition can be used. The major problem here is to be sure that the clean solvent is actually clean. It is also necessary to realize that the solvent may also dissolve material that is not a contaminant. Careful solvent selection is needed. Along with the problems of use, the problem of waste disposal must also be considered, particularly for toxic or hazardous materials.

A number of new methods are being investigated for cleaning surfaces in the electronic industries. The use of halocarbon solvents has been widespread for many years, especially in this industry for surface cleaning. The trichlorotrifluoroethanes (CFCs) have several advantages for use as a cleaning solvent. They are not flammable; they have low viscosities (< 1 centipoise), low surface tension (19 dynes/per cm), and a relatively low vapor pressure (7–10 mm Hg); and they have a low level of toxicity, with most threshold limit values in the range of 1,000 ppm. They are also useful for the electronic industry because their conductivity is very low. However, their effect on the environment is of great concern. As of March 1989, the U.S. Senate ratified the Montreal Protocol, which calls for a freeze in CFC production at the 1986 level, reduction to 80% of that level in the middle of 1993 and to 50% of that level in the middle of 1998. These reductions will be sought to protect the stratospheric ozone layer. For that reason, a search for new surface cleaning methods is going on. New surfactant/water systems are being investigated (Ko 1989). A water cleaning system using ultrasonic and spray cleaning was found feasible to replace CFC cleaning for many disk drive parts. Operating costs increase, but the cleaning operation is environmentally acceptable.

Several methods using combinations of cleaning procedures have been investigated. Combining centrifugal force with hydrodynamic drag has been found quite effective for removal of spherical particles larger than 1–2 μm from smooth surfaces (Kurz, Busnaina, and Kern 1989). It was shown that wet rinsing at rotational speeds over 3,000 RPM removed 2-μm particles with efficiency well above 90%. Combinations of surfactants and ultrasonic energy are being investigated with several solvent systems. Thermophoretic fields are being considered to keep surfaces cleaner where very small particles are of concern. Brushing with electret elements has been considered. It has been found that a stream of frozen CO_2 particles obtained from a container of liquid CO_2 by adiabatic expansion of the liquid emitted from a nozzle can be very effective in cleaning surfaces (Hoenig 1986). It is believed that the solid CO_2 ice particles moving over the surface drive a gaseous CO_2 film along the surface to penetrate the static air boundary layer. In this way particles that may have been embedded in that boundary layer are removed. This procedure has been used in the past for large-substrate cleaning. It has also worked very well for removing chemical films from some substrates.

Many other methods are being studied. An electrostatic dry cleaning method has shown some promise for removal of particles as small as 1 μm

from surfaces (Cooper et al. 1990). The surface to be cleaned is electrically grounded, and a thin insulating film is placed on that surface. A counter electrode is placed on the film with sufficient direct-current voltage to produce a field of several kilovolts per centimeter. When the film is stripped from the surface, micrometer-sized particles tend to become attached to the film and are removed with it.

Chemical cleaning processes are being studied for semiconductor wafer cleaning. A study of nine different chemical cleaning processes to reduce oxidation-induced stacking fault formation on wafer surfaces found that use of dilute hydrofluoric acid in combination with a mixture of choline [trimethyl (2-hydroxyethyl)ammonium hydroxide] and surfactant was most effective in reducing fault formation and reducing heavy metal deposits on wafer surfaces (Hariri and Hockett 1989). It is interesting that the cleaning mechanism involved with this combination of materials is not well understood. Another study to find chemical cleaning methods has shown that choice of specific surfactant and water systems or of heated, sonicated acid baths was very effective in removing a broad range of particles from specific materials (Lorimer et al. 1989). Further work showed that care is needed for selection of the particular method for a specific cleaning problem.

Cleaned parts must be dried after wet cleaning and before use. Obviously, any dissolved ionic impurities in the chemical used for cleaning is concentrated if the part is dried simply by evaporation. Studies have been carried out to find a means of drying the cleaned parts. It has been shown that use of high-purity isopropanol allows cleaned parts to be dried with minimum residue deposits (Mishima et al. 1988).

Ultrasonic cleaning has been used for electronic and mechanical parts cleaning for many years (Kashkoush et al. 1990). For example, disk file memory devices are assembled from many parts manufactured under standard machine shop fabrication operations. The inherent contaminating nature of machining operations means that parts must be cleaned thoroughly before the disk file can operate with any reliability.

The mechanism of ultrasonic energy cleaning can be explained as one wherein the acoustic waves passing through a liquid produce alternating compression and rarefaction. Where the pressure is reduced, the liquid cavitates, creating small bubbles and concentrating the acoustic energy. At high energy levels, the bubbles implode and the high-velocity vapors scrub the surface placed within the liquid. Solvent cleaning with the aid of ultrasonic energy is commonly used in this area. Both organic and aqueous solvents can be used. Depending on the solvent, operating temperatures from 70 to 190° F should be used (O'Donoghue 1984). Halocarbon solvents are used at temperatures of 70–90° F, water at 130–150° F, and perchloroethylene at 180–190° F. When high-temperature cleaning is used, solvents with high vapor pressure require addition of solvent recovery systems to the cleaner as well as making sure that the object to be cleaned will not be damaged at those temperatures. Operating frequencies most commonly used range from 15 to 120 kHz. Good efficiency in removal of contaminants is reported, but care is required so that fragile metal parts are not eroded or etched by some aqueous solvents in high-energy acoustic fields (Phillips, Stone, and Baldwin 1984). It is also very important to keep the solvent clean by cyclic or continuous filtration. It has been reported that use of ultrasonic energy with ozonated water can be very effective for removing both oily and water-soluble contaminants, while resulting in minimum damage to delicate structures (Hoenig, Gael, and Carter 1989). Very-high-energy acoustic fields have been investigated for effects on small-particle removal (Montz, Beddow, and Butler 1988). Removal efficiency at 165 dB was better than 95% for particles greater than 30 μm in diameter. Use of higher energy levels was expected to result in effective removal for smaller particles.

Ultrasonic cleaners usually operate at a frequency of 20–25 or 45–90 kHz. These devices produce cavitational bubbles that produce sufficient energy during collapse to remove particles from surfaces. Another technique involving sonic energy is the use of megasonics. These devices operate at frequencies near 800 kHz. They produce short sonic pressure waves that physically force particles from surfaces. The effects of varying cleaning solutions with these devices have been studied on a variety of contaminant deposits (Menon, Clayton, and Donovan 1989). Judicious selection of megasonics power levels and cleaning solution chemistry was shown to permit wafer cleanliness levels, particularly for submicrometer particles, not obtainable with other techniques.

Where most of the particle soil has been removed from a surface, final cleaning can be done by special methods. Effective wet cleaning has been reported with ozone in water that has been exposed to 254-nm ultraviolet radiation. Peyton and Glaze (1988) show that photolysis of aqueous ozone produces hydrogen peroxide, followed by secondary reactions that produce hydroxyl radicals, the active species responsible for destruction of many organic compounds.

Dry cleaning methods are also being advanced rapidly. These methods are particularly applicable to removal of organic films as well as unwanted metal ion layers. Organic films can be removed by exposure to a strong oxidant gas, which produces gaseous reaction products that can evaporate almost completely from the wafer surface. Many metal oxides can be removed by exposure to halogen gas combinations, which produce volatile metallic compounds. For example, dry cleaning with a combination of

NO, HCl, and N_2 at elevated temperatures is reported to remove both metal and organic contaminants through formation of volatile nitrosyl compounds (Ruzyllo 1990). Exposing a dry surface to ultraviolet radiation of 185- and 254-nm wavelengths in the presence of 2–3 ppm ozone results in photooxidation of organic films followed by dissociation to CO_2 and H_2O within minutes. Xenon flashtube cleaning provides an intense, short-duration irradiation that can result in annealing, vaporization, pyrolysis, or melting of surface contaminants. Caution is required with these methods because some surfaces can be damaged by the high-energy effects.

In each contaminant-sensitive product area, there are special problems associated with specific product types. Some of these problems are pointed out here. Early in the development of aerospace technology, it became apparent that control of propellant system operation and of inaccessible orbiting vehicle instruments required long-term freedom from contaminants. In the 1960–1970 period, the tendency of contaminant materials to migrate from one surface to another became a problem in long-operating-life systems. After some study, it became apparent that protection of components from contaminants released by other components in these systems was almost impossible. Many aerospace products are operated in orbital environments, where maintenance is extremely difficult. Others are used in military areas, where long-term storage in less-than-ideal environments may be required. Contamination arises from material outgassing, particle migration, propulsion system exhaust interaction, waste venting, extraneous debris interaction, and interaction with residual ambient atmosphere components (Phillips and Maag 1984). The best procedure was simply to ensure that contamination on space vehicles was reduced to a minimum. Procedures found acceptable for cleaning large surfaces, such as those found in space vehicles, include solvent wiping, vacuum brushing, and jet stream spray flushing. These procedures were tested on complex structures, and only the solvent wiping procedure was found acceptable for cleaning to level 100 ($<$ 200 particles $\geq$ 15 μm per square feet of surface) of MIL-STD-1246B (Kwan, Tower, and Mason 1984).

Even though the cleanroom air is very carefully filtered, local source generation of particles and of other contaminants causes deposition within the cleanroom of large particles that would not normally penetrate a filter. Particle size distributions on surfaces in cleanrooms resulting from fallout due to gravity sedimentation of large particles has shown that a model based on Stokes settling might fit the measured data reasonably well (Hamberg and Shon 1984). Once a meaningful model has been developed, data that do not fit the model might show when specific contamination sources begin to affect the cleanroom. Action can then be taken to control such sources.

The need for clean liquids and gases, used in surface processing, requires not only that the fluids be cleaned but also that they remain clean during their handling processes. This apparently obvious statement is more often ignored than not. Fluid-handling components, particularly those such as tubing, piping, valve assemblies, and flow control elements, are fabricated, flushed, and assumed clean. Unfortunately, the fabrication process may result in fissures, cracks, folds, and other roughness that can entrap large quantities of contaminants (Villafrance and Zambrano 1985). Electropolishing is recommended for the surfaces of fluid-handling devices to minimize contaminant entrapment with possible later entrainment into otherwise clean fluids.

Silicon wafer cleanliness has long been a concern in semiconductor manufacturing. The wafers must be cleaned continuously during the semiconductor manufacturing process. Changes were examined in the degree of particulate contamination on wafers resulting from use of etching chemicals and deionized water cleaned to various levels (Bansal 1983). Effective filtration was required in minimizing particulate loading from the chemical materials. A serious problem exists in chemical cleaning and photoresist stripping from the wafers before diffusion operations. The chemicals normally used for these cleaning operations include solutions of sulfuric acid and hydrogen peroxide, of hydrofluoric acid, and of other acid and oxidizer mixtures. Many of these chemicals have been found to contain particles at the submicrometer level in concentrations up to thousands per milliliter (Bansal 1984), particularly for hydrofluoric acid solutions. Point-of-use filters for the chemicals in the spray-cleaning equipment have been shown effective in reducing these levels by several orders of magnitude. A similar problem is seen in the need to maintain cleanliness of the containers in which critical processing occurs. Low-pressure chemical vapor deposition is used for depositing a variety of films on wafer surfaces. Wafers are loaded into quartz holders and containers, heated, and exposed to process gases at low pressure. Obviously, some material also deposits on the quartz components, and after sufficient exposure the deposit grows in thickness to the point where flaking occurs. Unless the containers are cleaned at regular intervals, flakes can deposit on wafers in subsequent operations. Quartz process vessels have been cleaned by conventional wet chemical rinsing, but quartz tube handling, liquid cleanliness, and corrosive liquid handling problems arise. In situ cleaning with radio frequency plasma-excited nitrogen trifluoride appears to be an effective cleaning method with very few handling problems with reasonable care (Benzing 1986).

Virgin wafer-cleaning technology in the late 1980s included scrubbing with a brush and/or high-pressure jets for removal of particulate contaminants, along with chemical cleaning to remove chemically bonded ionic contaminants (Fong and Daszko 1985). For liquid cleaning agent removal after cleaning is completed, spin drying followed with exposure to heated dry gas streams is used. To reduce some of the problems of liquid handling and cleaning, consideration has been given to the possible development of dry techniques for wafer cleaning (Ruzyllo 1988). Some of the techniques considered include application of thermal energy, plasma energy, short-wavelength irradiation, and particle beam energy. Although each technique shows promise as a means of applying sufficient cleaning energy to the wafer surface, each one may also cause harmful effects. Application of sufficient thermal energy to vaporize contaminants may also volatilize some transition metals. The very active plasma can remove a variety of contaminant materials but may also expose the wafer to potentially harmful gas discharge products and materials removed from the reaction chamber walls. Short-wavelength radiation may damage the silicon material directly. Exposure to selected ion beams can be used to cause selective interaction with specific impurities. At this time, it appears that dry cleaning procedures are not completely adequate, but they may have some use as adjuncts to conventional wet cleaning procedures.

References

Bansal, I. K., 1983. Control of Surface Contamination of Silicon Wafers in the Semiconductor Industry. *Journal of Environmental Science* 26(4):20–24.

Bansal, I. K., 1984. Particle Contamination during Chemical Cleaning and Photoresist Stripping of Silicon Wafers. *Microcontamination* 2(4):35–40.

Benzing, D. W., 1986. Reducing Contamination by In Situ Plasma Cleaning of LPCVD Tubes. *Microcontamination* 4(5):71–76.

Cooper, D. W., et al., 1990. Surface Cleaning by Electrostatic Removal of Particles. *Aerosol Science and Technology* 13(1):116–123.

Fissan, H. J., & Turner, J. R., 1988. Control of Particle Flux to Prevent Surface Contamination. Proceedings of the 9th International Committee of Contamination Control Societies Conference, pp. 33–36, September 26, 1988, Los Angeles.

Fong, G., & Daszko, M., 1985. Wafer Cleaning Technology for the Eighties. *Microelectronics Manufacturing and Testing* 8(6):1.

Gutacker, A., 1991. Vacuum Cleaning. In *Fundamentals of Contamination Control Technology,* ed. A. Gutacker, 3.32. Webster, NY: ARGOT, Inc.

Hamberg, O., & Shon, E. M., 1984. Particle Size Distribution on Surfaces in Clean Rooms. Proceedings of the 31st Institute of Environmental Science Annual Technical Meeting, pp. 14–19, May 1, 1984, Orlando.

Hariri, A., & Hockett, H. S., 1989. Evaluate Wafer Cleaning Effectiveness. *Semiconductor International* 12(9):74–78.

Hoenig, S. A., 1986. Cleaning Surfaces with Dry Ice. *Compressed Air Magazine* 91(3):21–24.

Hoenig, S. A., Gael, S., & Carter, S. R., 1989. The Application of Ozonated Water and Ultrasonic Energy to Cleaning Processes in Place of Organic or Chlorinated Solvents. Proceedings of the 35th Institute of Environmental Science Annual Technical Meeting, p. 333, May 1989, Anaheim, CA.

Jackson, L. C., 1978. Contaminant Detection, Characterization and Removal Based on Solubility Parameters. In *Surface Contamination, Genesis, Detection and Control,* Vol. 2, K. L. Mittal. New York: Plenum Press.

Kashkoush, I., et al., 1990. Particle Removal Using Ultrasonic Cleaning. Proceedings of the 36th Institute of Environmental Science Annual Technical Meeting, pp. 407–413, April 1990, New Orleans.

Ko, M., 1989. Comparison of Freon with Water Cleaning Processes for Disk-Drive Parts. In *Particles on Surfaces II: Detection, Adhesion, Removal,* ed. K. Mittal. New York: Plenum Press.

Kurz, M. R., Busnaina, A., & Kern, F. W., 1989. Measurement of Detachment Forces of Particles on Silicon Substrate. Proceedings of the 35th Institute of Environmental Science Annual Technical Meeting, pp. 340–344, May 1989, Anaheim, CA.

Kwan, S. C., Tower, R. J., & Mason, K. D., 1984. Precision Cleaning of Large Complex Structures. *Journal of Environmental Science* 27(4):27–30.

Lorimer, D., et al., 1989. Aqueous-Based Parts Cleaning Strategies for Semiconductor Production Equipment Manufacturing. *Microelectronics Manufacturing and Testing* 12(10):54–56.

Menon, V. B., Clayton, A. C., & Donovan, R. P., 1989. Removing Particulate Contaminants from Silicon Wafers: A Critical Evaluation. *Microcontamination* 7(6):31–36.

Mishima, H., et al., 1988. High Purity Isopropanol and Its Application to Particle-Free Wafer Drying. Proceedings of the 9th International Committee of Contamination Control Societies Conference, pp. 445–456, September 26, 1988, Los Angeles.

Montz, K. W., Beddow, J. K., & Butler, P. B., 1988. Adhesion and Removal of Particulate Contaminants in a High-Decibel Acoustic Field. *Powder Technology* 55(2):133–140.

O'Donoghue, M., 1984. The Ultrasonic Cleaning Process. *Microcontamination* 2(5):63–67.

Peyton, G. R., & Glaze, W. H., 1988. Destruction of Pollutants in Water with Ozone in Combination with Ultraviolet Radiation: 3. Photolysis of Aqueous Ozone. *Environmental Science and Technology* 22(7):761–767.

Phillips, A., & Maag, C., 1984. Maintenance of Contamination Sensitive Surfaces on Board Long-Term Space Vehicles. *Journal of Environmental Science* 27(4):19–21.

Phillips, Q. T., Stone, G. T., & Baldwin, J. M., 1984. Parts Cleaning: An Evaluation of Ultrasonic Systems. Proceedings of the 7th International Committee of Contamination Control Societies Conference, September 18, 1984, Paris.

Ranade, M. B., 1987. Adhesion and Removal of Fine Particles on Surfaces. *Aerosol Science and Technology* 7(2):161–176.

Ruzyllo, J., 1988. Evaluating the Feasibility of Dry Cleaning of Silicon Wafers. *Microcontamination* 6(3):39–43.

Ruzyllo, J., 1990. Issues in Dry Cleaning of Silicon Wafers. *Solid State Technology* 33(3):S1–S4.

Stowers, I. F., 1978. Advances in Cleaning Metal and Glass Surfaces to Micron-level Cleanliness. *Journal of Vacuum Science and Technology* 15(2):751–754.

Villafrance, J., & Zambrano, E. M., 1985. Optimization of Cleanability. *Pharmaceutical Engineering* 5(6):28–30.

12

Standards for Contamination Control Areas

The objective of standards and specifications used for contamination control is to establish controls and definitions that will allow satisfactory cleanroom construction and good product fabrication within cleanrooms. Application of the standards permits exchange of definitions and conditions so that a cleanroom design, performance, and operation can be understood in any area as long as it is defined in accordance with accepted standards. The standards and specifications apply to the design, construction, performance, and operation of the cleanroom. The standards are necessary because design engineers, contractors, equipment manufacturers, materials suppliers, process tool producers, cleanroom owners, quality control managers, and product purchasers are all involved with cleanroom operation and must understand one another. Contamination control document standards must define clearly and acceptably:

- Cleanliness levels for the cleanroom and the product
- Conditions when measurements are to be made
- Design and construction requirements for the cleanroom and for components used within the cleanroom
- Airflow and environmental conditions within the room
- Materials and tools allowed within the cleanroom for processing and manufacturing
- Test methods for definition of cleanroom quality
- Personnel training and activity requirements

The primary standard for use in cleanrooms in the United States is federal standard 209 (FS209). This standard, released by the U.S. General Services Administration (GSA), is presently approved at the D level and was available in final format in 1988. In late 1988, the GSA approved the start of work by the Institute of Environmental Sciences (IES) on the next version (FS209E) of FS209. Work was completed in 1991 and the revision was approved in 1992. The IES has distributed several Recommended Practices (RPs) covering procedures for testing and operating cleanrooms and related systems, including RP-6, *Testing Clean Rooms,* and RP-13, *Equipment Calibration or Validation Procedures*. The IES is in process of preparing additional cleanroom RPs, which are listed in Table 12-1. These recommended practices are in various stages of completion. As with other voluntary standards groups, the IES examines standard methods and practices at regular intervals for validity and timeliness. For a more complete listing of cleanroom standards, the IES has released IES-CC-009-84, which lists 126 documents concerned with contamination control in a format that includes a description of the document, a definition of the particular area of contamination control that is considered, and a source for the document. Documents from most of the countries with policies on contamination control are included. In addition to documented standards for cleanroom operation, there are some unofficial cleanliness requirements derived from FS209.

TABLE 12-1 IES Recommended Practices, 1990.

WG-001	HEPA Filters
WG-002	Laminar Flow Clean Air Devices
WG-003	Garments Required in Cleanrooms and Controlled Areas
WG-004	Wipers Used in Cleanrooms and Controlled Areas
WG-005	Cleanroom Gloves and Finger Cots
WG-006	Testing and Certification of Cleanrooms
WG-007	Testing ULPA Filters
WG-008	Gas-Phase Adsorber Cells
WG-009	Compendium of Cleanroom Standards, Practices, Methods
WG-011	Glossary of Cleanroom Terms and Definitions
WG-012	Considerations for Cleanroom Design
WG-013	Procedures for Equipment Calibration or Validation
WG-014	Calibration of Particle Counters
WG-015	Installation of Cleanroom Production Equipment
WG-016	Deposition of Nonvolatile Residue in Cleanrooms
WG-017	Ultrapure Water: Contamination Analysis and Control
WG-018	Cleanroom Housekeeping: Operation and Monitoring
WG-019	Qualifying Cleanroom Test Agencies and Personnel
WG-020	Documentation Substrates and Forms in Cleanrooms
WG-021	Testing HEPA and ULPA Filter Media
WG-022	Electrostatic Charge Control and Testing in Cleanrooms
WG-023	Viable Particles in Cleanrooms
WG-024	Vibration Measurement/Reporting in Semiconductor Areas
WG-025	Evaluation of Swabs used in Cleanrooms
WG-026	Ancillary Operations in Cleanrooms
WG-027	Personnel in Cleanrooms

New working groups are formed as a need for activity in new contamination control areas arises.

The American Society for Testing and Materials (ASTM) has released ASTM F50, *Continuous Sizing and Counting of Airborne Particles in Dust-Controlled Areas Using Instruments Based on Light-Scattering Principles,* which is an example of a large number of standards and specifications that apply to cleanroom operation. Many of the older standard methods were derived from work done by various government agencies interested in the performance of devices developed for protection against chemical, biological, or radiological warfare agents. Some of these standards are described later in this chapter. This background area also reflects much of the overall cleanroom technology used for even the most modern applications. Although many of these standard methods were prepared in the 1960s period, these standards are still applicable for most cleanroom use.

Although the preceding paragraphs have discussed cleanroom standards developed in the United States, this does not mean that the only standards available for cleanroom operation are American ones. Many standards for cleanrooms that have been developed in other countries are widely used and compatible with U.S. standards. Some Japanese semiconductor manufacturing areas are reported to meet a "0.1 μm-class 10" condition (Edmark and Quackenbos 1984). This condition is defined simply as one where the airborne particle concentration does not exceed ten particles $\geq$0.1 μm in diameter per cubic foot. No information on other environmental conditions or on the confidence limits of the measurement is provided. Obviously the area is very clean because good product yield is reported. Australia has developed a series of standards for cleanrooms and for filters that date back to 1973. They include AS 1132, *Methods of Test for Air Filters in Air Conditioning and General Ventilation,* and AS 1807, *Methods of Test for Cleanrooms, Work Stations and Their Accessories,* which are used in that country to good purpose. Great Britain has been providing standard methods for filter testing (BS3928, *Method for Sodium Flame Test for Air Filters*) and for cleanroom operation (BS 55295, *Environmental Cleanliness in Enclosed Spaces*) from the British Standards Institution for many years. Two French organizations, ASPEC and AFNOR, have been releasing cleanroom and filtration test standards since 1972 and have been using them even before that time. German Engineering Societies have also been releasing standards for many years in this area. Guideline VDI 2083, *Cleanroom Technique,* has eight sections covering:

Part 1. Definitions and determination of cleanliness classes
Part 2. Construction, operation, and maintenance of cleanrooms
Part 3. Clean air measurement methods
Part 4. Surface cleanliness

Part 5. Personnel comfort criteria
Part 6. Personnel activities
Part 7. Process media cleanliness (water, gases, etc.)
Part 8. Process and equipment cleanliness compatibility

Several contamination control groups in Scandinavia have been releasing standards for cleanroom design, operation, and testing that are widely used. Schicht (1989) recently summarized European cleanroom standards, primarily for pharmaceutical production areas. He emphasizes the need for better international harmonization as international market areas are broadened. This need is not unique to pharmaceutical areas alone. Worldwide production facilities located in almost any country, as well as the shipment of products throughout the world, has made harmonization of all standards a necessary consideration in their preparation.

Another major source for standard methods has been Japan. At this time, a great deal of interest in Japan has been directed to cleanroom requirements for electronic product production. Coordination among the several contamination control standards has been promoted widely by the biannual meetings of the International Committee of Contamination Control Societies. Although this group does not produce standard documents, the coordination has been invaluable in aiding international transfer of contamination control technology.

FS209D is used so widely in cleanroom operations that it is discussed here in more detail. Application of chosen portions of this standard are used throughout the world, are referenced in many product designs, and are used in many foreign cleanroom standards.

FS209D defines particulate air cleanliness levels; the standard includes methods for certifying that the cleanroom can operate at those levels and methods of monitoring to ensure that the room maintains its classification. The cleanroom classes apply only to air cleanliness levels and not to room cleanliness, and components, surfaces, and process materials in the room are not classified. Only the air within the cleanroom is defined.

Cleanliness levels are defined for as-built, at-rest, or operational areas. An *as-built* cleanroom is complete and ready for operation, except for production equipment or personnel in the room. An *at-rest* cleanroom is one that is complete, with tools and production equipment operational, but no personnel working in the room. An *operational* room is one in which normal manufacturing procedures are in progress. Cleanliness classes are stated as classes 1; 10; 100; 1,000; 10,000; and 100,000. The numerical value of the class refers to the maximum number of particles $\geq 0.5\ \mu m$ per cubic foot of air. Classification levels can be defined for any operating status as specified and agreed upon by both the buyer and seller of the room or its product. The classification level for an operational cleanroom is about one order of magnitude greater than that for an at-rest room, which is about one order of magnitude greater than that for an as-built room.

FS209D defines particle size ranges and sample volumes to be measured for classification purposes. At this time, FS209D is expressed primarily in terms of English units, with some metric units in use as well; for example, concentration is specified in terms of "particle (sized in micrometers) number per cubic foot," and cleanroom area is defined in terms of square feet. Plans are in progress to convert the standard to either metric or SI units as soon as possible. Table 12-2 shows the class limits in terms of the number of particles per cubic foot that cannot be exceeded for a number of particle sizes that can be measured to establish the class level. Note that for any class, the air classification level is established by particle counting in accordance with valid sampling and measurement plans. At any class, the particle count data are obtained at one or more particle sizes. The allowable particle sizes that can be measured to define a class are those shown in Table 12-2 for that class. Measurements need not be made at all the levels shown for a class unless the contract to procure or operate the cleanroom so specifies. Figure 12-1 shows the information in Table 12-2 in graphical form. Intermediate class limits can be defined within the levels in this figure, but no extrapolation out of these limits is allowed; that is, class 5 or even class 7 can be stated, but any level outside those given in Table 12-2, such as class 150,000 or class 0.5 is not allowed.

TABLE 12-2 Federal Standard 209D Cleanliness Class Levels

Class limits in particles per cubic root of size equal to or greater than particle sizes shown*

	Measured particle size (micrometers)				
Class	0.1	0.2	0.3	0.5	5.0
1	35	7.5	3	1	NA
10	350	75	30	10	NA
100	NA	750	300	100	NA
1000	NA	NA	NA	1000	7
10000	NA	NA	NA	10000	70
100000	NA	NA	NA	100000	7000

(NA - not applicable)

* The class limit particle concentration shown in Table 12-1 and Figure 12-1 are defined for class purposes only and do not necessarily represent the size distribution to be found in any particular situation.

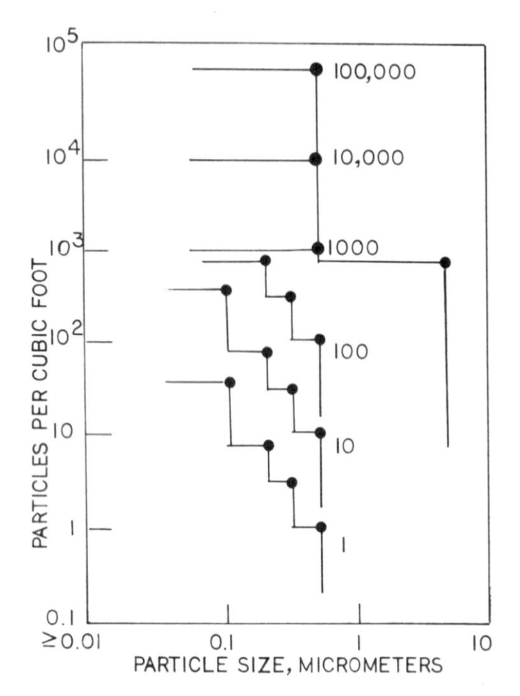

FIGURE 12-1. Federal Standard 209D cleanliness class levels. The points shown come from the values in Table 12-2, and the lines limit the maximum concentrations for intermediate particle sizes.

Air classification levels and monitoring data are established by particle counting at each sample point. For those particles ≥ 0.1 μm, an optical particle counter is used. For particles ≥ 5 μm, a sizing and counting method using membrane filter collection and manual or automated microscopic counting may be used. As air cleanliness increases, particularly for semiconductor processing, measurement specifications for particles smaller than 0.1 μm will be required.

Sampling and measurement plans are required for both verifying and monitoring cleanliness. Good sample acquisition procedures should be used. The plans should be designed to produce statistically valid information based on adequate data. Table 12-3 shows the air sample volume that must be measured for any class at each of the particle size levels that can be used to certify that class. When the precepts of Tables 12-2 and 12-3 are followed, each sample should provide a particle count of approximately 20 particles if the cleanroom particulate contamination level is at or near the classification point. At that data level, it is possible to state the classification with 95% confidence level after a reasonable sample time.

TABLE 12-3 Minimum Air Sample Volumes for Federal Standard 209D Class Levels

Minimum volume (in cubic feet) per sample for the air cleanliness class and measured particle size shown

Class	Measured particle size (micrometers) 0.1	0.2	0.3	0.5	5.0
1	0.6	3.0	7.0	20.0	NA
10	0.1	0.3	0.7	2.0	NA
100	NA	0.1	0.1	0.2	NA
1000	NA	NA	NA	0.1	3.0
10000	NA	NA	NA	0.1	0.3
100000	NA	NA	NA	0.1	0.3

Minimum volume in cubic feet per sample for the air cleanliness class and measured particle size shown. These volumes were chosen to allow collection of sufficient particle count data in each sample to justify the statistical requirements for defining the air cleanliness classification.

Area classifications are verified if the average count at each sample point location meets the class limit and if the upper 95% confidence limit of the mean value of the averages is below the class limit. The calculation procedures required to certify a cleanroom are given in FS209D. Because the variability and the relative standard deviation of data decrease with the quantity of data, acquisition of more data in minimum time aids in improving data quality. Tables 12-2 and 12-3 show that more data can be acquired in less time if measurements are made at the smallest particle size that can be defined with available instruments. Once the cleanroom classification is initially determined, the facility classification level must be reverified at periodic intervals, when major changes occur in the facility operation, or when routine monitoring data indicate that problems may be present. The reverification intervals vary from one facility to another. For the most critical areas, an interval as short as 6 months can be used, but most areas are reverified at intervals of several years if no major changes in the facility have been made.

Observation of particles in any single size range may permit cleanroom classification, but cleanroom performance may not be adequately defined. By observing only submicrometer particles, local source generation of large particles is not detected; the particle number is so much larger in the smaller size range that the few large particles (which can still cause product damage) may not be noticed in the total particle count information. Similarly, observation of only large particles that can cause harm to a product may ignore early failure process or tool problems that generate primarily small particles. Electrical component overheating is one example of a problem that can produce large numbers of very small particles with essentially no particles larger than 1 μm produced until after catastrophic failure occurs.

The number and location of sampling points for classification are specified on the basis of class levels, facility size, and airflow patterns. The older terms *laminar* and *turbulent* have been replaced by *unidirectional* and *nonunidirectional,* respectively. The objective in specifying number and location of sample points for certifying an area is to ensure that the entire cleanroom or specified portions of the cleanroom meet the classification level. For a unidirectional airflow cleanroom, the required number of sample points is "the lesser of (a) the area of the entrance plane perpendicular to the airflow (in square feet) divided by 25, or (b) the area of the entrance plane (in square feet) divided by the square root of the airborne particulate cleanliness classification." For nonunidirectional airflow rooms, the number of sample locations is equal to the square feet of floor area of the clean zone divided by the square root of the airborne particulate cleanliness class designation. The sample size and number of samples are specified to permit definition of the cleanroom classification level within an upper confidence limit of 95%. The required sample number specifications can lead to a high number of samples required to verify a cleanroom class. Table 12-4 shows the minimum number of sample points required to verify cleanrooms of varying sizes at several classification levels. The number and location of sampling points for facility classification are specified on the basis of class levels, facility size, and airflow

TABLE 12-4 Sample Point Number Requirements for Federal Standard 209D Class Levels and Various Cleanroom Sizes

Area Square Feet	Classes 1,10,100	Class 1000	Class 10k	Class 100k
100	4	3	2	2
200	8	6	2	2
400	16	13	4	2
1000	40	32	10	3
2000	80	63	20	6
4000	160	126	40	13
10000	400	316	100	32

type. In order to classify a clean area, the number of required sample measurement points can be very large; however, the number of sample points for monitoring are not specified.

Classification levels are verified by simple statistical procedures; the average particle count at each sample location must meet the class level, and the upper 95% confidence level of the mean value of the averages is below the stated class limit. It is assumed that the particle distribution in the cleanroom follows a Student's t-test distribution and that the standard deviation for any mean value multiplied by the appropriate factor defines the maximum concentration that can be expected within the required 95% confidence limit. Procedures, along with an example for defining the class level, are given on the basis of particle count data, sample size, and number of sample points.

The individual sections of FS209D are as follows:

1. *Scope and limitations*. Classes of air cleanliness for particulate levels are established with methods prescribed for class verification. The classifications do not apply to equipment or supplies for the room.
2. *Referenced documents*. A 1964 text on statistics for engineering is referenced to aid in definition of the Student's t distribution.
3. *Definitions*. Terms specific to FS209D are defined here, including *class, calibration, clean zone, cleanroom, as-built, at-rest, operational, unidirectional, nonunidirectional, condensation nucleus counter, optical particle counter, particle, particle size, concentration, Student's t distribution,* and *upper confidence limit*.
4. *Airborne particulate cleanliness classes*. The class limits and particle sizes that can be used to specify each class are tabulated and plotted.
5. *Verification and monitoring of airborne particulate cleanliness class*. Measurement procedures are given here for both unidirectional and nonunidirectional flow facilities. Locations and number of sample points and sample volumes are stated for the several facility classes and sizes. Table 12-4 shows the minimum number of sample points that might be required for air class verification in rooms of varying size. Following verification of the class, implementation of a detailed monitoring plan during cleanroom operation is required. Particles larger than 5 μm are sized and counted by manual counting and sizing methods. Particles greater than 0.1 μm are sized and counted by an optical particle counter. Limitations and the need for calibration are stated for both methods. The statistical analytical procedures and their bases are stated. Procedures for calculating average concentrations, standard deviations, standard error of the averages, and the 95% upper confidence limit are given.

Appendix A gives the manual procedures for membrane filter sampling and microscopic counting of particles over 5 μm. Appendix B discusses operation, use, and testing of optical particle counters for particles over 0.1 μm. Both appendices are based on ASTM methods and include procedures for validation and instrument calibration. Appendix C gives a sample calculation to illustrate the procedures used to obtain the values for the statistical parameters defined in Section 5 of FS209C. Appendix D gives some sources for supplemental information that may be helpful in preparing documents related to design, testing, acceptance, and maintenance of clean facilities. It is of some interest to note that within a year of the issue date for FS209D, the particle size sensitivity limits for the specified measurement methods have been exceeded both by capabilities of optical particle counters developed in 1987 and by adding the use of the condensation nucleus counter (CNC). The optical instruments are counting and sizing particles at the 0.05 μm level and the CNCs are being used to define particles smaller than 0.01 μm. As of 1991, documented procedures for validating and calibrating the CNC for use in cleanrooms were not available.

In early 1989, the Institute of Environmental Sciences agreed to begin work on preparation of the next version of Federal Standard 209. There are several reasons for revising FS209D. Several areas in FS209D have led to some confusion and a desire for changes. For example, the definition of required sample point locations and frequency requires clarification. The increasing need for work with products that require cleanrooms much cleaner than the present FS209D class 1 makes it necesary to consider measurement at smaller particle sizes than the present low size range of 0.1 μm. The rapid advances in technology require that any cleanroom standard be written so that new technology in cleanroom design, operation, and measurement be capable of definition with a standard that does not become obsolete in too short a time. The widespread adaptation of this standard to European and Asian countries that use metric rather than English units of measurement requires that the English units in FS209D be revised to allow work with either metric or SI (international system of units) units. The updated FS209E should be approved by the end of 1992.

Most of the other U.S. standards sources for cleanroom operation are derived from government agency work. For example, when NASA was developing vehicles for extraterrestrial exploration, much cleanroom effort was carried on in order to increase the reliability of the operational and guidance systems in these vehicles and also to maintain sterility of the vehicle so as to avoid possible transfer of viable terrestrial organisms. Several NASA standards were developed for cleanrooms. In 1963, NASA Marshall Space Flight Center Standard 246, *Design and Operating Criteria of Controlled Environment Areas, Standard for,* was released. This standard defined four cleanroom classes, with control over particle concentration, temperature, and relative humidity. Anticipated work types for each cleanroom class were mentioned in the standard. As an indication of cleanliness, microscopic measurement of particles over 5 μm was required for the cleanest room.

The U.S. Department of Defense has developed many military standards and specifications. One of the first standards having to do with cleanrooms was Air Force Technical Order 00-25-203, *Contamination Control of Aerospace Facilities, U.S. Air Force.* This standard, released in 1960, described cleanroom designs and classifications for specific operations. The most widely used Defense Department standard even today in cleanroom areas is MIL-STD-1246B, *Product Cleanliness Levels and Contamination Control Program.* Even though this standard defines particle loadings in fluids and on surfaces for particles of 5 μm up, the problems of maintaining and monitoring liquid and surface cleanliness are still important at that size range. Both MIL-STD-1246B and TO-00-25-203 are overdue for revision. Revision of the former standard is in progress and work is considered for the latter standard, but no definite plans have been made as of 1991.

The standard methods mentioned above are concerned mainly with cleanrooms for electronic and mechanical technology devices. However, there is great interest in clean device operation from groups concerned with contamination-sensitive operations other than electronic, optical or mechanical production. Food processors and pharmaceutical and biological products manufacturers are concerned with cleanroom performance for both appearance and health reasons. Most of the groups concerned with these products rely on FS209 to define cleanroom operation, but devices such as biological safety hoods require rigid specifications for both inert and viable particles. FS209D states specifically that viable particles are not considered in its definitions. Airflow requirements for positive assurance of hood integrity and containment of viable or hazardous materials in any circumstance are also required. Standards for measurement and control of viable bacteria, spores, and viruses are used to control operation of biological hoods and work where these materials are used (NASA 1967). Some consideration was given to adding microbiological standards to FS209D in its early stages (Seeger 1983), but it was decided then that only inert particles would be considered.

Standards for cleanrooms and other areas sensitive to contamination may involve parameters other than just the ambient air particle content of FS209. Other standards dealing with environmental cleanliness are discussed next.

Very few effective standards exist to define cleanliness of compressed gases. The Current Good Manufacturing Practices required for production of pharmaceutical products state that any compressed gas used in the area, as for container drying, should be as clean as is required for FS209D class 100. This statement is rather vague and does not specify measurement conditions rigidly enough. It is accepted that the cleanliness definition applies to atmospheric pressure gases, even if the gas cleanliness is measured at line pressure. However, this implicit assumption is not specifically defined.

The semiconductor industry uses a wide range of different compressed gases for processing operations. Some gases are obtained from large-quantity vessels and flow lines at pressures of no more than 10 to 15 bars. Other gases are provided in cylinders at pressures of up to 150 bars. As of early 1992, the cleanliness standards for these gases are those being prepared by Semiconductor Equipment and Materials International (SEMI). At that time, specifications for the maximum particle content of inert gases were stated at no more than 10 particles $\geq$0.02 μm per standard cubic foot of gas, with a requirement that at least 8 cubic feet of gas be measured. In addition to defining the allowable threshold concentration, a reference procedure must be identified.

Definition of working standards is a particular problem for cylinder gases because the procedure must also include consideration of the cylinder treatment during use. This includes the effects of handling, shock, and vibration as well as the fact that the cylinder pressure changes as the gas content is depleted. In addition, cylinder gases are often dispensed through convoluted tubing and flow control systems that can vary widely from one installation to another. Because gas cylinders are normally refilled after use, the question of methodology for cleaning procedures before filling always arises in terms of defining the refilled cylinder gas cleanliness specifications. Time of storage also affects particle concentration in cylinder gas because deposition and retention on walls vary with time. It has been stated (Kasper, Wen, and Wang 1989) that for cylinder gases "standards writers are presented with several quite different approaches to selecting analytical procedures and selecting concentration limits for particles." It is therefore doubtful that any single universal standard will be developed to define all cylinder gas cleanliness.

Perhaps the most widely used standard for water cleanliness is the ASTM reagent water specification D1193 (ASTM 1983). This standard has been used widely for assuring satisfactory water cleanliness for test reagent preparation. It covers requirements for water suitable for chemical testing and defines maximum quantities of total solids, electrical conductivity, pH, organics content, soluble silica level, and total bacteria count. Unfortunately for purposes of semiconductor manufacturing, the allowable levels for any of the specified four grades are too high for use in that industry. Therefore, SEMI has been considering more rigorous standards for water to be used in semiconductor manufacturing (Balazs and Poirier 1984). Purity levels orders of magnitude greater than those of ASTM D1193 are considered necessary (Couture and Capaccio 1984). Where D1193 allows a total solids load of up to 0.1 mg per liter, particle counts in operating fabrication area process water show as few as 10 particles per liter larger than 0.5 μm. A solids loading of 0.1 mg per liter of particles in the micrometer size range may result in a particle concentration of several hundred thousand per liter of liquid.

The major specification for manufacturing pharmaceutical materials in the United States is the Current Good Manufacturing Practice (CGMP) for drug products (Title 21, Code of Federal Regulations, parts 210 and 211), which specifies practices for pharmaceutical manufacturing area operations, controls, validation, documentation, and other matters. The CGMP is verified by inspection by auditors from the FDA, who have authority to require rejection and disposal of pharmaceutical product batches that do not meet the CGMP requirements. Details on required practices and recommendations for compliance have been distributed by the FDA (1987). The guideline describes building and facility requirements, including critical and controlled areas. It describes the establishment of written procedures for storage, handling, testing, and approval of components, as well as validation of production and process control measures, including filtration and sterilization processes. The U.S. Pharmacopeia has prepared standards for monitoring and controlling particle concentrations in large-volume (1983) and in small-volume (1986) parenteral fluids. *Large-volume* refers to containers larger than 100 ml and *small-volume* to any container less than 100 ml. The large-volume standard limits inert particles to no more than 50 particles per ml greater than 10 μm and no more than 5 particles per ml greater than 25 μm in diameter. The small-volume particle standard limits inert particle concentration to no more than 10,000 $\geq$ 10 μm and 1000 $\geq$ 25 μm per container. In order to meet these requirements, the CGMP requires that pharmaceutical product filling lines be operated in class 100 conditions.

Other standards for control of bacteriological hazards include NASA procedures to ensure sterility of space-qualified equipment (Marshall Space Flight Center 1980). A family of standards has been prepared to control operation of biohazard control cabinets. NSF-49 (National Sanitation Foundation 1976) for low-risk systems is one example of this family.

This standard describes requirements for containment of biohazard materials within the cabinet to ensure operator safety and environmental protection and to prevent contamination of the material within the cabinet.

References

American Society of Testing and Materials, 1983. *Standard Specification for Reagent Water,* ASTM D 1193-77. Annual Book of ASTM Standards, Vol. 11.01. Philadelphia, PA: ASTM.

Balazs, M. A., & Poirier, S. J., 1984. Those Confusing Pure Water "Specifications": Setting the Record Straight. *Microelectronics Manufacturing and Testing* 7(2):22–23.

Couture, S. D., & Capaccio, R. S., 1984. High-Purity Process Water Treatment for a Microelectronic Device Fabrication Facility. *Microcontamination* 2(2):45–49.

Edmark, K. W., & Quackenbos, G., 1984. An American Assessment of Japanese Contamination Control Technology. Proceedings of the 30th Institute of Environmental Science Annual Technical Meeting, pp. 24–31, May 1984, Anaheim, CA.

Food and Drug Administration, 1987. *Guideline on Sterile Drug Products Produced by Aseptic Processing.* Rockville, MD: Center for Drugs and Biologics and Office of Regulatory Affairs, FDA.

Institute of Environmental Science, 1984. *Compendium of Standards, Practices, Methods and Similar Documents Relating to Contamination Control.* IES-CC-009-84. Mt. Prospect, IL: IES.

Kasper, G., Wen, H. Y., & Wang, H. C., 1989. Developing Particle Standards for Cylinder Gases. *Microcontamination* 7(1):18–26.

Marshall Space Flight Center, 1980. NASA Standard Procedures for the Microbio logical Examination of Space Hardware, NHB 53401B.

National Aeronautics and Space Administration, 1967. *NASA Standard for Clean Room and Work Stations for Microbially Controlled Environments.* NASA Publication NHB 5340.2, Washington, DC: NASA.

National Sanitation Foundation, 1976. *Class II (Laminar Flow) Biohazard Cabinetry.* NSF STD #49, Ann Arbor, MI: NSF.

Schicht, H. H., 1989. Contamination Control Standards and Recommended Practices for Pharmaceutical Production in Europe. *Swiss Contamination Control* 2(6):11–19.

Seeger, G. A., 1983. Microbiological Standards for Clean Rooms. Proceedings of the 29th Institute of Environmental Sciences Annual Technical Meeting, pp. 297–300, May 1984, Dallas, TX.

United States Pharmacopeia, Vol. 21, 1985. <788> Particulate Matter in Injections; Large Volume Parenterals, p. 1287. Washington, DC: United States Pharmacopeial Convention.

United States Pharmacopeia, Vol. 22, 1989. <788> Particulate Matter in Injections: Small Volume Injections, p. 1596. Washington, DC: United States Pharmacopeial Convention.

13

Verification and Monitoring: Requirements and Procedures

When a cleanroom is first installed or any significant changes take place in its operation, the air cleanliness level should be verified. Once work operations begin in the room, monitoring is required to observe the status of a number of conditions in the room that are related to contamination generation and control. Many of the procedures for cleanroom monitoring are described in IES RP-006, *Recommended Practice for Testing Clean Rooms* (1984). The procedures for verification of the cleanroom classification, given in FS209D, have been discussed in the previous chapter.

The reason for cleanroom verification is to establish the air classification level. This activity can be carried out by the cleanroom operator or by an independent third party. Cleanroom monitoring should be the responsibility of the cleanroom operator. Monitoring is carried out only in operational areas, whereas verification can be carried out in as-built or in at-rest areas. A major objective of cleanroom monitoring is to ensure that the cleanroom performance is maintained without significant change during normal operation. Monitoring is carried out to observe environmental trend data within the cleanroom and also to make sure that critical operations are not subjected to atypical contamination generation. Another reason for monitoring is to observe the effects of a specific tool or operation on environmental conditions. Even though the purpose of monitoring in the cleanroom is different than the purpose of verification, the procedures and results obtained from verification should be kept in mind when either establishing or carrying out details of a monitoring plan.

Cleanroom verification can be carried out for as-built, at-rest, or operational areas. These terms for activity have been defined previously in the discussion of FS209D. Once the cleanroom class has been defined, then routine monitoring is usually carried out. Sampling and measurement plans are required for both verification and monitoring. The plans are designed to produce statistically valid data based on good measurement procedures. The number and location of sampling points in a cleanroom are defined when verification is carried out. The number and location of sampling points are specified in FS209D for various classification levels and cleanroom area. Each sample point is designed to represent the particle concentration within an area of interest. The smallest area of interest is 25 square feet, assuming that this is the area within which any single tool will operate. Unfortunately, in a unidirectional flow regime, the sampler draws air from only a very small area. In a downflow area with air moving in true laminar flow at 90 feet per minute, a sampler drawing air at 1 cfm samples from a cross-sectional area of about 0.01 square feet. This area is about 1/2,000 of the 25 square feet being sampled. Even with the divergence of flow seen in an actual cleanroom, the source of any particles seen by a 1 cfm sampler is much smaller than the area that the sample is supposed to represent.

The objective of verification is to establish the classification level for the room in its specified operating condition. Verification procedures allow definition of the classification level for all parts of the cleanroom where that level exists. Using these procedures, a room may be subdivided into areas with different classification levels. A schedule for reverification is established on an acceptable time base. As mentioned previously, the interval can vary from months to years. The data produced during verification may be used as a basis to modify cleanroom components or procedures controlling airflow or work flow. If significant changes are made in components or procedures in the room or if the monitoring data show unjustified changes in the room performance, the cleanroom should be reverified. As an example of the amount of effort that may be required for verification of a small class 10 cleanroom approximately 8 meters by 5 meters, particle concentration measurements were made at 15 locations at 5-foot center points, using two optical particle counters (Helander 1987). The total time for this process was 3 hours, not including setup and calculation time. For larger rooms, the time required for verification can be quite long. Even so, when the total cost of the room and its components is considered, the additional cost of a few days of measurement to verify the room classification level is not excessive.

Cleanroom verification can be handled by a variety of organizations, including cleanroom construction or design groups, equipment manufacturers, cleanroom operators, and independent certification and testing companies. The last group mentioned has certain independent third-party qualifications that may make it the most attractive for an independent certification. If such a group is used, however, the cleanroom operator should be quite certain that the testing company is well qualified and widely experienced. Aside from the standard procedures for buyer evaluation, there are a few programs available that can certify the cleanroom testing organizations. Cleanroom and contamination control short courses are presented by Arizona State University, Harvard, the University of Minnesota, the University of Rochester, and a number of private training organizations. The National Environmental Balancing Bureau (NEBB) has developed a certification program based on a detailed manual for cleanroom testing. The program also specifies a series of qualification requirements for personnel, for verifying instrument operation and calibration status, and for preparation of acceptable test procedures based upon standard methods. NEBB has also set up a qualification testing program and has qualified a number of independent cleanroom testing companies.

FS209D states that a monitoring plan should be established following verification of the room class level. As part of the monitoring plan, a specific location or locations may be selected as representative of the cleanroom area. Monitoring data are obtained as desired. The data can be procured continuously, on a defined schedule, or when specific operations occur that should be observed closely. If anomalous data appear, then either the data must be justified or the room must be reverified.

A monitoring plan might also include requirements for scheduled inspection of the temperature, relative humidity, and differential pressure data for the room. The location of sensors for these parameters can be chosen to indicate overall conditions or specific point conditions. Differential pressure indicators can show pressure differences between the cleanroom and any ancillary area servicing that cleanroom. A weekly program of monitoring cleanroom structural integrity should be set up to evaluate door seals, pass-throughs, windows, wall joints, and any other penetrations where service lines for air, liquids, or vacuum enter the cleanroom. Garment storage locker integrity and correct handling of new and used garments should also be checked weekly. Integrity of point-of-use filters and ion generators should be checked weekly. Visual inspection of painted or tile cleanroom surfaces should be made weekly. A more complete monthly monitoring procedure is suggested. Pressure drop across selected prefilters in the HVAC system should be verified to ensure adequate airflow. Air shower filter integrity should be verified, and the ducts to the air shower vents should be inspected. If carbon adsorbers are used, samples should be obtained from the bed interior and checked to ensure that contaminant vapor breakthrough is not imminent. If sealed cartridges are used for vapor removal, a sample cartridge should be analyzed for this purpose.

An excellent summary of cleanroom test procedures for both verification and monitoring in a variety of cleanroom classes is provided in IES RP-006 (1984), which is summarized in Table 13-1. After several years of testing, this recommended practice is in process of revision, primarily for recommended test sequences. The following series of tests is described in this document.

Airflow velocity and uniformity testing is recommended for uniform flow areas. This test verifies that the airflow is effective in moving contaminants out of the cleanroom with minimum recirculation. Anemometer measurements are made at selected locations within the room to verify uniformity. A square grid pattern with 4- or 10-foot centers is used for the measurement points. Visualization of airflow patterns has been measured by several methods. Motion picture photography of smoke from point sources has been used where cleanup of deposited material is no problem. Zero-buoyancy helium-filled bubbles have been used, but deposition of the bubbles and spreading of the film used to form bubbles may be a problem. A recent report (Lawless and Donovan 1988) describes use of larger

TABLE 13-1 Recommended Test Sequence from IES Recommended Practice for Testing Clean Rooms, IES-RP-CC-006-84-T

Section	Tests	Unidirectional Airflow	Nonunidirectional Airflow	Mixed Airflow
4	Velocity/Volume/Uniformity	1	N/A *	0
5	Filter Leak	1	1	1
6	Particle Count	1,2,3	1,2,3	1,2,3
7	Pressurization	1,2,3	1,2,3	1,2,3
8	Parallelism	1,2	N/A	0(1,2 only)
9	Integrity	1,2	1,2	1,2
10	Recovery	1,2	1,2,3	1,2
11	Particle Fallout	1,2,3	1,2,3	1,2,3
12	Lighting Level	1	1	1
13	Noise Level	1,2	1,2	1,2
14.2	Temperature	1,2,3	1,2,3	1,2,3
14.3	Moisture	0	0	0
15	Vibration	0	0	0

* N/A refers to the test not being applicable to the room.

Key:

0 Test optional, depending on process requirements
1 Test suited to as-built phase
2 Test suited to at-rest phase
3 Test suited to operating phase

Source: (Courtesy Institute of Environmental Sciences.)

"particles" as tracers. Helium-filled Mylar balloons of 1-foot diameter were used with good results. As an alternate means of providing flow visualization aids that should not cause any contamination, the use of pure water droplet clouds has been investigated and found useful (Ramsey, Liu, and Gallo 1988). The droplets are produced by low-temperature condensation of vapor produced by heating extremely clean water. The droplet clouds evaporate after a short time, leaving no harmful residues.

A filter leak test is recommended for all cleanroom types. This test ensures that the air entering the cleanroom will not contaminate any surfaces over which it passes. Although most filters installed in cleanrooms have already been tested by the manufacturer, handling and installation can cause damage to the filter media that result in leaks. The in-place leak tests are made by providing an upstream challenge aerosol to the filter and scanning the filter face and edges for leaks with a photometer or particle counter. The formal test procedures require a challenge aerosol of dioctyl phthalate (DOP) or of dioctyl sebacate (DOS) submicrometer droplets in a concentration range of 10 to 100 μg per liter of air for HEPA filter testing. Table 13-2 shows the number concentration that corresponds

TABLE 13-2 DOP Particle Concentrations versus Mass Loadings

Monosized Particles

Mass Conc'n µgrams/Liter	Equiv. Eff.	Particles per cubic foot at unit density >0.5 µm	>0.3 µm
100	0%	4.32×10^{10}	2.01×10^{11}
10	90%	4.32×10^{9}	2.01×10^{10}
1	99%	4.32×10^{8}	2.01×10^{9}
0.1	99.9%	4.32×10^{7}	2.01×10^{8}
0.01	99.99%	4.32×10^{6}	2.01×10^{7}
0.001	99.999%	4.32×10^{5}	2.01×10^{6}
0.0001	99.9999%	4.32×10^{4}	2.01×10^{5}

ANSI N101.1 Particle Size Distribution

Mass Conc'n µgrams/Liter	Equiv. Eff.	Particles per cubic foot at unit density >0.5 µm	>0.3 µm
100	0%	8.5×10^{9}	2.5×10^{10}
10	90%	8.5×10^{8}	2.5×10^{9}
1	99%	8.5×10^{7}	2.5×10^{8}
0.1	99.9%	8.5×10^{6}	2.5×10^{7}
0.01	99.99%	8.5×10^{5}	2.5×10^{6}
0.001	99.999%	$8,5 \times 10^{4}$	2.5×10^{5}
0.0001	99.9999%	8.5×10^{3}	2.5×10^{4}

The upper portion of this table was calculated assuming that the particles were all either 0.5 μm or 0.3 μm diameter. The lower portion of the table uses the particle size distribution shown in ANSI N101.1 and in ANSI N510 for "cold" DOP aerosols.

to this mass loading when a standard Laskin pneumatic nozzle is used to generate this test aerosol. Where DOP or DOS vapors may be a problem to products, some testing has been carried out by bypassing the prefilter and using atmospheric particles as a challenge. This procedure is effective only if the atmospheric particle concentration is sufficient to produce a statistically valid number of particles in a reasonable measurement time after passing through a small leak in a filter.

Airflow parallelism testing is recommended for unidirectional flow rooms to ensure that any contaminants generated within the cleanroom are not dispersed widely before removal. This test is carried out by generating a defined smoke stream and measuring its offset path dimension from the axis of the airflow streamline in the area.

Cleanroom recovery time testing illustrates how fast the cleanroom recovers from any introduction of contaminants. It is recommended for all operational cleanroom types. This test is carried out by generating a relatively dense smoke cloud and defining the time interval for the airflow in the room to reduce smoke concentration to a level below 100 0.5 μm particles per cubic foot in areas downstream of the smoke source.

Particle count testing defines the cleanroom class and illustrates if any changes in cleanliness have occurred since the last measurements were made. It is also recommended for all cleanroom types. This test is carried out by measuring airborne particle concentrations with either a manual filter and microscope measurement method for large particles or a particle counter for small particles. The measurement point locations are usually selected in accordance with FS209D specifications. A procedure for selecting optimal sample point locations can be based upon criteria for a nearly square aspect ratio for sampled area shapes and for minimizing the number of samples required for characterizing the area (Cooper et al. 1990).

Particle fallout testing is an optional test that does not affect the cleanroom air cleanliness determination. However, fallout testing data are definitely allied to the quality of product produced in any cleanroom. Witness plates are placed at critical locations and the number of particles per unit area per hour that have settled on or deposited on the plates are measured. The witness plates can be oriented vertically or horizontally. Figure 13-1 (Gutacker 1991) shows some witness plate configurations that have been used in cleanrooms. Deposits on the plates can be measured by bright field or dark field light microscopes, by dark field photography, by electron microscopes, or by surface scanners. There is a relationship between airborne and settled particle concentration. This relationship is not firm because it depends on many factors, even for a single area, as shown in Figure 13-2 (Hamberg and Shon 1984). Note the wide spread of

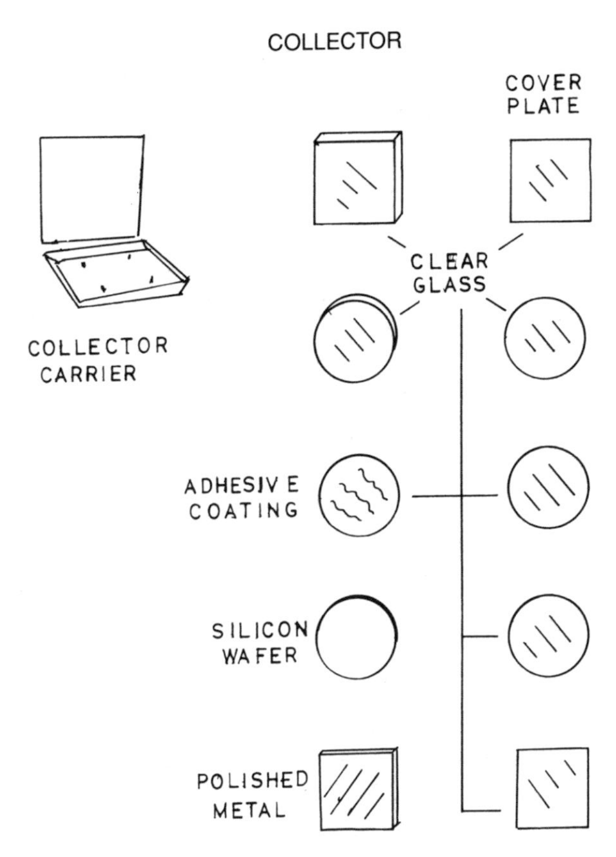

FIGURE 13-1. Some witness plate configurations for cleanroom use. Flexible metal foil plates can also be used for deposited particle collection. The surrogate surfaces should have essentially the same electrical charge as the surface of concern. (Courtesy A. Gutacker, ARGOT, Inc.)

the data relating airborne and settled particle concentrations. Accelerated deposition of larger particles on a clean silicon wafer used as a witness plate has been obtained by passing cleanroom air through a single impaction nozzle directed at the test wafer. Particle deposition is determined by using a surface particle detector. It has been reported (Berger and Bahr 1988) that this procedure is about 30 times more efficient for large-particle detection than a passive witness plate.

Even though RP-006 (1984 version) refers to the fallout test as optional, I strongly recommend that deposition testing be made in *any* cleanroom because local source contamination may deposit on the product without entering the airstream. Selection of witness plate location and configura-

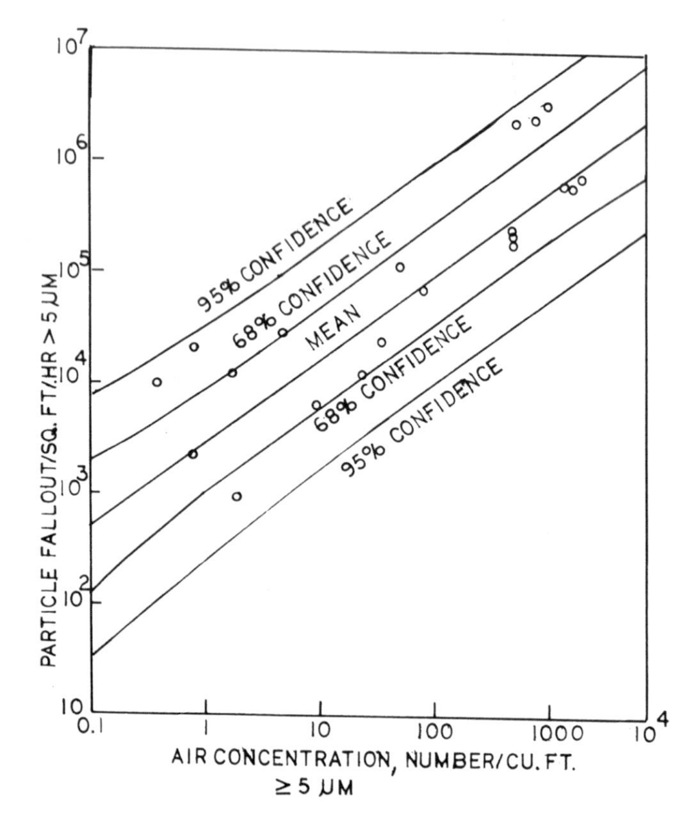

FIGURE 13-2. Fallout rates versus airborne particle concentrations in cleanroom areas. Even though there is a relationship between airborne particle concentration and fallout of particles ≥ 5 μm, the variability of data shows that many factors affect the fallout in addition to the airborne particle concentration. Local source activity, airflow patterns, movement in the area, and other factors all modify the deposition rates, resulting in minimal correlation between airborne particle concentration and deposition rates. (From Hamberg, O., and Shon, E., 1984. Particle Size Distribution on Surfaces in Clean Rooms. Proceedings of the 33rd Institute of Environmental Sciences Annual Technical Meeting, May 1984, Orlando, FL.)

tion should consider the facts that particles can be deposited upon vertical surfaces as well as horizontal ones and that the witness plate conductivity and charge accumulation should simulate the same effects upon critical product surfaces. Tests (Beeson and Weintz 1988) on surface fallout particle size distribution in a class 10,000 area showed that particles in the size range 10 to 100 μm followed a lognormal distribution, as would be expected for aged airborne particles. However, larger-sized particles followed a Poisson distribution, indicating a separate source and/or physical phenomena affecting these particles.

Contaminant induction testing shows any contaminant that is introduced through leaks at any wall or ceiling joints or through doors. This test is recommended for all cleanrooms because contaminant introduction at a joint, door, or other opening near a work area may cause product damage by particles that have not entered the monitored cleanroom air. This test is carried out by measuring particle concentration and size distribution external to the area and comparing that data with particle data taken along possible leak sites.

Pressurization testing is carried out to make sure that the cleanroom air pressure is greater than that in surrounding areas. Differential pressure of 0.1 to 0.2 inches of water, gauge pressure (w.g.) is usually adequate. If any leaks are present, air passes from the cleanroom to the external environment, rather than from the probably dirtier external air to the cleanroom. Differential pressure between the cleanroom and the next external area is defined to the closest 0.01 inch w.g.

Air supply capacity testing defines the capability of the HVAC blowers and airflow system to provide adequate airflow even if filters begin to load. This is normally an as-built room test for any room type. The main air supply volume to the cleanroom, makeup air volume, and reserve capacities of the air handling system are measured. This test is used mainly to ensure that the HVAC system has sufficient excess capacity to handle the airflow requirements if filters load after use or if changes occur in the cleanroom operation.

Lighting level and uniformity and noise level testing are as-received tests for all room types. These are advisable to ensure that work conditions are acceptable to personnel. Light levels for certain operations (photolithography, for example) may be outside preferred levels for comfort. The noise level in most cleanrooms is higher than preferred because of the need for large airflow to the cleanroom. A standard portable photoelectric illumination meter is used at normal operating conditions for light levels and a standard calibrated sound level meter is used at specified locations for noise level determinations.

Temperature and humidity levels and uniformity tests are carried out in all room types to ensure that products that are sensitive to variations in these conditions are not adversely affected. The temperature and humidity sensors can be permanently installed, and continuous measurement data are recorded if so desired.

Testing of vibration levels within the cleanroom structure and components is an optional procedure that may be a vital test for precise manufacturing operations. Location of critical operations may be moved from one point to another, depending on room and building structure, as well as external activities within the building. A standard accelerometer is used at prescribed locations within the room.

Although the documents summarized here illustrate present thinking on cleanroom verification and monitoring, there have been extensive efforts in the past on recommended procedures for both verification and monitoring. One of the first discussions on cleanroom monitoring (Lieberman 1957) recommended that airborne particle count monitoring be carried out even in the uncategorized cleanrooms of that era. Measurements of particle content in some clean areas were shown to indicate the value of routine measurements even at that time. Lieberman (1979) presented a survey of monitoring procedures, primarily for use in pharmaceutical manufacturing areas. However, the methods presented were applicable to measurement of inert particles, rather than for biological materials, making them useful for almost any area at that time. Even before FS209D was released, the value of periodic testing and recertification of cleanrooms was known. One cleanroom manufacturer recommended reverification at 6-month intervals and suggested continuous surveillance of the area where rigorous requirements for products exist (Nelson 1983). Almost all of the tests indicated in IES RP-006 are suggested to be repeated for recertification. These requirements may be excessive for many cleanrooms. Reverification at 1- to 2-year intervals is adequate for most needs.

After the B version of FS209 became widely distributed, several cleanroom monitoring and verification protocols were described for a number of applications. Depending on the nature of the cleanroom to be characterized, the sophistication of the measurement procedure can vary over a wide range. If cost of monitoring is a primary concern and the cleanroom is to be used for products affected mainly by large particles, then simple manual sample collection and measurement systems can be adequate (Madden 1986). A membrane filter, a vacuum pump, and a low-power optical microscope can serve to define particles larger than 5 μm in size. Although capital costs are low with this type of monitoring system, labor costs may be high and manual monitoring procedures can be used effectively for rooms of class 10,000 or worse. For areas at class 1,000 or better, optical particle counters are required for effective particle measurement. In either case, the sample acquisition procedures must be followed to minimize losses or artifact generation during sample collection or handling. FS209D has considered the sample handling requirements to minimize these problems (Lieberman 1988a). Both the sampling inlet efficiency and line loss problems are considered in stating the sample requirements for valid data accumulation in FS209D.

One of the problems in sampling cleanroom air has always been the difficulty in obtaining data for large particles that may be generated locally when measuring devices with low sample collection flow rates are used to monitor very clean areas. Based upon the anticipated particle count in a class 100 area, each cubic foot of air contains 21 particles at 1 μm, 5 particles at 2 μm, and 0.6 particles at 3 μm. Time for acquisition of meaningful data is excessive. A virtual impactor concentrator has been suggested for particles in the size range of 0.65 to 2.5 μm in diameter (Wu, Cooper, and Miller 1986). An impactor with a high airflow is used to collect and concentrate larger particles for examination. After calibration, a large-particle concentration ratio of nearly 10 was achieved for this system. Although used widely in air pollution management, this type of device has not found acceptance in the cleanroom industry.

Although most new cleanrooms are considered for extremely clean applications and production of small devices as used in the semiconductor industry, there are still many areas where large mechanical and optical devices are constructed that must also be kept clean. An example of such a requirement is the area where the Hubble Space Telescope assembly took place. The monitoring program for this area included procedures for monitoring in a very large area. Both floor space and room height were extreme to permit handling the large optical system. Airborne and fallout particles were observed, along with gas sampling for hydrocarbon gases and surface sampling for organic film deposits (McClellan 1986). The large vertical assembly area incorporated areas with classification levels below 1,000 as well as some with levels as high as 10,000.

In characterizing particle content of class 100 and better cleanrooms, FS209D requires use of optical particle counters for measurement of particles smaller than 5 μm in diameter. In addition to conventional instruments that observe the particles directly, use of condensation nuclei counter (CNC) instruments has become popular for extremely clean areas. The CNC is capable of counting—but not sizing—particles larger than approximately 0.01 μm (Fisher 1987). For local source particle generation, the capability to detect these small particles may be important (Liu et al. 1987). However, depending on the nature of the cleanroom and the air recirculation rate through the filters, the concentration of particles below 0.1 μm may be extremely low because the removal efficiency of filters increases significantly for these small particles. Figure 13-3 (Liu et al. 1987) shows the size distribution in clean room air. A similar study carried out in 1985 to define the particle size distribution of aerosols in a cleanroom, using differential mobility and a CNC led to the conclusion that "Due to the cumbersome nature of diffusion battery operation . . . and the long sampling times . . . it is not recommended that the diffusion battery be

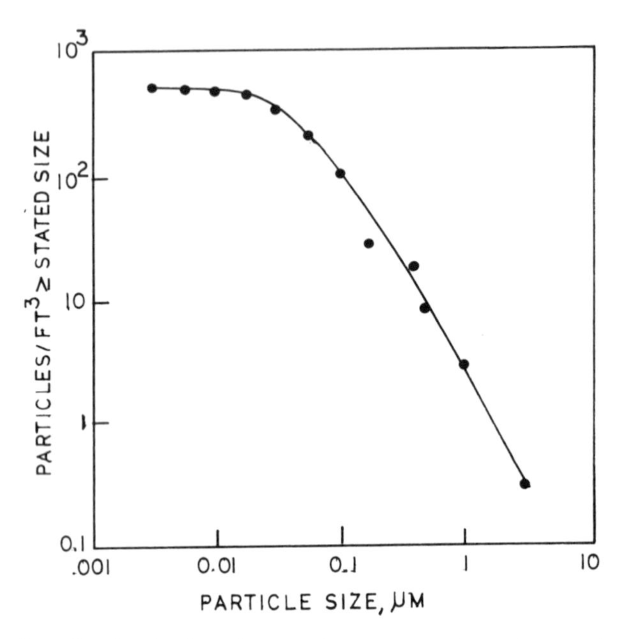

FIGURE 13-3. Size distribution of aerosol in cleanroom. As can be seen from the very low number of particles smaller than approximately 0.01 μm, there was essentially zero activity in the room at the time of measurement. (From Liu, B. Y. H., et al., 1987. Performance of a Laboratory Clean Room. *Journal of Environmental Sciences* 30(5): 22–25.)

used for routine cleanroom operations'' (Locke et al. 1985). In the meantime, optical particle counters with capability of producing particle size data with good resolution have been developed with capability for routinely sizing and counting particles smaller than 0.1 μm (Murray 1986).

Measurements made in a ''superclean'' area at levels with a mean concentration of less than 1 particle per cubic foot for particles smaller than 0.1 μm led to the conclusion that the most critical factor in the characterization of such areas is the background counts of the particle counter used for measurement (Suzuki et al. 1988). As of 1988, an optical particle counter was available that can count and size 0.05 μm particles at a sample rate of 0.1 cfm with good counting accuracy and very low false count data production (Lieberman 1988b).

Characterization of a cleanroom used for manufacturing very large surface integration (VLSI) semiconductor devices requires definition of temperature, relative humidity, and airflow patterns, as well as particle concentration information, to produce the desired cleanliness level for acceptable yield of high-precision devices (Gupta and Sheng 1985).

A robot system for monitoring the high-quality cleanrooms required for products of this type has been described (Ohsakaya 1988). A wheeled, bar code–guided robot is moved throughout a cleanroom to be monitored via a computer control system. Time and labor costs for manual operation of monitoring systems would have been prohibitive. The robot assembly contains instruments for observing airborne particle content, temperature, relative humidity, and air velocity. Mobile manipulator elements permit use of the robot to measure these parameters at specific locations in the cleanroom; the robot is designed and constructed to stay clean.

In some instances, semiconductor cleanrooms have been monitored for particle content with multiplexed systems where either a number of sensors send data to a single data analyzer (Sem et al. 1985) or an air sample manifold system sequentially provides air samples to a single particle counter (Hope 1987). The two methods have advantages and problems (Rossi 1988). In the multiple-sensor multiplexing system, it is possible to monitor a number of locations simultaneously; complexity, variability between sensors, and sample size may produce minimal data. In the air sample multiplexing, single-sensor system, larger sample flow rates are possible, cost is reduced, and sensor variability is eliminated, but all the sample points are not observed at all times. In some instances, combinations of the multiplexing system must be used. This can occur in a large installation, especially if a diversity of equipment, long-term system expansion, and real-time control of data choices must be available (Horne 1988). If a choice must be made, the decision should be based on a cost-benefit analysis that considers initial and operating costs as well as monitoring requirements.

In characterizing cleanroom class levels, it is necessary that room cleanliness be stated with a 95% confidence level. The statistical procedures for ensuring this situation are laid out in FS209D and permit assurance of the level if followed carefully. Each sample measurement can be assumed to be reasonably representative of the entire cleanroom population. The experimental procedures require collection of particle count data in the cleanroom with good data collection procedures. If the cleanroom particle concentration is so high that several hundred or several thousand particles are counted each time, then the statistical problems are minimal. If the cleanroom is class 100 or better, then any individual sample may consist of very sparse data. In that situation, careful examination of the data and consideration of the statistical limitations are required. To reduce the problem to a simple analogy, consider a class 100 room with no more than 100 0.5 μm particles per cubic foot and a sampling system with 0.01 cubic foot capacity at that size. The mean sample population is one particle per sample. If two sequential samples are procured, one may have zero particles and the next may have two particles. If the technician uses only single samples, the first sample ensures that the room meets the criterion, but the second sample means that the room fails the requirement. Thus it can be seen that both the number and size of samples must be adequate.

For more information on the subject, the reader is referred to recent papers by Cooper (1988a) and Bzik (1986). Cooper shows how statistical analyses can aid in helping a cleanroom operator to meet the criteria of FS209D. Suggested procedures include reducing true mean concentrations, reducing sample-to-sample variability, and optimizing the sampling plan. Bzik points out the need to make sure that data management decisions are based on reliable data for knowing the nature of the variability type represented by the data; that is, are the data arriving in truly random manner, in bursts, or with a steady change rate? Even if the mean concentration level is well below the desired class concentration limit and some of the concentration values are close to zero, then the standard deviation of the data can be so large that the requirement for a 95% confidence limit cannot be met (Mielke and Shaughnessy 1988). If the frequency distribution of count data is found to be highly skewed, a normal distribution cannot be used to fit that data; when the formula for the Poisson distribution is used to define the cleanroom class, the large relative standard deviation value results in the room classification assignment to be significantly higher than the mean value.

Use of sequential sampling statistics has been suggested as a means of reducing the quantity of samples required for ensuring cleanroom quality assurance (Cooper 1988b). Sequential sampling involves sequentially measuring a series of samples and recording the number of samples measured and the number failing. These values are then compared with values preset by the sequential analysis statistics. If the number failing is excessive, then the entire lot is rejected; if the number failing is acceptably low, then the entire lot is accepted. Although an exact cleanroom class level cannot be defined by this technique, the probability of assurance that the desired classification level has been met is quite good. The technique is very good if the true cleanroom level is well above or well below the desired class level but may be marginal for borderline cases.

To this point, the considerations for monitoring in cleanrooms have been directed mainly toward examination of the air cleanliness. The next areas to be considered are the tools and product surfaces that must be kept free of contaminants for good yield. Surface cleanliness levels are usually defined on the basis of the classes stated in MIL-STD 1246B (1987) as shown in Figure 13-4. Desirable product monitoring capabilities (Bowling, Larrabee, and Fisher 1988) for manufacturing semiconductors include

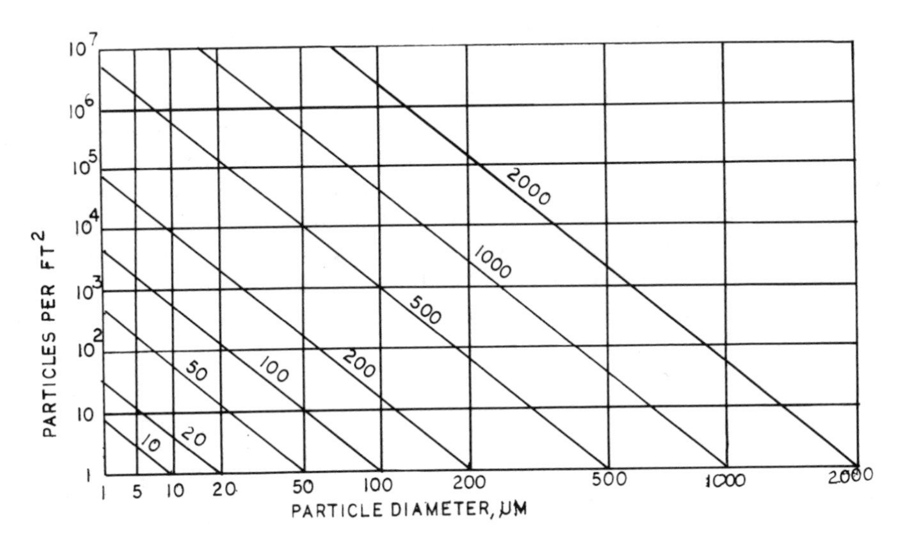

FIGURE 13-4. MIL-STD 1246B chart of cleanliness levels. Although not stated in the document, the cleanliness level chart of particle concentration versus particle size does not represent an actual size distribution but rather shows class levels based on a concept of no more than one particle larger than a specific size (the class level) per square foot of area and a convenient, if unrealistic, particle size distribution. Larger concentrations of smaller particles can be anticipated.

real-time observation of a large area of the in-process product surface with sizing sensitivity to less than 0.1 μm. Measurement of particles within tools operating under vacuum with real-time data production would also be helpful. For this purpose, a particle flux monitor has been described (Borden and Knodle 1988). The sensor head of the monitor can be placed in a vacuum process tool, and relative particle flux data are reported for particles 0.5 μm in diameter and larger. Some questions as to adequacy of sampling efficiency have been raised regarding this device. In many semiconductor manufacturing operations—and in sterilization of pharmaceutical product containers—high-temperature gases must be used and must be kept clean. Use of a quartz isokinetic sample probe within a furnace, followed by electropolished stainless-steel sample lines, permitted measurement of particles from 0.02 μm and larger within a processing system (Cheung and Hope 1988). This system allows the operator to pinpoint sources of contamination within a working system and to implement remedial measures for control of contamination. Contaminant sources from both system component actuation and material processing were identified.

Analysis of contaminant material can be extremely helpful when it is necessary to identify and control a source. Sometimes the procedures require rather subtle analytical techniques and careful study. A metal oxide semiconductor device was found to show a very high failure rate, and pinhole defects in a gate dielectric were seen. Careful monitoring of the failed parts finally showed that contaminant particles had deposited on the dielectric. After identification of the contaminant particles, the problem was controlled (Canfield and Klein 1986). A simpler problem is illustrated in analyzing contamination produced by personnel on computer memory disks (Brar and Narayan 1988). The particular contaminant material on the disk surface was analyzed and found to be primarily spit marks and fingerprints, which were produced by careless personnel procedures. In this situation, the contaminant is easily traced to personnel training problems.

For detailed analysis of contaminants deposited upon surfaces, several very sensitive analytical techniques can be used. These are mainly based upon exposing a sample to ion, electron, or photon beams and detecting emitted particles caused by these beams. A detailed discussion of all of the available analytical methods is beyond the scope of this discussion. A summary of small-spot surface analysis methods, such as secondary ion mass spectrometry, Auger electron spectroscopy, and electron spectroscopy for chemical analysis, allows preliminary evaluation of potential testing methods for surface-deposited contaminant materials (Van der Wood, Bowers, and Gavrilovic 1988).

Several surface monitoring systems have been developed for use in defining contamination on wafer surfaces (Burggraaf 1988). Most of these systems use scattered laser light to define particulate contamination on virgin wafer surfaces. They are capable of detecting individual particles as small as 0.1 μm in diameter on the wafer surface. The surface monitoring systems can produce both particle concentration and particle size distribution data for the deposited particles. Knowing that particulate material deposited on wafers can have a signature particle size distribution that can be identified as having come from specific sources permits definition has made the surface monitoring system a tool to aid in controlling those contamination sources (Granger 1990). It was shown that specific sources could be identified by the change in particle size distribution on wafers that had passed through those specific tools.

A major need still exists for rapid and accurate wafer cleanliness definition on the surface of patterned wafers. Present systems for inspecting patterned wafers are based on computer-controlled pattern matching or holographic analysis systems. They are based on high-magnification optical microscopy and are relatively slow, but are very sensitive in terms of the complexity of the image-processing algorithms used to determine acceptable patterns.

Some product fabrication and medical procedures have required the use of specialized work stations and monitoring procedures. In many surgical operating rooms where large area operations are carried out, positive unidirectional flow of filtered air is used to minimize the possibility of infection during surgery (Franco et al. 1977). Similarly, persons at high risk of infection within hospital areas are kept within such areas used as isolation units. In such areas, it is advisable that both inert airborne particles and airborne bacteria be monitored. One study (Newman and Schimpff 1982) showed that using a clean air supply kept a patient care room at a level equivalent to class 10 or better, whereas concentration of viable organisms was maintained below one colony-forming unit per 200 cubic feet of air.

Certain biological products must be produced in work areas that are free of contaminants but must also prevent any possible emission of potentially hazardous biological materials into the environment. Standards for biohazard control cabinets released by the National Sanitation Foundation (1976) require very thorough control of airflow from these devices. Adequate safety systems are required, including exhaust airflow and pressure-monitoring devices, along with reliable alarming systems when materials such as biological agents, hazardous viruses, or toxic materials are studied. Sterile dry powder antibiotic materials are produced and filled in special facilities. The facility for filling containers for shipment includes a sterile, high-temperature handling tunnel for washing and sterilization of containers, as well as a powder-filling unit under unidirectional sterile airflow with a contained vacuum exhaust to control any possible powder emission into the working space, and all within a ''conventional'' cleanroom facility. Routine monitoring of airborne and settled particulate and biological contamination is required at all stages of the operation. Thus a degree of control on viable particle contamination of sensitive materials is obtained. However, as the need for better controls arises, as the sensitivity of biological products increases as new materials are generated, and as toxicity of new compounds may increase, the problem of viable material monitoring will become more serious. The problem is associated with the lack of universal standards for viable material controls and with the need for more sensitive real-time monitoring systems. As of 1991, there were no completely satisfactory solutions to these requirements. Even so, application of presently available bacteriological monitoring methodology can aid in identifying microbiological material sources and correcting problem areas (Galson 1988).

References

Beeson, R. D., and Weintz, W. W., 1988. Cleanroom Facility Analysis of Fallout Particles for Quality Control Applications. Proceedings of the 9th International Committee of Contamination Control Societies Conference, pp. 628–631, September 26, 1988, Los Angeles.

Berger, J., & Bahr, D., 1988. Surface Particle Detector Monitors Work Station Air Quality. *Microelectronics Manufacturing and Testing* 11(2):19–21.

Borden, P., & Knodle, W., 1988. Monitoring Particles in Vacuum Equipment. Proceedings of the 9th International Committee of Contamination Control Societies Conference, pp. 204–207, September 26, 1988, Los Angeles.

Bowling, R. A., Larrabee, G. B., & Fisher, W. G., 1988. Status and Needs of In-Situ Real-Time Process Particle Detection. Proceedings of the 34th Institute of Environmental Science Annual Technical Meeting, pp. 508–516, May 1988, King of Prussia, PA.

Brar, A. S., & Narayan, P. B., 1988. Analyzing Spittle Mark Contamination on Computer Disks. *Microcontamination* 6(9):67–71.

Burggraaf, P., 1988. Auto Wafer Inspection: Tools for Your Process Problems. *Semiconductor International* 11(13):54–61.

Bzik, T. J., 1986. Statistical Management and Analysis of Particle Count Data in Ultraclean Environments. *Microcontamination* 4(5):59–65; 4(6):35–41.

Canfield, E., & Klein, J. C., 1986. Determination of Particulate Contamination in Gate Dielectric: A Failure Analysis Case Study. Proceedings of the 4th Millipore SEMI Symposium, May 19, 1986, San Mateo, CA.

Cheung, S. D., & Hope, D. A., 1988. Monitor Particles in Real Time to Sleuth Contamination Sources. *Semiconductor International* 11(11):98–102.

Cooper, D. W., 1988a. Statistical Analysis Relating to Recent Federal Standard 209 (Cleanroom) Revisions. *Journal of Environmental Science* 31(2):32–36.

Cooper, D. W., 1988b. Sequential Sampling Statistics for Evaluating Low Concentrations. *Journal of Environmental Science* 31(5):33–36.

Cooper, D. W., et al. 1990. Selecting Nearly Optimal Sampling Locations Throughout an Area. *Journal of Environmental Science* 33(3):46–53.

Fisher, W. G., 1987. Particle Monitoring in Clean Room Air with the TSI 3020 Condensation Nucleus Counter. *TSI Journal of Particle Instrumentation* 2(1):3–16.

Franco, J. A., et al., 1977. Airborne Contamination in Orthopedic Surgery. *Clinical Orthopedics and Related Research* 122:231–243.

Galson, E. L., 1988. Controlling Microbiological Contamination: Standards, Techniques and a Case Study. *Medical Device and Diagnostic Industry* 10(2):34–40.

Granger, G. F., 1990. The Analysis of Surface Inspection Particle Data Using Computer Generated Probability Plots. Proceedings of the 36th Institute of Environmental Science Annual Technical Meeting, pp. 173–177, April 1990, New Orleans.

Gupta, P., & Sheng, D., 1985. Environmental Characterization of a VLSI Cleanroom. *Microcontamination* 3(10):62–68.

Gutacker, A. R., 1991. Contamination Measurement and Characterization. In

Contamination Control Technologist Handbook, ed. A. Gutacker, Chapter 15. Webster, NY: ARGOT.

Hamberg, O., & Shon, E., 1984. Particle Size Distribution on Surfaces in Clean Rooms. Proceedings of the 33rd Institute of Environmental Sciences Annual Technical Meeting, May 1987, Orlando, FL.

Helander, R. D., 1987. Certifying a Class 10 Cleanroom Using Federal Standard 209C. *Microcontamination* 5(9):45–51.

Hope, D. A., 1987. Automated Contamination Monitoring. Proceedings of the 33rd Institute of Environmental Science Annual Technical Meeting, May 1987, Orlando, FL.

Horne, K. D., 1988. Facility Monitoring in a VLSI Wafer Fab. Proceedings of the 34th Institute of Environmental Science Annual Technical Meeting, pp. 455–457, May 1989, King of Prussia, PA.

Institute of Environmental Science, 1984. *Recommended Practice, Tentative, TESTING CLEAN ROOMS,* RP-006, Mt. Prospect, IL: Institute of Environmental Science.

Lawless, P. A., & Donovan, R. P., 1988. Visualizing Airflow Patterns with Helium-Filled Balloons. *Microcontamination* 6(9):72–75.

Lieberman, A., 1957. Contamination Control in Critical Component Assembly Areas. Proceedings of the Joint Military-Industrial Guided Missile Reliability Symposium, November 5, 1957, Pt. Mugu, CA.

Lieberman, A., 1979. Free Air Monitoring of Nonviable Aerosol Particles. *Pharmaceutical Technology* 3(2):71–80; 3(3):61–67.

Lieberman, A., 1988a. Sampling the Air to Monitor and Validate a Room under Federal Standard 209C. *CleanRooms* 2(6):16–17.

Lieberman, A., 1988b. A New 0.05 μm, 0.1 CFM Optical Particle Counter. Proceedings of the 20th DOE/NRC Air Cleaning Conference, August 1988, Boston.

Liu, B. Y. H., et al., 1987. Performance of a Laboratory Clean Room. *Journal of Environmental Science* 30(5):22–25.

Locke, B. R., et al., 1985. Assessment of the Diffusion Battery for Determining Low Concentration Submicron Aerosol Distributions in Microelectronics Clean Rooms. *Journal of Environmental Science* 28(6):26–29.

Madden, P. G., 1986. Low-Cost Analysis for an Effective Contamination Control Program. *Microcontamination* 4(10):20–25.

McClellan, M. S., 1986. Monitoring Contamination and Component Cleaning for the Hubble Space Telescope. *Journal of Environmental Science* 29(1):41–44.

Mielke, R. L., & Shaughnessy, G. J., 1988. A Study of the Statistical Distribution of Airborne Particles in Class 100, 1000, and 25,000 Cleanrooms and Clean Areas. Proceedings of the 9th International Committee of Contamination Control Societies Conference, September 26, 1988, Los Angeles.

MIL-STD 1246B, 1987. Product Cleanliness Levels and Contamination Control Program. Washington, D.C.: U.S. Department of Defense.

Murray, C., 1986. Airborne Particle Monitoring Approaches 0.1 μm. *Semiconductor International* 9(2):60–65.

National Sanitation Foundation, 1976. *Class II (Laminar Flow) Biohazard Cabinetry,* NSF STD #49, Ann Arbor, MI: National Sanitation Foundation.
Nelson, L., 1983. Periodic Testing and Recertification of Cleanrooms. *Medical Device and Diagnostic Industry* 5(1):33–36.
Newman, K. A., & Schimpff, S. C., 1982. Microbiological Evaluation of Laminar Airflow Cleanrooms. Proceedings of the 28th Institute of Environmental Sciences Annual Technical Meeting, May 1982, Atlanta.
Ohsakaya, A., 1988. Development and Performance of a Fully Automatic Clean Room Measuring System. Proceedings of the 9th International Committee of Contamination Control Societies Conference, pp. 110–114, September 26, 1988, Los Angeles.
Ramsey, J. T., Liu, B. Y. H., & Gallo, E., 1988. A Non-Contaminating Fog Generator for Flow Visualization in Clean Rooms. Proceedings of the 9th International Committee of Contamination Control Societies Conference, pp. 632–636, September 26, 1988, Los Angeles.
Rossi, P., 1988. Continuous Multipoint Sampling and Counting in Critical Areas. *Microcontamination* 6(11):64–69.
Sem, G. J., et al., 1985. New System for Continuous Clean Room Particle Monitoring. Proceedings of the 31st Institute of Environmental Sciences Annual Technical Meeting, pp. 18–21, May 1985, Las Vegas, NM.
Suzuki, Y., et al., 1988. Super Cleanliness Evaluation Method. Proceedings of the 9th International Committee of Contamination Control Societies Conference, pp. 171–174, September 26, 1988, Los Angeles.
Van der Wood, T. B., Bowers, C. B., & Gavrilovic, J., 1988. Put Surface Analysis to Work—On Small Areas. *Semiconductor International* 11(13):106–110.
Wu, J. J., Cooper, D. W., & Miller, R. J., 1986. Virtual Impactor Aerosol Concentrator for Cleanroom Monitoring. *Journal of Environmental Science* 32(4): 52–56.

14

Standards for Contamination Control Testing Systems

When a cleanroom is constructed and its classification level is verified, assurance of continued good operation is still needed. Product quality records are the most meaningful indication of satisfactory cleanroom operation, but some production operations may continue for long periods before final product delivery and assurance of acceptable quality. For that reason, real-time definition of cleanroom acceptability is needed. A variety of testing methods and devices are used to verify satisfactory cleanroom operation. Airborne and settled particles are characterized, the particle concentrations in and on process materials and products are measured, freedom from miscellaneous contamination in addition to particles is proven, and cleanroom environmental conditions are verified as acceptable. Many of the tests require the use of sophisticated and delicate instruments, whose performance and accuracy must also be verified. A number of standard methods and test materials are available to verify correct operation and status of these testing instruments. Although many of these methods are discussed, no detailed references are given in this section. All of the methods are referenced elsewhere in other discussion areas.

Many of the tests discussed in the widely used National Environmental Balancing Bureau cleanroom testing manual and in the Institute of Environmental Sciences Recommended Practice RP-006, *Testing a Cleanroom,* are performed with commonly used measurement devices, including instruments to measure airflow, temperature, relative humidity, static pressure, electrical charge, vibration, and light and sound levels. In addition, instruments to measure particulate and gaseous contaminant concentrations are also used. Because the measurements must be reasonably accurate for good cleanroom operation, standard methods and known materials must be used to verify instrument operation. Instruments are used that respond mainly to the parameter being measured. This response situation is achieved for many but not all of the instruments used. The problems of possible anomalous response is discussed in more detail, particularly for particle measurement devices, in chapter 27.

Air velocity can be measured in cleanroom applications by any of several instruments. Airflow in open space is usually measured with a rotating vane or a thermal anemometer. Air velocities can also be defined by using a Pitot tube or static pressure probe to indicate air velocity and recording pressure with a sensitive indicator, such as an inclined micromanometer capable of reading gas pressures accurately to approximately 7.5 Pa (0.05 inches of water). For very low gas flows, a hook gauge can be used to indicate pressures below 1 Pa. Devices of this type are usable for airflows greater than approximately 75 feet per minute. For lower flows, more sensitive manometers must be used. An electronic manometer can be used to define pressures as low as 0.05 Pa. A rotating vane anemometer or a thermal anemometer is sensitive to airflows as small as 30 feet per minute. Procedures for calibration and correct use of air velocity measuring devices are discussed in more detail in the American Society of Heating, Refrigeration, and Air Conditioning *Engineers Guide and Data Book*.

The same pressure indicators used to define air velocity can be used to indicate differential pressures from one area to another. Static pressures in any area can be measured with any of the pressure-indicating devices used to show air velocity with a Pitot tube or static pressure tube. Because most cleanroom differential pressures for contiguous areas are set for 0.01 to 0.05 inches of water (2.5 Pa to 12.5 Pa), an inclined manometer is acceptable for an immediate indicating device. For recording, data transmission, or both, a diaphragm gauge or a digital electronic manometer should be used.

Temperature and relative humidity measurements are usually acquired at the same location. Although mercury in glass thermometers may be used to calibrate other systems, they are manual observation devices with a very slow reaction time and are not normally seen in a cleanroom operation. Cleanroom temperature measurements are often made with resistance temperature devices or bimetallic thermocouples, which are capable of indicating temperatures to within 0.1 Celsius degree. Relative humidity data are obtained with thermohygrometers that may use thin film capacitance sensors to show percent of relative humidity directly. The use of wet and dry bulb temperature measurement to determine relative humidity has almost completely disappeared from cleanroom practice. Dew point indicators can also be used to show moisture content of the air.

Illumination levels in the cleanroom are measured with a photovoltaic cell calibrated in footcandles, as described in the *Lighting Handbook,* Application Volume, prepared by the Illumination Engineering Society of North America. Several acceptable field measurement instruments are capable of indicating illumination levels that can range from 30 to 500 footcandles with ±7.5% accuracy.

Noise levels in the cleanroom are important not only for operator comfort but also because excessive noise levels may result in sufficient vibration transmitted through the air to degrade stability of some sensitive systems. Sound levels are measured with a combination sound level meter and octave band analyzer. Sound level testing and meter calibration methods are defined in the National Environmental Balancing Bureau (NEBB) document *Procedural Standards for Measuring Sound and Vibration.* Some test groups have also been using an American National Standards Institute proposed method for noise measurement, ANSI S12.2 (Proposed), *Procedure for Measuring and Rating Steady-State Room Noise,* for convenience.

The procedures for using and calibrating accelerometers used for vibration testing are given in ANSI document S2.17, *Techniques of Machinery Vibration Measurement.* Additional information is also available in the NEBB manual, *Procedural Standards for Measuring Sound and Vibration.* Vibration data reports must include both vibration intensity and frequency to indicate whether a specific system may be affected.

Many of the parameters that must be defined in cleanrooms are measured by instruments that respond to more than one facet of the measured system. The major example of this problem area is the measurement of particle size and concentration. The optical particle counters and condensation nuclei counters commonly used for this measurement respond not only to particle size but also to one or more particle physical or chemical property. Further, because neither these instruments nor microscopic measurements respond at all to particles below their sensitivity threshold, size measurement error at the minimum size also results in incorrect particle concentration data for the reported particle size. For this reason, standard calibration methods have been established to verify sizing accuracy of these instruments.

ASTM F25, *Standard Method for Sizing and Counting Airborne Particulate Contamination in Clean Rooms and Other Dust-Controlled Areas Designed for Electronics and Similar Applications,* describes procedures for sample acquisition on membrane filters and for microscopic counting and sizing of particles larger than 5 μm in cleanrooms. This method discusses collection of airborne particle samples upon a membrane filter and procedures for handling the filter for microscopic measurement. It describes procedures to reduce sample handling errors and states the maximum measurement error allowed in the microscopic observation of the collected particles. ASTM F328, *Standard Practice for Determining Counting and Sizing Accuracy of an Airborne Particle Counter Using Near-Monodisperse Spherical Particulate Materials,* and ASTM F649, *Standard Practice for Secondary Calibration of Airborne Particle Counters Using Comparison Procedures,* define procedures for calibrating optical airborne particle counters. The first method, F328, is used to calibrate the particle size response of the particle counter to several sizes of monosized latex spheres and to verify that the counter produces the same concentration information as produced by a referee method when measuring a latex sphere aerosol. In this way, reproducible particle size data are obtained, and the operator is assured that the particle counter will report all particles producing a signal at least as large as the one produced by the minimum-size latex particle used for calibration. Because optical particle counters of different optical design can produce different data for particles of varying refractive index, F649 describes a method to permit establishing correlation between different counters when observing atmospheric aerosols. Aliquot samples are passed through a "standard" counter and the test counter. Response of the test counter is adjusted until it matches that of the standard counter acceptably.

ASTM F658, *Defining Size Calibration, Resolution and Counting Accuracy of a Liquid-Borne Particle Counter Using Near-Monodisperse Spherical Particulate Material,* and NFPA T.2.9.6, *Method for Calibration of Liquid Automatic Particle Counters, Using "AC" Fine Test Dust,* are the primary standard methods to verify operation of the instruments used in counting and sizing particles in liquids. The NFPA method is in the process of replacement by a method based upon ASTM F658; it uses monodisperse latex spheres that are traceable to National Institute of Science and Technology (NIST) for calibration, but suspends the latex spheres (in a protective surfactant layer) in hydraulic oil and modifies the concentration verification methodology. A more detailed discussion of liquidborne counter calibration requirements and methods has been presented elsewhere (Lieberman 1988).

Calibration of optical particle counters is carried out with one of the standard methods previously mentioned. Calibration with a standardized reference material is necessary because the optical particle counting instruments respond to particle parameters more than just the particle size. For this reason, standard particle materials are available for use with the standard methods. These materials are isotropic polystyrene latex spheres. They are available as a dry powder or in water suspension with sufficient surface-active material added to minimize agglomeration during storage and with sufficient bactericide to control growth of bacteria or fungi that may enter a container during use. The spheres can be procured with median diameters ranging from approximately 0.05 μm to 100 μm and with relative standard deviations of no more than 5%. Figure 14-1 is a photomicrograph of some typical latex calibration particles showing the uniformity of this material.

As well as the latex spheres can be characterized, they have one problem when used as a calibration standard: Their morphology is quite differ-

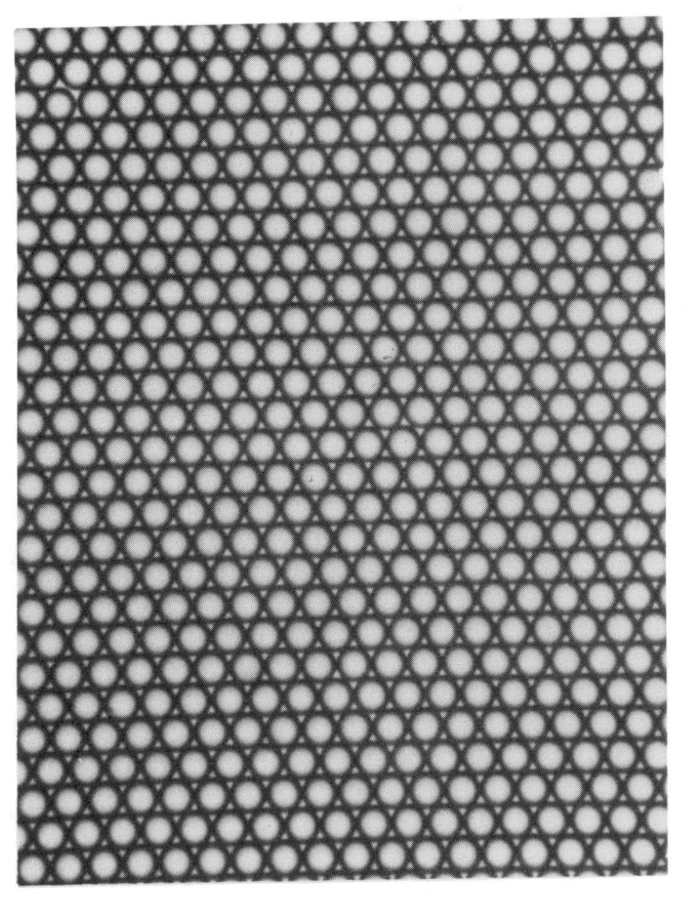

FIGURE 14-1. Monodisperse latex sphere photomicrograph. The particles shown here were very carefully prepared to serve as a standard calibration material with a median diameter of 10.0 μm and a relative standard deviation of 0.8%. (Courtesy Duke Scientific Co.)

ent from that of normal contaminating particles. The latex is spherical, whereas contaminants are irregular; it is transparent, with refractive index 1.6, whereas contaminants may have a refractive index with an imaginary coefficient greater than zero, indicating some light absorption by the material. For these reasons, some liquidborne particle counter calibration is carried out with Air Cleaner Fine Test Dust (ACFTD) particles using the NFPA T2.9.6 method. These are shown in Figure 14-2. This material is quite similar to many actual contaminants because it is composed of granite dust particles that have been elutriated to a known mass distribution. Unfortunately, batch-to-batch variation and dispersion problems have made this material impossible to use consistently as a calibration standard material. For this reason, the National Fluid Power Association, which prepared and distributed the original calibration procedure using ACFTD, has prepared a new calibration procedure for liquidborne particle counters used with hydraulic fluids. This procedure is based on instrument response to latex spheres suspended in hydraulic oil with sufficient anionic surfactant to produce a critical micelle layer of surfactant to protect the latex from attack by the organic liquid.

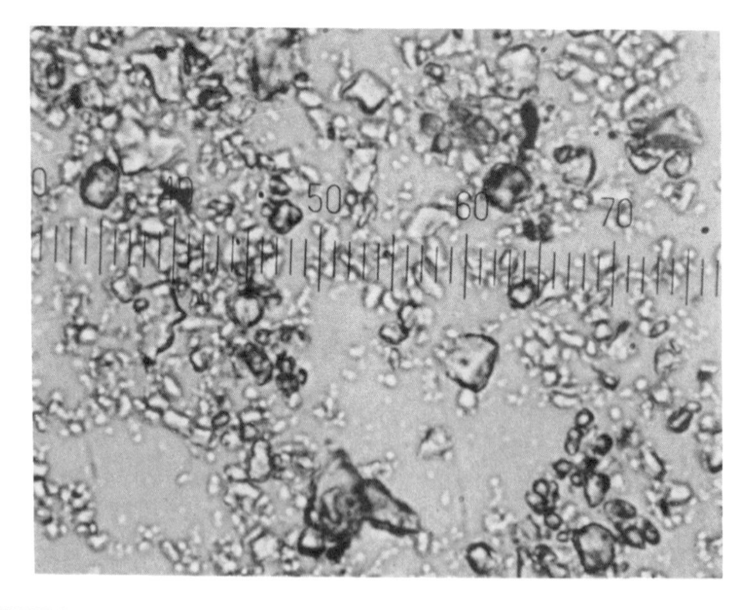

FIGURE 14-2. AC Fine Test Dust photomicrograph. Each scale division represents 2 μm. A significant difference in both shape and appearance can be seen between small and large ACFTD particles, making this material awkward to use as a standard test dust over a wide size range.

For further details on calibration and use of equipment in cleanrooms, the Institute of Environmental Sciences has prepared a recommended practice for cleanroom equipment calibration or validation procedures. This recommended practice (RP-013) covers definitions and procedures for calibrating instruments used for testing cleanrooms and clean air devices and for determining intervals of calibration. The practice references many of the standard test procedures that are listed in this chapter. Although RP-013 does address many of the problems of cleanroom test device calibration methods, generally accepted specifications for calibration frequency are not available. Most of the calibration interval specifications reference the instrument manufacturer's recommendations. These recommendations may not be acceptable in some cleanroom areas and may be too severe for others. For this reason, the calibration interval should be adjustable. A suggested procedure for selecting this interval would be to begin with a time interval that is about half of the manufacturer's recommendation. At the end of that time, if the instrument performance has not changed, then extend the time period by 25%. If the instrument performance has changed, then reduce the time period by 25%. Continue with the adjusted time period until stable operation is assured. Once that time period has been established, then reduce it by an additional 10% or so as a safety factor.

Standard documented calibration methods for size specification of particles by microscopy are available at this time only for use with manual microscopic observation of particles larger than approximately 1 μm in diameter. The microscopic techniques have been in use for many years (Loveland 1952) and are discussed here only briefly. Bright field illumination is used with overhead viewing of the surface. Either manual microscopic measurements of the particle size and concentration can be carried out, or an image analyzer can be used to automate the process. Although documented methods for use of image analyzers are not widely accepted, these instruments have been shown to be fast, accurate, and reproducible for particle size measurement. They are also capable of a wide range of image manipulation and data processing operations.

Automated methods of observing particles as small as 0.1 μm in diameter on polished silicon wafer surfaces are also in use in the semiconductor industry. The instruments are reliable and widely used. However, historical experience with standard specifications and methods for verifying instrument operation has not been developed. Some surface scanner manufacturers have provided usable methods and materials for evaluating and verifying their products, but universally accepted methods have not yet become available as of 1991.

Test wafers have been manufactured with etched pits in specific patterns for checking operation of surface scanners. Some attempts have been made to use bare test wafers upon which known quantities of latex particles of specific size have been deposited in a specific pattern. Unfortunately, continued use of these wafers has not been satisfactory because handling can modify the latex deposits and deposit additional contamination on the wafers. In addition, preparation and handling of standard wafers as clean as needed to simulate actual manufacturing conditions is difficult, especially if statistically reproducible test wafers are desired for use in round-robin testing procedures. Transport and handling of the wafers from one location to another without adding contaminant particles to the wafer surface are very difficult unless very careful procedures are used.

References

Lieberman, A., 1988. Calibration Requirements and Methods for Liquid-borne Particle Counters. *Journal of Environmental Science* 31(3):34–36.

Loveland, R. P., 1952. Methods of Particle Size Analysis. In *Symposium on Light Microscopy,* ASTM STP 143, p. 94ff. Philadelphia: American Society for Testing and Materials.

15

Cleanroom Design Bases and Allowable Components

Early cleanrooms were constructed by modifying building space in existing factory areas. Some of the 1930s factory areas were located in New England and had been used for fabric or furniture manufacturing. The cleanliness of these building interiors was very poor, and structural components were almost impossible to clean adequately, no matter how much effort was applied. These modified cleanroom areas were modeled after medical and surgical rooms and were used mainly for reducing the deleterious effects of visible contamination. The products manufactured in these rooms were primarily mechanical devices, some pharmaceutical products, optical systems, and some electrical devices, such as precision resistors and high-voltage capacitors. A more serious approach to the conventional cleanroom performance requirements resulted from military requirements for gyroscopic guidance and navigation systems used for long-range vehicles and for control and optical elements used in high-altitude bomb-aiming systems. The guidance systems used floated gyroscopes that could be unbalanced by deposited particles in the size range of 50 μm. The Norden bombsight used during World War II contained fine threaded position controls for the optical elements that required extremely smooth operation for good control and operation in the aircraft.

Much early pharmaceutical product manufacturing was carried out in buildings that were modified from older mechanical manufacturing operations. Cleaning those manufacturing areas to a level equivalent to a present-day class 100,000 was very difficult to achieve. Requirements for control and performance in these early cleanrooms were satisfied much more easily than those for present-day products. Product requirements in the future will probably impose still greater demands on cleanroom performance than those in effect at this time, and it appears that the same increase in severity of demand will continue in the future.

This chapter describes present-day thinking on cleanroom designs and the types of components allowable in those areas. It has been reported (Baillargeon 1988) that the fraction of clean space in the class 10 to class 100 range would increase from less than 25% to more than 50% after 1988. The rooms were predicted to decrease in size from an average of 8,900 square feet to less than 6,000 square feet, and their costs to pass the level of $2,000 per square foot. In considering cost figures for cleanrooms, the cleanroom condition must be defined. Costs for the same cleanroom in the as-built, at-rest, and operational conditions of FS209D might vary by a factor of 20 or more. The added costs for production tools and personnel facilities can be very great in an operating cleanroom. In addition to these items, an operating cleanroom also generates cost items such as off-shift cleaning costs, waste management systems, and peripheral service area maintenance.

As products become more sensitive to contaminants, particularly for present and future semiconductor manufacturing, cleanroom costs will continue to increase. There are several reasons for this situation. More space in the cleanroom is required to move the larger process equipment needed for new products into the clean area. The building volume increases as greater volumes of clean fluids (filtered air, deionized water, superclean compressed gases and chemicals) must be processed and moved into the cleanroom, along with removing the used processing materials following a production operation. In 1970, a fluids handling and HVAC systems facility support area at least equal in area to the space for initial processing was required. In 1987, the ratio was 3 to 1; by 1995, the ratio will probably be at least 5 to 1 (Faure and Thebault 1987).

A very large number of present-day cleanrooms are designed for manufacture of precision mechanical devices and electronic components. The overall requirements for the cleanroom design and construction for these systems are basically similar, differing mainly in terms of the room dimensions and the cleanliness requirements for the two product types. Particularly for semiconductor devices, greater cleanliness is needed, and the manufacturing tools are becoming larger every year. Another category of products sensitive to contaminants in general includes those materials that may also be affected by biological contaminants. The cleanrooms required for these products differ from the other cleanroom types in that bacterial control capability is required. This category is discussed separately.

Aside from some specialized cleanroom designs for specific operations, present-day cleanrooms can be described as one of three types. Depending on the cleanroom class level, the airflow system may be unidirectional or nonunidirectional. In general, cleanrooms used for more critical products use unidirectional airflow, and nonunidirectional airflow is used for rooms that can handle less critical work. Further discussion of the meaning of these terms is provided later. Figure 15-1 shows a standard nonunidirectional (mixed flow) area. Mixed-flow cleanrooms are typically rated at class 10,000 and greater. Figure 15-2 shows a unidirectional (horizontal flow) area, and Figure 15-3 shows a unidirectional (vertical flow) area. Horizontal flow rooms are capable of operation at class 100 to class 1,000, whereas vertical flow areas are used where operation as low as class 1 is required. At levels below class 10, small clean areas are normally used within a larger, less clean space.

Consider normal practice in standard factory areas. Most component fabrication takes place on site with material cutting, shaping, trimming, joining, and the like. The workers often leave tools in place at shift end, smoke, eat, and leave debris on and in components during and at the end of the construction job. If products must be clean, then they are cleaned after construction has been completed.

Products built in a cleanroom require much greater cleanliness to oper-

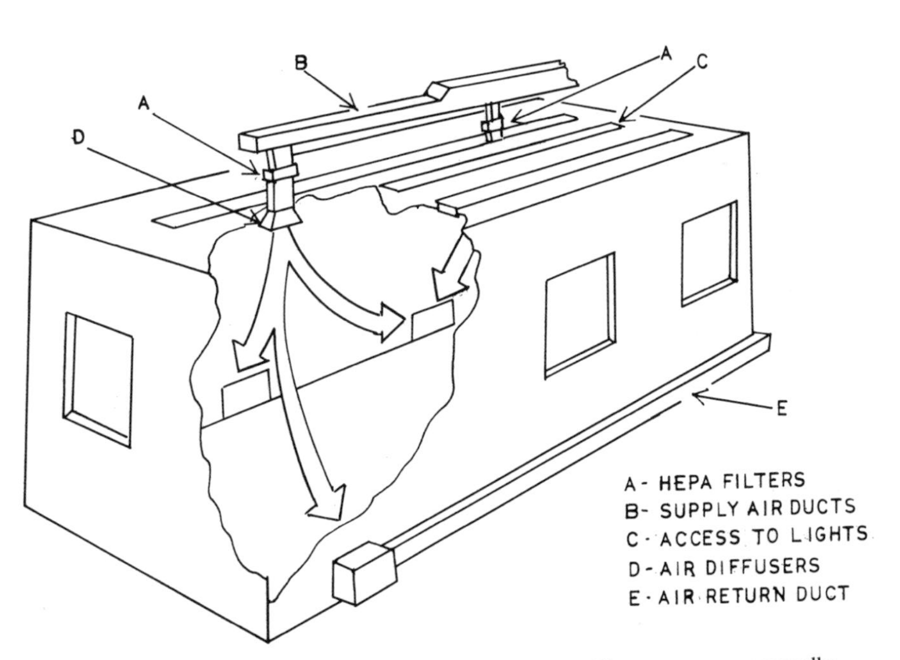

FIGURE 15-1. Nonunidirectional, mixed-flow cleanroom. These rooms are generally used for operations requiring cleanliness classes of 10,000 or more. (Courtesy A. Gutacker, ARGOT, Inc.)

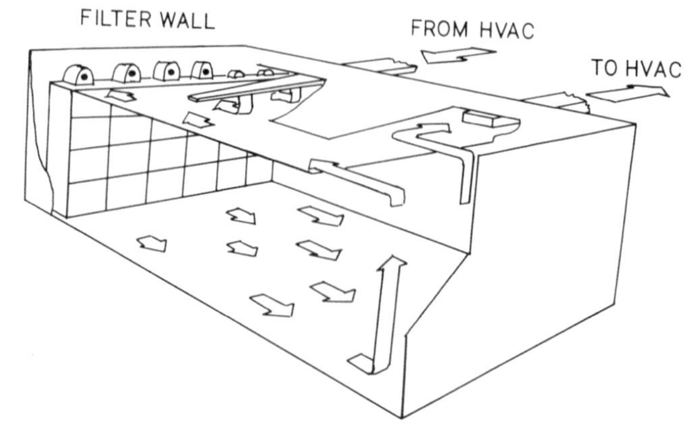

FIGURE 15-2. Unidirectional (horizontal flow) cleanroom. Air cleanliness close to the filter exhaust area can be class 100 or better. The air becomes more contaminated as the air moves to the exhaust wall. (Courtesy A. Gutacker, ARGOT, Inc.)

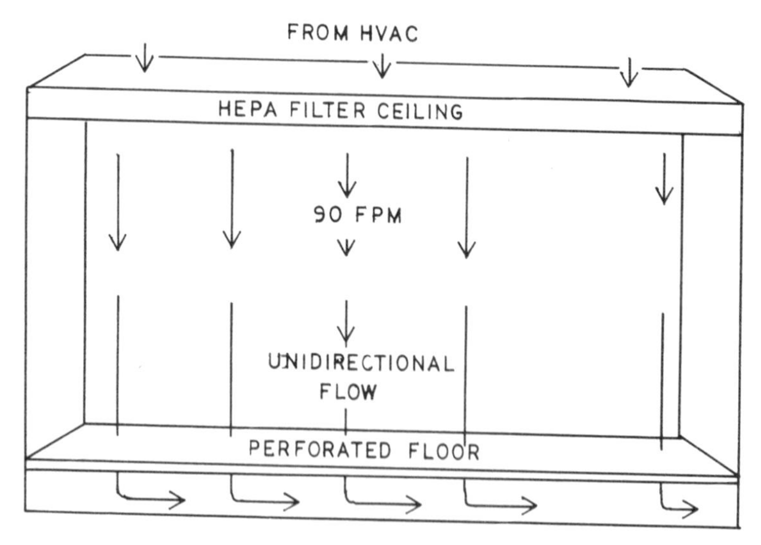

FIGURE 15-3. Unidirectional (vertical downflow) cleanroom. Rooms of this type are used for the most critical operations. Air cleanliness classes better than class 1 can be obtained in this type of room. (Courtesy A. Gutacker, ARGOT, Inc.)

ate acceptably, and their components must be kept clean throughout the manufacturing process. The fabrication area—the cleanroom—must be kept clean both before and during product fabrication. Cleanroom construction require continuous cleaning and restricted access. Workers must be continually reminded to clean all items brought into the area. Materials and tools must be kept out of the clean area and cleaned before entry. Special materials and techniques may be required to prevent contaminant generation within the cleanroom support structures both before and after the cleanroom has been constructed; otherwise, contaminants are generated and subsequently released into the clean areas. Even though all construction and manufacturing processes are inherently generators of contamination, a protocol should be established for production procedures to minimize this problem (Schneider 1990). Procedures for clean construction should be laid out during the manufacturing process, including specifications for the area to be kept clean during construction, the cleaning processes to be used on components and tools used in construction, and guidelines for contamination control responsibilities in the construction operation. If the support structure is steel or concrete, the surfaces must be protected from oxidation or spalling. Wall panels, doors, hardware, and glazing should have interior finishes that are not contaminating. Metal-faced and high-pressure laminated interior walls may be required. Wall panel edge joints must be sealed to prevent air leakage and uncontrolled air flow at those points. In semiconductor production areas, conductive floors may be required; in class 10 to class 100 rooms, raised, perforated floors may be required to optimize airflow paths. For rooms at this cleanliness level, from 20% to 100% of the ceiling area can consist of HEPA filters.

The means of mounting the filters, which varies with filter type, is discussed later. The primary requirement, no matter what the filter-mounting technique may be, is to make sure that there is no leakage at the filter edges. Electrical lines to overhead light fixtures must be sealed to ensure no leakage at that point. The same requirement exists for any wall fixtures, switch panels, and communication connectors. Many of the construction considerations specific for cleanroom construction are seldom needed for construction in other areas. The building structure may require design and construction to minimize vibration transmission. Special clean and sterile surfaces may be required in the cleanroom. Wall panel structure material is selected that does not release either particulate or vapor contaminants. Power and utility fittings may require seals to control possible contaminant transfer to the cleanroom from possibly contaminated wall interior spaces. The heating and ventilation systems are usually much larger and more complex for cleanroom operation than for an equivalent normal manufacturing operation.

Present cleanroom design bases are aimed at providing an environment clean enough for the product. Overkill in design must be avoided, or construction costs will become excessive. Compared to the cost of a standard manufacturing area, a class 10,000 cleanroom is about 4 times as expensive; a class 100 room about 5 times as expensive, and a class 10 area for the same space may be 8 to 10 times more costly. Modular designs, where the working part of an entire class 10,000 cleanroom is kept at class 100, have been shown to be far more economical than building an entire facility at class 100 (Stokes 1982). The design base must consider the initial requirements for the product: Where class 10,000 is adequate, a class 100 facility should not be installed. Product requirements are paramount in selecting the cleanroom designs and materials.

Always provide adequate observation ports for visitors; otherwise, careless activity by untrained visiting dignitaries who wish to observe the cleanroom operations at close view within the room may cause disastrous results. Auxiliary furnishings, such as work tables, chairs, and storage facilities, are also needed in the cleanroom. They should not produce contaminants when they are used or moved about in the cleanroom. Their design should provide maximum work efficiency and permit personnel to work in comfort. Height adjustment and padding are required for chairs. Chair pads can pump air when workers move about on them. Both fiber and foam pads can also emit large quantities of particles when they are compressed. The pads should be vented through high-efficiency filter ports. Material selection and system design for these elements are discussed later.

More and more, the concept of the "integrated" facility is considered in cleanroom design. This type of facility provides for control of airborne contaminant interchange between the room and its process tool operation, particularly at exposed work locations. Such locations are minimized by containing exposed products within either protection containers or within the tools as much as possible. Localized process control systems such as the SMIF systems and cluster tool production facilities are becoming more and more a part of modern semiconductor production needs.

A number of factors are important in cleanroom design. The enclosure containment system requires careful selection of adequate wall, floor, and ceiling material to minimize contaminant emission from and/or passage through the materials used for these surfaces. The lighting fixtures and power lines must be sealed and designed so that they do not produce or pass contaminants into the cleanroom. The filters must be installed and sealed so as to keep clean air flowing correctly into the area. Processing equipment must be selected so that process contaminants are not released either to the product or into the cleanroom. Finally, structural or vibration-resistant supports for the equipment must not result in any problems. The use of large process tools and equipment has made it necessary to consider tool locations and installation needs as one of the cleanroom design steps. This may require consideration of movable structures if product modifications may occur. The installation of robot handling systems and controls for manufacturing operations may be part of the cleanroom design needs for either extremely sensitive products or for products where hazardous materials or operations may be involved. Selection of the location of large manufacturing tools should also consider the effects of the tool on the cleanroom airflow patterns and the potential for generation of recirculating eddies at the tool edges.

Experience has shown the need for careful planning in order to provide a good cleanroom, whether it is class 100,000 or class 1. The design plan must consider such aspects as work and personnel flow, room access, and the human factors for work efficiency, all at minimum cost. Lighting must fit the needs of the product and personnel in terms of illumination type and intensity. Wall, floor, and ceiling material and design must fit initial design and performance requirements. Processing equipment and materials, product components, services, tools, personnel, and storage must fit conveniently in the cleanroom. Access for maintenance must be provided in such a way as to avoid contaminating the work area. A recent design plan (Bangtsson, Larsson, and Neikter 1988) for a cleanroom that may produce an environment almost completely free of contaminants is based on five specific guidelines.

1. Cleanliness requirements for all fluids are defined at the point of use.
2. Critical areas are identified and isolated from other areas that do not need the extent of control.
3. All contamination sources in the environment are identified and controlled at the source.
4. Any obstacle, device, or other cause of nonunidirectional fluid flow around the product is removed.
5. Product exposure to the environment is minimized.

In addition to reducing contamination in the cleanroom, implementation of these guidelines reduced airflow requirements by 75%, resulting in significant energy savings. Cost savings are an important factor in semiconductor cleanroom operation. Care in system design and operation can reduce costs significantly. Reducing airflow requirements by modifying duct return path lengths, minimizing external air volume intakes, and reducing thermal loads in humidity and temperature control systems can be extremely effective in cost control. Cost reduction as great as by a factor of 4 has been reported (Takenami, Inaba, and Ohmi 1989).

The first step in cleanroom design is definition of the product requirements and an understanding of how the design and construction plans must be handled to ensure that these requirements are met. For example, semiconductor manufacture requires the ability to reproduce optical images with definitions that may be less than one micrometer. Therefore, processing equipment must minimize image motion, and vibration frequency and amplitude of that equipment must be very well controlled. Analytical techniques are used in cleanroom design so as to permit that control. The maximum floor vibration can be stated to which a particular tool can be exposed before excessive relative motion between critical elements occurs. Vibration limits are typically required to be no more than 5 μm per second in the frequency range of 50 or 100 Hz. Good design is necessary to avoid exceeding these limits. Poorly designed structures can exceed these limits even when heavy personnel walk rapidly too close to critical devices (Deaves and Malam 1985). Measurements were reported (Silver and Szymkowiak 1988) of component displacements and vibration data for several floor types. Displacement of up to 20 microinches was found for a main floor at locations far from vertical supports; an external pad with vertical support showed displacement of 0.5 microinch for similar stresses. As in all advancing technology, cleanroom designs will have to keep up with the requirements for both improved production capability and for control of production costs. One approach suggested to accomplish these ends is to standardize manufacturing equipment to the utmost (Ohmi 1990).

When optical products are to be manufactured, deposition of molecular contamination must be avoided, particularly if the optical device is to be used in an aerospace application where molecular film contaminants may migrate to critical surfaces. This application may involve measurements at wavelengths ranging from the far ultraviolet to the far infrared. Absorption of energy within the wavelength range from 120 nm to the millimeter range by organic contaminants on optical surfaces can degrade data severely. The plastic components that may be used in many cleanrooms can outgas, and organic vapors can deposit on critical surfaces. Potential problem materials that can affect some processes have been described briefly (McClellan 1985). They include dioctyl phthalate (DOP), used as test material for HEPA filters. Because DOP is also used as a plasticizer for polyvinyl chloride, this tubing type should also be avoided in those areas. The same point is noted for many conductive materials used in static control and for impact protection. In general, polymers with weak cross-link bonds may be prone to sloughing and should be avoided. A procedure for testing materials for use in areas where deposition of molecular contaminants may be a problem has been described (Keilson 1985). The testing cycle included soaking a sample of the material at controlled temperature and time in a solvent that should dissolve a suspect material. It is followed by measurement of the added nonvolatile residue content of the solvent and species identification of the residues by infrared spectroscopy. Particulate emissions from the materials into the solvent were quantified by sampling and measurement methods documented in ASTM standards.

Many pharmaceutical products and processes may be hazardous to the operator during the processing operations. The type of work considered here includes sterility testing, preparation of low-level radioactive materials, and handling of chemotherapeutic agents for cancer therapy. Operator exposure to these materials must be minimized but can be effectively controlled in class II safety cabinets. For this reason, most small-scale developmental work is frequently carried out in safety cabinets. Careful air velocity control and HEPA filters that have been previously checked for leaks for any recirculated air is normally adequate for these areas (Avis 1987; Brader 1988). A sufficient amount of background data and experience exists to show that these precautions are adequate for routine control in these areas. Needless to say, hazard control equipment and safety systems are not enough in themselves. Well-trained operators are probably the most important means of assuring that hazardous materials are not released to the environment in any operating area.

Manufacture of some products may dictate more specific cleanroom design and components. An example is the production needs for some pharmaceutical and bioprocessed materials. Safety controls both for production personnel and for environmental safety are strict and are mandated by governmental control agencies (U.S. Food and Drug Administration 1982). Along with the physical hazards that may result from release of many materials, there is always a political and emotional aspect that must be considered. Operators must be trained and must understand both the hazards and the control methods that are required to protect against these hazards. For example, physical containment requirements for large-scale production of recombinant DNA are specified by National Institutes of Health guidelines. Some requirements are given by West and Snell (1987) as:

1. Viable cultures must be contained in a closed system or in primary containment equipment designed to prevent escape of the product. No more than 10 liters of material can be handled outside a closed system.
2. Cultures must be inactivated by a validated procedure before removal from the closed system.

3. Sample collection, material addition, or transfer from one system to another must be done so that aerosol release and contamination of exposed surfaces are minimal.
4. Exhaust gases from a closed system must be filtered or incinerated to minimize release of viable organisms.
5. A closed system must be sterilized using a validated procedure before it is opened for maintenance or any other reason.
6. Emergency plans must be in place, including procedures for handling inadvertent release of cultures.
7. The containment system integrity must be tested using the host culture before the DNA culture is introduced and after any essential containment features have been modified.

In many medical device manufacturing operations, a class 10,000 cleanroom is adequate. The product requirements are frequently not as critical as those for parenteral liquids; even for devices that are to be used internally, terminal sterilization can be used. For this type of product area, airflow, and filtration requirements can be relaxed somewhat. Good manufacturing procedures normally keep essentially all inert particles larger than 2 or 3 μm from depositing upon the surface of the device during production, and these particles will not cause problems even for most internal devices. Production personnel should understand that medical devices must be kept free of major contamination so that patient end use problems do not occur and that the devices are not a hazard during production. Work flow and personnel access control is usually not as severe for these devices as for the class 100 filling areas. A large-area, mixed airflow system is acceptable, with partial overhead filtered air entry and side wall return air grilles (Brenn 1983).

The additional components required to ensure the safety and acceptability of the bioengineering cleanroom mean installation of correct fluid flow systems for processing the process materials. These systems may include both process liquids and flushing liquids, as well as compressed gases used for product preparation. In these areas, it may also be necessary that wastes are completely sterile before disposal. In addition, materials are required that will withstand the potentially corrosive and toxic fluids that may be required for sterilization of that area. For example, cleanrooms used for pharmaceutical manufacturing may use epoxy terrazzo floors in aseptic cleanrooms with high traffic conditions (Thompson 1985) because this material has excellent chemical resistance to sterilant liquids and can be cleaned relatively easily. Wall, floor, and ceiling joints should be monolithic so as to minimize potential leaks into or out of the room. Chrome-plated components should seldom be used in pharmaceutical areas because these rooms are washed down frequently with chemicals that can attack the chrome plate and cause flaking. The metal flakes are capable of depositing on critical surfaces and resulting in excessive quantities of inert particulate contamination in the area. The use of prefabricated metal sheet panels with foamed cores has been shown to produce very stable wall systems with a rigid and cavity-free structure as long as the metal sheet surface is chosen to be resistant to the sterilizing wash materials used regularly in typical pharmaceutical manufacturing areas (Walti 1983).

A contamination-sensitive area related to pharmaceutical manufacturing is the problem of design and construction of surgical suites where long-term, large-area surgery may occur. The need to maintain a sterile air environment in an area where infectious agents may be present is the same in both areas. The desirability of using unidirectional filtered airflow to aid in maintaining sterility of the air in areas where hip joint replacement and other massive wound surgery has been known for many years (Bernard et al. 1965). One major difference in activity in the surgical suite and in a pharmaceutical processing area is the number of personnel and the level of activity in the areas. Even where unidirectional flow of filtered air has been established in an operating room, personnel loading can be very high, and movements of personnel can occur that interfere with filtered airflow. In some situations, there may be some question of control and priorities for activity in the surgical area. Obviously, the surgeon has primary responsibility; however, some procedures that may produce contamination may be required and cannot be avoided. In addition, the possibility of infectious material released from surgical procedures is sometimes quite high. Some tests have shown that the cleanroom type of surgical theater is less a requirement than more careful practices and disciplines on the part of the surgical suite personnel (Belkin 1984). In fact, some workers have found that an overhead airflow with very little containment of the airflow has been satisfactory in maintaining sterility in some operating theaters (Howorth 1987). An arrangement of air withdrawal inlets located so that any entrained contaminants are withdrawn at the periphery of the clean zone was found adequate to maintain sterility and provide maximum freedom of movement for personnel in the area.

The requirements for manufacturing cleanroom design where sterility of the product and of the room during production is a primary requirement can be summarized, once it is realized that the need to protect the personnel from product components is as important as the need to protect the product from contamination due to the environment and from any personnel emissions (Wetzel 1988). Among other factors, it is necessary that the room design and production tool installation be laid out to provide access to room and exterior tool surfaces easily so that disinfectant materials can be applied as needed; the surfaces must withstand repeated applications of such materials without degradation; capability is needed for moving large quantities of liquids in a sterile manner into and out of the rooms and using those liquids where needed; personnel protection and safety must be assured, while simultaneously protecting products from personnel emissions. In some cases, negative pressure devices within the rooms may be needed for positive protection of personnel working there; in other cases the entire room may require operation at negative pressure to protect the environment. Walls and floors must be constructed so that a continuous surface is present with no seams. If wall return-air grilles are used, they must be removable and capable of being sterilized.

Powder processing operations can result in some very special air handling requirements. In some situations, the product of concern is the powder material. It is necessary to be sure that one batch of powder is not contaminated by another material and that no production residue remains to contaminate a future batch. In other situations, the powder may be a processing material that is used for another product, and the amount and nature of the applied powder must be very carefuly controlled. The powder must be protected from contamination generated by the environment and/or personnel. The product can be exposed to the environment during many processing steps. It must be protected from contamination from the environment, and it must be prevented from entering the general airstream; personnel exposure should be avoided. Powder-grinding and capsule-filling machines are contained within minimum-volume areas with low-turbulence airflow patterns (Schicht 1988). Any dust released in the process is carried as directly as possible to a wall return and duct leading to a safe collection and disposal system that prevents any product powders from entering the return air system. Ceiling panels must also be cleanable, especially if less than 100% filter ceilings are used. Any seals at doors or windows should be smooth rubber or neoprene for easier sterilization.

Modular cleanrooms are constructed for a wide range of applications. This type of room may be designed, assembled, tested, and verified after erection at the manufacturer's plant for later installation at the final facility (Crumley 1987). There are many advantages to use of modular cleanrooms. Standard elements are used to save on costs and to reduce delivery time. They are most useful for standard products with minimum sensitivity to contaminants that may be in a modular cleanroom. Such contaminants may include vapors from filter testing or may be plasticizers or fillers from any plastics used in the modules. The cleanroom user should know which contaminants may be present and whether they will affect products. Modular cleanroom specifications allow a choice of wall, window, and door materials, along with a selection of finishes. Modular components offer a cost-effective alternative to a one-of-a-kind facility in almost any situation. The vendors of such facilities can acquire construction materials for these rooms with better economy of scale than the prospective builder of a single cleanroom, and they can use experience gained in previous construction to improve performance of later cleanrooms.

Modular construction is also advantageous if changes in the room layout or production may become necessary a few months or years after construction has been completed. Many cleanrooms are found to be overdesigned or underdesigned for the desired operation. Product modifications may also dictate a change in the cleanroom. Where design changes are required, an economic necessity for retrofitting or upgrading an existing cleanroom may occur. If the products being manufactured may change, then the designer may retrofit to allow changes in room layout or to use different tools by using movable rather than rigid walls. The use of modular designs is specially valuable in this situation. The air supply must be fitted to each movable module. Care in design is needed to maintain the required air cleanliness levels in changed configurations. A recent development similar to modular cleanrooms is based on intensive application of the standard mechanical interface (SMIF) system, along with use of isolation technology (Hughes, Moslehi, and Castel 1988). The process is described as "open-area SMIF isolation site (OASIS)" methodology. The product components are contained in clean chambers during all processing operations and during transfer from one tool to any other tool. Containment is maintained until the manufacturing process is complete and the product is ready for shipment. This concept is similar to the localized area control or "cluster tool" approach, which was first used in the late 1980s. In the OASIS approach, an isolated, clean environment is created for the product alone, rather than for an entire cleanroom containing processing equipment, personnel, and cleaning devices. The approach is reported to be effective in reducing particle count; it is also helpful in separating process contamination from environmentally caused contamination. Implementation of the OASIS procedure requires cleaning and modifying process equipment to handle the isolation canopies and automatic product-handling mechanisms, testing the installation, and reviewing all of the possible effects on operator productivity. Once the initial system has been installed, significant cost savings over and above equivalent cleanroom costs are expected. The absence of strict personnel control needs should also increase productivity and decrease operator stress.

Previous comments on cleanroom materials have discussed general requirements for desirable material characteristics. Summaries of specific properties of various materials have been prepared by several investigators. Scott (1987) presented an example of that type of study. His criteria for cleanroom suitability included abrasion, corrosion, and wear resistance; low particle-generation characteristics; cleanability; magnetic, radio frequency, and electrostatic properties; stability and life expectancy; material costs and availability; ease of fabrication; structural strength; and surface treatment needs. He rated several metals, both ferrous and nonferrous, a number of thermoplastic and thermosetting polymers, and ceramic building materials for suitability in cleanroom applications.

In designing a cleanroom and selecting materials and components, cost is very important. Cleanroom costs are normally well above those for normal construction. Major costs are due to the air treatment systems required for good cleanroom operation. The air must be filtered with high-efficiency filters that have more flow resistance than normal air conditioning filters. In many cleanrooms, the air change rate in the room occurs several times per minute, rather than a few times per hour, as in most other industrial areas. Moving large quantities of conditioned air at pressures even slightly greater than ambient results in costly air handling and HVAC systems. The HEPA or ULPA filters and their mounting systems are more expensive than standard air conditioning filter systems. The walls, ceiling, floor, windows, and doors for a cleanroom are costly because of the need for special materials and finishes and the need to ensure against leaks. For some cleanroom needs, special seismic supports may be required for cleanroom components such as heavy tools. Fire safety systems are extremely costly for cleanrooms. Very large airflows are present, and many fluid lines that may contain flammable solvents are not easily accessible. Safety controls are always a problem in cleanrooms because personnel access is controlled and doors closed in normal operation. For these reasons, the cleanroom designer must work within the building and fire codes and regulations that are applicable for the product and cleanroom type, as well as for the municipal agencies in the area where the cleanroom is built (LeClair 1990). The high airflow normally present in the cleanroom, where there may be flammable materials and hazardous production, mandates planning and designing safety systems into the cleanroom before any construction begins.

The need for careful consideration of the required cleanroom classification is closely related to cost. Table 15-1 shows relative cost requirements for some class 10,000 and class 100 cleanroom components. Even with the optimistic assumptions that the same wall and floor materials can be used for both rooms, the class 100 materials cost 35% more than those for a class 10,000 room. In addition to those costs, the energy costs for operating the HVAC system for the higher airflow needed in the class 100

TABLE 15-1 Class 10,000 and Class 100 Cleanroom Costs

	Class 10,000		Class 100		
	$/ft	(%)	$/ft	(%)	▲
Walls	$14.25	(5.4)	$14.25	(4.0)	
Floor	22.39	(8.5)	22.39	(6.4)	
HVAC					
Equipment	48.45	(18.4)	70.27	(20.1)	21.82
Piping	17.88	(6.8)	25.96	(7.4)	8.08
Controls	59.63	(22.6)	86.61	(24.8)	26.98
Duct work	44.22	(16.8)	64.31	(18.4)	20.09
HEPAs installed	21.29	(8.1)	30.99	(8.8)	9.70
Electrical with switch gear	23.66	(9.0)	23.66	(6.8)	
Miscellaneous	11.53	(4.4)	11.53	(3.3)	
	262.94	(100)	349.43	(100)	86.67

The same layouts are used, but airflow control and operations are much more closely controlled, requiring different components in the class 100 area.

room continue during operation of the rooms. Generally speaking, a class 100 area requires nearly 10 times as much air-moving costs as a class 10,000 area. Ceiling panel grids for HEPA filter installation can also be more costly for class 100 installations when gel seals and special coatings are used for these panels. Costs can vary by nearly a factor of 2 for these components. Another area where costs must be considered is that of floor materials and coatings. In most cleanrooms rated at class 1,000 and greater, solid floors are used. Cleanrooms rated at class 100 and below frequently use grated floors. In either case, the floor material is coated with materials that resist contaminant generation and are easily cleaned. Acrylic, latex, or baked enamel paints frequently develop cracks, especially after protracted cleaning operations, and should not be used. A 3-mil coat of polyurethane paint applied over a good primer coating is quite good for most applications. Epoxy paints have been used for solid floors. For materials, the cost for a 3-mil application of epoxy paint is equivalent to the cost for similar polyurethane paint. When a prime and two finish coats of antistatic nonskid epoxy paint are applied, a layer up to 50-mil thick is produced at a cost nearly 10 times the paint cost alone.

In considering cost and production requirements, the cleanroom designer should look at the possibility of restricting high-performance, high-

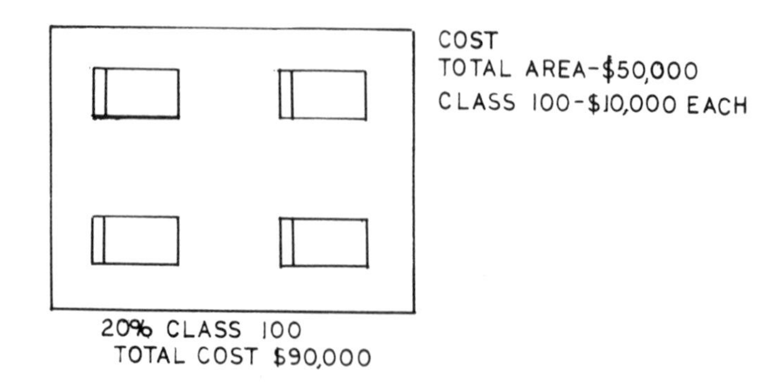

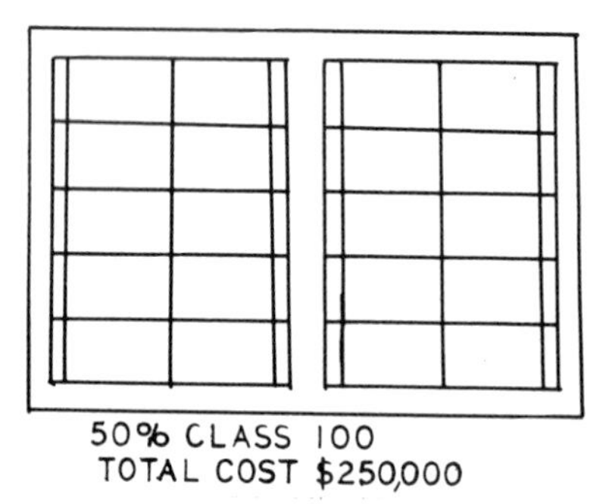

FIGURE 15-4. Cost-effective cleanroom modules. Not all areas in a cleanroom require great cleanliness levels. Access areas and passageways can be left at a lesser cleanliness class level with significant construction and operation savings. Good operating control is required to keep the working module as clean as required. (Courtesy A. Gutacker, ARGOT, Inc.)

cost cleanroom systems to only that portion of the overall cleanroom that requires that level of contamination control. Some of the approaches discussed in the section on modular cleanrooms can result in significant cost advantages. Figure 15-4 shows how the use of high-performance module installation for specific applications in a large, lower performance cleanroom can result in significant savings.

References

Avis, K. E., 1987. Appropriate Laminar-Flow Cabinets for Pharmaceutical Applications. *Pharmaceutical Technology* 11(11):28–33.

Baillargeon, P., 1988. Trend toward Class 10-100 Cleanroom Challenges Construction Contractors. *Microcontamination* 6(7):26–32.

Bangtsson, B., Larsson, B., & Neikter, K., 1988. Design of Clean Room for Optimal Contamination Control. Proceedings of the 9th International Committee of Contamination Control Societies Conference, pp. 489–493, September 26, 1988, Los Angeles.

Belkin, N. L., 1984. Clean Room Technology and Aseptic Practices in the Surgical Suite. *Journal of Environmental Science* 27(3):30–32.

Bernard, H. R., et al., 1965. Airborne Bacterial Contamination. *Archives of Surgery* 91 (Sept.).

Brader, P. E., 1988. Contamination Control in Biological Safety Assessment Facilities. Proceedings of the 9th International Committee of Contamination Control Societies Conference, pp. 567–571, September 26, 1988, Los Angeles.

Brenn, H. J., 1983. Planning and Installation of a Class 10,000 Cleanroom. *P & MC Industry* 2(3):26–32.

Crumley, R., 1987. Modular Clean Room Design Reflects Changes in the Use of Contamination-Controlled Environments, Part 1. *Microelectronic Manufacturing and Testing* 10(10):23–25.

Deaves, D. M., & Malam, D., 1985. Advanced Analysis Techniques for the Optimum Design of Clean Rooms. *Journal of Environmental Science* 28(5):17–20.

Faure, L.-P., & Thebault, H., 1987. Perspectives on Contamination Control: Part II. *Microcontamination* 5(4):10–16.

Howorth, F. H., 1987. A Clean Zone without Walls. *Journal of the Society of Environmental Engineering* March:25–27,

Hughes, R. A., Moslehi, G. B., & Castel, E. D., 1988. Eliminating the Cleanroom: Experiences with an Open Area SMIF Isolation Site (OASIS). *Microcontamination* 6(4):31–37.

Keilson, S. E., 1985. Evaluating Clean Room Products for Aerospace Applications. *Journal of Environmental Science* 29(4):19–22.

LeClair, D., 1990. The Impact of Building and Fire Codes on Cleanroom Design and Safety. *CleanRooms* 4(5):14–18, 63.

McClellan, M. S., 1985. Clean Room Considerations for Avoiding Molecular Contamination. *Journal of Environmental Science* 28(5):21–22.

Ohmi, T., 1990. Closed System Essential for High-Quality Processing in Advanced Semiconductor Lines. *Microcontamination* 8(6):27–32, 106–107.

Schicht, H. H., 1988. Clean Room Installations for Contamination Control during the Processing of Pharmaceutical Solids. *Swiss Contamination Control* 1(2): 45–52.

Schneider, R. K., 1990. Developing and Implementing a Cleanroom Construction Protocol. *Microcontamination* 8(8):35–38.

Scott, C. M., 1987. Material Selection for Cleanroom Compatibility. *Microcontamination* 5(4):18–28.

Silver, W., & Szymkowiak, E. A., 1988. Vibration on Various Floor Types in Microelectronic Fabrication Facilities. Proceedings of 9th International Committee of Contamination Control Societies Conference, pp. 187–191, September 26, 1988, Los Angeles.

Stokes, K. H., 1982. Personal communication.

Takenami, T., Inaba, H., & Ohmi, T., 1989. Total System Cost Effectiveness Must Keep Pace with Submicron Manufacturing. *Microcontamination* 7(8):25–34.

Thompson, R. G., 1985. The Architectural Design and Detailing of Pharmaceutical Clean Rooms. *Pharmaceutical Engineering* 5(3):19–24.

U.S. Food and Drug Administration, 1982. *Status of Current Good Manufacturing Practice Regulations,* 21 CFR Section 210.1. Washington, DC: U.S. Government Printing Office.

Walti, H., 1983. Sterile Rooms in Prefabricated Element Construction. *Swiss Pharma* 5(11a):9–14.

West, J. M., & Snell, J. T., 1987. Bioprocess Facility Design. *Pharmaceutical Technology* 11(3):24–29.

Wetzel, L. E., 1988. Clean Room Design for Sterile Manufacturing. Proceedings of 9th International Committee of Contamination Control Societies Conference, pp. 521–526, September 26, 1988, Los Angeles.

16

HVAC Filter and Flow Control Systems for Cleanrooms

In any cleanroom, the HVAC (heating, ventilating, and air conditioning) system, the air filters, and the airflow controls are crucial to good operation. The HVAC system must be capable of providing a uniform flow of conditioned air at sufficient velocity and pressure to overcome the airflow resistance of the air ducts, air conditioning components, and filters. Very large airflow is required for modern-day cleanrooms. Even for a mixed-flow class 100,000 cleanroom, sufficient airflow to provide 20 to 30 air changes per hour may be required. A mixed-flow cleanroom that is 50 feet wide and 100 feet deep with a 10-foot ceiling requires a blower system capable of providing at least 17,000 cubic feet per minute airflow for this air change rate. The airflow should be provided at a pressure at least 0.5 inches water above ambient to assure maintenance of positive pressure in the cleanroom and to overcome the pressure drop of the filters and the HVAC components. A vertical unidirectional flow cleanroom of the same size requires a much larger blower system. For a 5,000-square-foot room with 100% filter ceiling, the airflow requirement is 450,000 cubic feet per minute for the 90-feet-per-minute air velocity specified for such cleanrooms. Energy costs just to operate the blowers to provide this airflow make up an important part of cleanroom costs.

Heating and cooling coils are needed to maintain air temperature and relative humidity as required for the particular cleanroom. Pharmaceutical manufacturing areas can be a special problem in that some product components must be washed with water, and regular room and component sterilization is usually a wet procedure. Walls and floors may require mopping with bactericide, or the room contents may be exposed to moist disinfectant gas. In some pharmaceutical manufacturing areas, it has been decided that an ideal operating temperature is 66° F ± 2° F (Cattaneo 1984). Operators using sterile gowns perspire and emit excessive particles at temperatures much above 68° F but are uncomfortable when working in cleanroom garments at temperatures below 64° F. Relative humidity limits ranging from 45% to 55% have been found satisfactory to control mold growth in these rooms, with reasonable worker comfort.

Depending on the cleanroom location and the time of the year, external ambient conditions may require use of either humidification or drying to keep the cleanroom moisture content at the desired level. Some process equipment operations may generate excessive heat in the cleanroom, resulting in a requirement for air cooling. To maintain the desired relative humidity levels, it may be necessary to remove moisture added by processes and personnel in the cleanroom. The HVAC system should be laid out to minimize duct pressure loss; otherwise, power costs can soar as pressure drop imposes additional load on the air-moving system. The ducts should be large enough within the limits of allowable space so that high air velocity does not cause excessive noise. Stainless-steel ducts minimize particle emission, but galvanized iron ducts can be used in many temperate zone areas with suitable precautions. The HVAC system must be designed for intake of enough makeup air to keep gaseous contaminants at an acceptable level, as they may build up in normal operations. Typical cleanroom makeup air quantity is 10 to 20% of total airflow. Higher quantities of makeup air result in excessive costs to filter and heat or cool large quantities of external air. A typical cleanroom HVAC system is shown in Figure 16-1. There are some possible choices for flow control; for example, return air can be vented through a raised floor or side wall return air grilles. Supply air can be directed through individual plenums containing individual or banked HEPA filters or to an overall ceiling filter system for an entire cleanroom. The cleanroom HVAC system may also include components in addition to filters that clean the air of gaseous and particulate contamination. Ionizers may also be required to reduce charge levels, both for prevention of electrostatic discharge hazard to sensitive electronic products and for reduction of particle deposition in any area.

The HVAC system performance requires a variety of components in order to condition the air and to move it properly. The major elements include fans and blowers to move the air to the cleanroom, ducting and control dampers to control its motion, grilles and vents to emit it into or remove it from the cleanroom, humidifiers, heating and cooling coils to control its temperature and moisture content, filters to remove particulate contaminants, adsorbing beds to control gaseous contaminants, and instruments to verify that each of these elements is operating correctly.

Some of the requirements and operating limits for these elements are

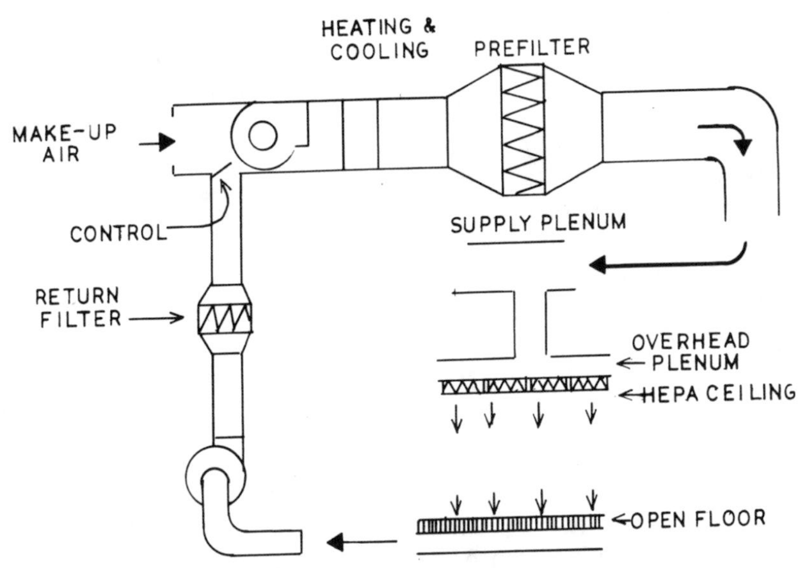

FIGURE 16-1. Cleanroom HVAC system. The general requirements are shown. Location and layout can be varied as required for specific needs. (Courtesy A. Gutacker, ARGOT, Inc.)

mentioned here briefly. In this discussion, it is assumed that the HVAC system is operating under optimum conditions. The reader is reminded that HVAC systems are similar to other mechanical or electrical systems. They can be subject to a variety of failure modes, ranging from gradual performance degradation to catastrophic failure. In normal use, blower capacity can be decreased if airfoils collect large amounts of adherent dust deposits; heat transfer coefficients can be reduced as surfaces are fouled by dust or crystal deposits; excursion of external ambient conditions beyond design limits can degrade HVAC system performance, and unanticipated component wear effects can cause performance loss.

HVAC air movers for cleanrooms are either axial or centrifugal blowers. Axial impellers are more efficient when high airflow at low pressure is needed, whereas centrifugal impellers are used where higher pressure is required. Because the required pressure to be developed is seldom more than 10 to 20 millibars (1–2 inches water head), either forward or backward curved blades can be used. The impeller choice depends on the cleanroom airflow requirement and its variability. For situations where the fan loading may change, then backward curved blades may be preferred. Typical blower sizes may range from one with capability of a few thousand cfm to ones with capability of up to 20,000 cfm.

As power costs increase, the question is frequently raised as to the advisability of using variable-speed motors in order to reduce operating costs. As tempting as this option may be, it should be used only if the cleanroom classification is at a level greater than class 1,000. The possibility of particle deposition on cleanroom surfaces always exists during periods when the blowers are operating at reduced rates. In this situation, the unidirectional flow patterns can be lost, with eddy formation resulting. The deposited particles may be reentrained and moved to hazardous locations when the fan speed is brought back to normal.

As shown in Figure 16-1, airflow from the HVAC system to and from the cleanroom is contained within ducts and plenum chambers. Both preconditioned and treated air flows through these components. Because the air exhausted from the cleanroom may contain corrosive air pollutants, the HVAC ducting should be resistant to the pollutant gases that may be produced or be present in the particular cleanroom area. In most parts of the world where normal relative humidity is relatively low, galvanized iron ducts can be used with few problems. In moist climates, pinhole failures in the galvanized coatings can occur and produce a release of corrosion product to the air in the duct. In cleanrooms where any particulate material corrosion products may cause problems, stainless steel ducts should be used. In cleanrooms located where acid rain is known to be present, galvanized iron ducts may not be satisfactory for *any* clean manufacturing process. Most aluminum ducts should also be avoided because aluminum oxide dust is produced very easily and can be released with flexing or vibration of the metal. When the ducts require insulation, as may be the case in areas where humidity and temperature of ambient air may be a problem, Fiberglas insulation is effective but should be applied to the duct exterior. Some commercial insulated duct material has a layer of Fiberglas and adhesive applied to the duct interior. This duct type should never be used. Sleeve joints should be oriented with the male end of a duct element at the element outlet. Duct joints should be sealed with clean, noncontaminating materials using good commercial practice. Sealing with adhesive tape alone is seldom acceptable for duct components upstream of conditioning elements unless very skilled assembly personnel do the work. Access ports in the HVAC ducting can be very useful if damper or turning vane adjustments may be required after the HVAC system is installed. Spring-loaded port closures with gasket seals can be used.

In any cleanroom, flow testing, balancing, and adjusting (TBA) of the HVAC system is almost always needed after design, construction, and installation. Variations can occur between calculated design load requirements and standard equipment capacities, and between original cleanroom flow conditions and conditions after prolonged use and/or process changes. A complete TBA program is concerned with measurement and control of airflow rates, temperature, relative humidity, pressure gradients, and other factors within the cleanroom. Because the objective of TBA is to ensure that the correct environmental conditions are produced by the HVAC system within the cleanroom and that the point-to-point parametric variations of flow rate, temperature, and relative humidity are within the specified ranges for that cleanroom, several environmental parameters are measured in accordance with specified procedures. Measurement procedures have been discussed in chapter 13 and are described in more detail in the IES Recommended Practice RP-006, *Testing Clean Rooms*.

Control of the range of cleanroom environmental airflow parameter variation is usually fine-tuned by adjusting damper position for diffuser and return air grilles in mixed-flow cleanrooms. The dampers are used to control both the quantity and direction of airflow into the cleanroom from a diffuser. Air inlet diffusers are not used, and return air grilles may not be used in unidirectional flow cleanrooms. Air inlets are provided through ceiling or wall-mounted filters, and returns may be through grated floors or through wall-mounted grilles. Effective TBA for such areas usually requires damper and flow control adjustment in duct sections outside the cleanroom. Figure 16-2 illustrates the use of damper control to balance air flow in a vertical flow cleanroom. The dampers are typically sheet metal elements that can be adjusted to restrict flow in an air line. They are essentially adjustable butterfly valves in the air line. If dampers are located

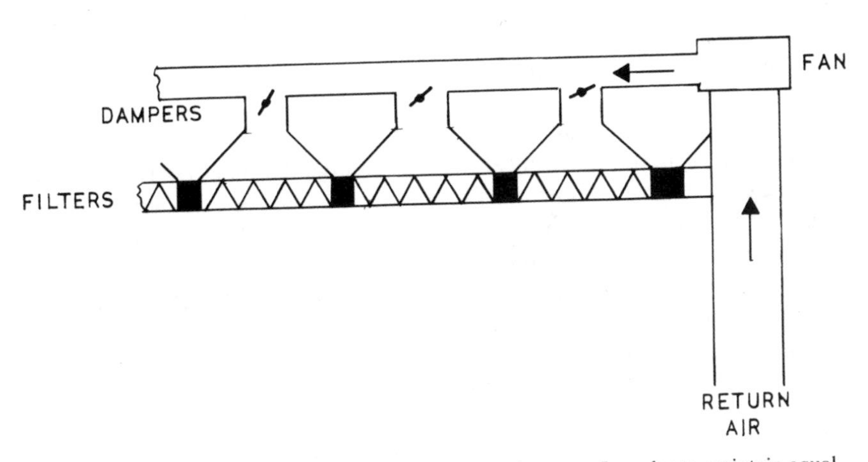

FIGURE 16-2. Cleanroom flow control by plenum dampers. In order to maintain equal air velocity at the plenum inlet and exit sides of the room, it may be necessary to vary flow to filters at each side by damper controls.

downstream of filters, then both the dampers and their bearing surfaces must be designed so that contaminant generation is minimized. Metal or plastic bearings or hinges should be considered, and the damper material and surfaces must be similar to the duct materials. Perez (1986) has described the advantages of well-planned and effective TBA operations for better cleanroom performance as well as for cost control.

Air coolers in industrial HVAC systems typically use chilled water or direct expansion refrigerant systems. Air temperature at the cooling coils is close to 50° F or so. Following the dehumidification process, it is necessary to restore the air temperature above the 50° F point. The cooling coil temperature must be controlled so that it does not drift much below 32° F, at which point condensate on the coil begins to freeze, blocking airflow through the system. Refrigeration-type dehumidifiers are not often used to reduce relative humidity below 35 to 40% at 72° C (Delli Paoli 1986). Reduction of relative humidity below this level is seldom required in any cleanroom operation. The procedure for controlling relative humidity frequently requires changes in air temperature. In warm, arid environments it may be necessary to add moisture to the air before sending it to the cleanroom. Water sprays are used for this purpose. Thus, the control system may involve spray nozzles, along with cooling and heating coils or panels. Atmospheric moisture and condensible gases can deposit on cooling elements with later release; particle emission can occur from heating panels with the wrong surfaces; water used for humidification must be filtered to remove suspended particles that can remain in the air after the sprayed droplets have evaporated. Components of the HVAC system can be exposed to mildly corrosive aqueous solutions during condensation and evaporation processes. The materials must be selected so that exposure to this material does not produce corrosion products that can result in contaminant emissions.

Along with the normal HVAC functions, many semiconductor and pharmaceutical manufacturing areas have installed ionizers to reduce electrostatic charge levels on products and to minimize particle deposition on grounded surfaces in the cleanroom. The latter capability justifies installation of ionizers in many cleanroom areas where electrostatic discharge is not a problem. Deposition of inert particles on any substrates that may become charged can be reduced sufficiently by use of ionizers, which are becoming more often used in pharmaceutical, optical, and mechanical assembly cleanrooms. Even though ionizers are not normally considered to be HVAC system components, their widespread and increasing use indicates the need for better understanding of their interactions with that system. In operation, ion emitters are placed within the filtered airflow and spaced on 20- to 40-cm centers located 10 to 20 cm below the filter ceiling.

Ionizer materials are selected to minimize erosion and particle emission when in operation. The ionizers are usually operated in a DC mode with alternate elements emitting oppositely charged ions. Modern ionizers appear to provide control of atmospheric charge levels with good efficiency and with very little contamination addition to the cleanroom atmosphere.

Developments in ionizer improvements open another possible area for cleanroom air control. For many years, the possibility of electrostatically enhanced filtration has been considered as a means of providing better particle removal with more open filter structures. The energy for both particle charging and electrostatic field establishment in the filter body is less than that required for an equivalent increase in filtration efficiency by use of filters with higher efficiency. One system that has been considered is simply addition of a corona source upstream of the filter, followed by application of an electrical field at the filter (Cucu and Lippold 1984). Although the concept has been shown in the laboratory to be capable of improving filter efficiency by as much as an order of magnitude or more, with pressure drop reduced by half, it has not found widespread use. The excellent performance of modern-day HEPA and ULPA filters, in combination with reasonable energy costs, has not justified development and testing of commercial electrostatically enhanced filter systems for cleanrooms. As class 1 areas come into use, the increased airflow needs may result in further investigation of very high filtration efficiency with very low pressure drop. At that time, developments now occurring in new ionizer material and electrical control technology may find application in electrostatically enhanced filtration as well. The two systems may well be operated as synergistic devices both to clean the air and to reduce particle deposition.

The cleanroom filter system should be designed with the room requirements in mind. For class 100 or better areas, vertical downflow rooms may be required with 100% ceiling coverage of HEPA or ULPA filters. For rooms up to class 10,000, vertical flow rooms with ceiling filter coverage of 40 to 80% may be adequate; horizontal flow rooms with one filter wall can be effective at this level for several rows of work areas before air cleanliness drops below the required level for the room class. As room cleanliness requirements decrease, a mixed-flow room with filters located in the duct system and nonunidirectional flow in the room may be adequate. In any case, the air going to the HEPA or ULPA filters should have been prefiltered with a 90 to 99% prefilter. This unit should be mounted in a duct so that any required replacement can be handled with minimal interruption of cleanroom activities, even though many prefilters are located just ahead of the final filter. It is emphasized that HEPA or ULPA filter mounting must include adequate sealing and that the filter installation must be verified to ensure that no filter damage has occurred during installation. As an additional air-cleaning operation, many cleanrooms contain clean benches with filtered airflow of up to 1,000 cfm. Depending on the operation and bench type, as much as 20% of the clean air may be emitted into the cleanroom from the bench. These benches can act as local sources of very clean air within a class 10,000 cleanroom.

In most cleanroom operations, the product area may be immediately exposed to the cleanroom airflow. In this situation, the air is filtered only once if any contaminants have been added in the cleanroom or if the air has been passed through the HVAC system only once. In some areas, however, the product may be in a system where air is continuously recirculated through a filter system at a reasonably high exchange rate. A good example of this situation is the Winchester disk drive airflow system. A nearly completely closed system is operated with internal air recirculation through a filter; external makeup air is a very small fraction of the total recirculating airflow. In this system, the time to achieve the required cleanliness level is related to the product of the particle removal efficiency of the recirculation filter and the airflow rate through that filter. Modeling and experimental tests showed that, with sufficient air recirculation, satisfactory cleanup could be obtained in reasonable time by sufficient recirculation through a filter of even indifferent efficiency (Japuntich 1985). The same conclusions would apply to a full-size cleanroom, except that the higher pressure and larger airflow requirements would probably result in excessive power costs.

Some special requirements for filter performance are present in pharmaceutical applications. These requirements result from the need to ensure sterility of components in these areas and to ensure that any toxic materials that may be present in the production process are not a hazard to operating personnel during normal cleanroom operation or maintenance. The need for aseptic or sterile conditions may require that the filters are occasionally exposed to high temperatures or to moist corrosive environments. Toxic material containment may require special components when a loaded filter must be handled. HEPA or ULPA filters are usually composed of glass fiber and adhesive media with aluminum or wood frame and spacer components. These materials are capable of operating at temperatures well above that required for sterilization. Most HEPA filters can operate reasonably well at temperatures to 250° C. A special filter system has been described that can operate at temperatures up to 400° C (Otake and Watanabe 1988). The organic binder is replaced with a special inorganic binder, the filter medium is retained in a stainless-steel mesh, and the frame is retained with a metallic seal.

If toxic materials may be collected on a HEPA filter in a pharmaceutical or radioactive material-handling operation, then procedures have been developed for handling such contaminated filters. The procedures ensure containment of the used, contaminated filter before and during the removal process, along with replacement with a new filter in such a way that emission of hazardous material from contaminated duct and filter retaining systems does not occur. The procedures are based on methods developed in the nuclear energy industry in the past (Burchsted et al. 1976). They require planning for components that allow filter changeout in suitable housings with sealable filter receivers, airtight dampers around the filter location, and suitable disposal sites for the contaminated filter package. Avery (1984) has described the procedures for safe filter change and disposal in some detail. The contaminated filters are enclosed in sealable bags and removed while a fresh filter is put in place with no leakage to the environment. A similar procedure (Melchert and Hogh 1985) can be used in pharmaceutical areas with personnel precautions during filter change, based on protective clothing and area sterilization after possible exposure to the filtered material. The filter to be removed is contained within a plastic bag that can be positively heat-sealed.

Filter life in a cleanroom operation is based on the time taken for the filter pressure drop to rise from the typical as-received level of 25 mm H_2O to a level of 50 mm H_2O. Tests (Asada 1988) indicate that the service life of a HEPA filter in normal cleanroom use should be 8 to 13 years before the pressure drop reaches 50 mm H_2O; at that time, the collection efficiency does not decrease and may even increase slightly.

Airflow control within the cleanroom is required because modern cleanroom filtration systems can remove essentially all particulate contaminants from the air entering the cleanroom. Most of the contamination encountered in the cleanroom is generated in that cleanroom and deposited on the products by turbulent flows within the room (Abuzeid, Busnaina, and Ahmadi 1990). The contamination may circulate within eddy flows caused by work surfaces projecting into an otherwise uniform airflow or by other flow interruptions within the cleanroom. It is important to prevent formation of local recirculating airflows that may concentrate contaminants near work areas. Airflow control begins with the needs for adequate contamination control; it includes assuring uniformity of airflow from a filter wall or ceiling for unidirectional areas; it considers the costs for maintaining that air cleanliness level and flow uniformity. As part of these needs, it includes layout of components that might interfere with controlled flow in any rooms. It has been shown that airflow uniformity in the supply plenum to an overhead filter system must be controlled (Sadjadi and Liu 1990). Nonuniform airflow distribution from the filters can arise as a result of airflow direction changes in the plenum upstream of the filter ceiling, with flow impingement directly at the face of the filters located at the direction change point. In downflow unidirectional areas, care is required in the design and location of components such as light fixtures in the air path. Otherwise, turbulence can be generated into the clean airflow, as shown in Figure 16-3.

The many theoretical and empirical studies made on cleanroom airflow patterns have aided in development of aerodynamically correct work surfaces and work stations (Schicht 1988). Sharp-edged corners that may cause flow separation and lateral mixing of air should be avoided. Surface configurations that may change airflows to the point where vortices can form often cause lateral contaminant transport. Air recirculating within a vortex can collect and retain particles generated by a source over which the air had originally passed. Some work (Oh, Lim, and Lee 1990) reports that the location of maximum particle concentration changes from the particle source to the recirculation zone illustrate this effect.

Because cleanroom airflow requirements are so high and stringent temperature and humidity controls may be required, these areas tend to use large amounts of energy. As energy costs increase, energy conservation

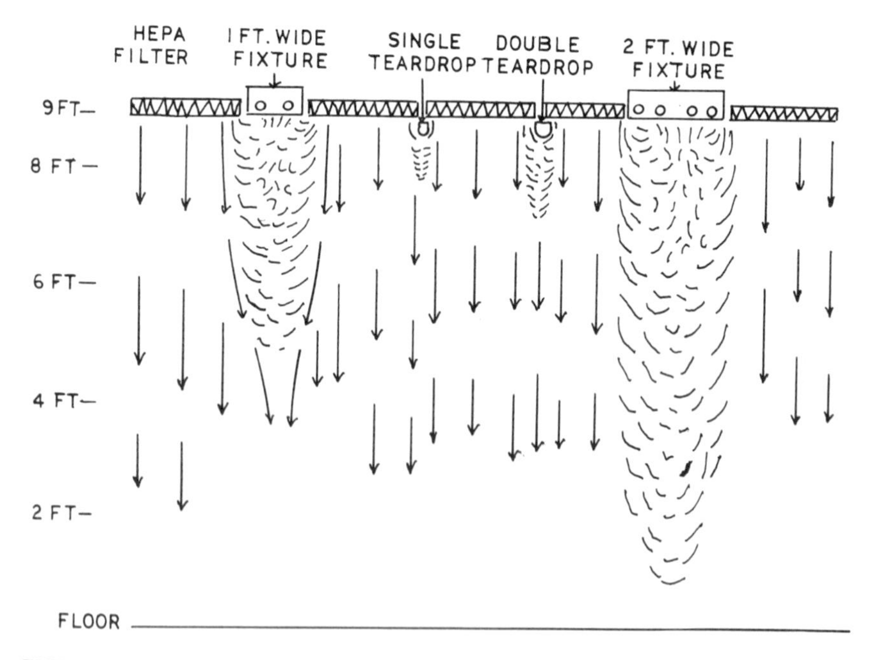

FIGURE 16-3. Turbulence generated by cleanroom overhead light installations. (Courtesy National Air Balancing Bureau.)

must be considered. As air cleanliness needs increase, the air change rates also increase. A class 100,000 room may require 20 air changes per hour; a class 10,000 room may require 75 per hour; and a class 100 room may require 500 per hour. The cleanroom designer and operator must examine the classification requirement carefully to minimize energy needs. In addition, the possibility of reducing energy consumption during off-hours should be examined. Reducing airflow during off-hour periods is one process that should be examined carefully. Although fan and blower energy is a major cost element, the cleanup time and the possible reentrainment of contaminants deposited during periods when flow is low may cause problems worse than cost. However, illumination level control during off-hours may allow significant savings (Schneider 1984).

The cleanroom designer must also realize that any interferences with minimum turbulence airflow not only will result in potential contamination effects in the cleanroom but also may cause greater load on the blowers supplying air to the cleanroom. Power costs for energy can quadruple if the designers are not aware of the potential loading on the blower system and if the users of the cleanroom are not careful as to equipment layout and operation (Gerbig 1984). For example, blower motors generate heat at the rate of 3.4 Btus per hour for each watt input to that motor. If the motor enclosure is in the airstream entering the cleanroom, additional cooling may be required to maintain desired cleanroom temperature.

Perhaps the greatest problem in maintaining unidirectional airflow in an operating cleanroom is the interference caused by the presence of both human operators and of functioning tools. Because of the high cost of cleanroom construction, the desire to obtain maximum use of expensive floor space can lead to installation of process tools in areas where interference with airflow cannot be avoided. Application of valid fluid mechanics models to cleanroom design has been a relatively recent addition to cleanroom technology. This unfortunate situation is evidenced by the continuing use of the term *laminar flow* to unidirectional flow areas. Even though many people will continue to use the incorrect terminology for many years, knowledge of the correct wording for cleanroom airflow patterns is very common.

Many studies using fluid mechanics theory have been made of the airflow in cleanrooms. Although details of the theory are beyond the scope of this discussion, some of the basics are summarized here, along with assumptions used in the calculations. The fundamental equations are the continuity equation for conservation of mass and the Navier-Stokes equation for conservation of momentum. Air properties are assumed to remain uniform with the exception of density. Particle transport within a cleanroom is affected by airflow. Because of the low particle concentra-

FIGURE 16-4. Air velocity vectors and particle trajectories. Any tool, cabinet, or other installation interferes and modifies the air velocity vectors. Particle trajectories diverge from the air velocity vectors, depending on the particle sizes in the area. (From Busnaina, A., et al., 1988. Numerical Modeling of Fluid Flow and Particle Transport in Clean Rooms. Proceedings of the 9th International Committee of Contamination Control Societies Conference, pp. 600–607, September 26, 1988, Los Angeles.)

tion, airflow is not influenced by particle content. Particle movement can be modeled on the basis of particle size and reasonable assumptions as to environmental conditions that affect particle movement; that is, small particles follow the airflow with negligible effects of inertial, electrostatic, or gravity forces. Diffusion has the main effect on movement of the small particles; Brownian motion is of concern in laminar flow, and turbulent eddy diffusion is of concern in turbulent flow. For larger particles, the effects on their motion of external energy gradient forces can be calculated, as discussed previously. More detailed models and discussions of airflow patterns in cleanrooms and around tools and other objects in these areas have been presented in the literature in the last few years (Kuehn 1988; Yamamoto, Donovan, and Ensor 1988; Gallo, Ramsey, and Liu 1988). Computer simulations of air and particle transport in cleanrooms describe flow path trajectories over and around obstacles that may be found in typical process tool and clean bench installations. Figure 16-4 (Busnaina, Abuzeid, and Sharif 1988) illustrates the model calculation results for both airflow patterns and particle trajectories in a cleanroom area. Turbulent flow patterns are generated at tool edges and at interfaces between tool and bulkhead surfaces.

Verifications of these model airflow calculations have been made using Schlieren observations (Settles and Via 1988) and observations of smoke flow around obstacles. It is observed that the theory does indeed predict the actual flow paths for particles in the airflow environments of cleanrooms. Kuehn et al. (1988) show nearly identical calculated airflow trajectories and measured particle streams. The effects on airflow and resulting particle motions caused by eddy and vortex formation behind an obstacle are shown quite clearly, even for relatively low-velocity cleanroom airflows (Ljungqvist 1988).

References

Abuzeid, S., Busnaina, A. A., & Ahmadi, G., 1990. Numerical Simulations of Particle Deposition from a Point Source in Turbulent Flow. Proceedings of the 36th Institute of Environmental Science Annual Technical Meeting, pp. 295–302, April 1990, New Orleans.

Asada, T., 1988. An Experimental Study on Deterioration Model of HEPA Filter in a Current Clean Room. Proceedings of the 9th International Committee of Contamination Control Societies Conference, pp. 156–159, September 26, 1988, Los Angeles.

Avery, R. H., 1984. Changing Contaminated HEPA Filters with Bag-in/Bag-out Housings. *Pharmaceutical Manufacturing* 1(7):18–21.

Burchsted, C. A., et al., editors, 1976. *Nuclear Air Cleaning Handbook,* Energy R & D Administration Publication 76-21. Oak Ridge, TN: Oak Ridge National Laboratories.

Busnaina, A. A., Abuzeid, S., & Sharif, M. A. R., 1988. Numerical Modeling of Fluid Flow and Particle Transport in Clean Rooms. Proceedings of the 9th International Committee of Contamination Control Societies Conference, pp. 600–607, September 26, 1988, Los Angeles.

Cattaneo, D. J., 1984. HVAC and the Clean Room. *Pharmaceutical Engineering* 4(6):42–45.

Cucu, D. D., & Lippold, H. J., 1984. Electrostatic HEPA-Filter for Particles 0.1 μm. Proceedings of the 30th Institute of Environmental Science Annual Technical Meeting, pp. 74–78, May 1984, Orlando, FL.

Delli Paoli, A., 1986. Control of Humidity in Cleanrooms. *Medical Device and Diagnostic Industry* 8(4):55–59.

Gallo, E. J., Ramsey, J. W., & Liu, B. H. Y., 1988. Flow Visualization Studies in Clean Rooms. Proceedings of the 9th International Committee of Contamination Control Societies Conference, pp. 637–645, September 26, 1988, Los Angeles.

Gerbig, F. T., 1984. Energy Consumption in Vertical Laminar-Flow Cleanrooms. *Microcontamination* 2(3):50–55.

Japuntich, D. A., 1985. Disk-Drive Recirculation Filtration System Evaluation. *Microcontamination* 3(1):52–56.

Kuehn, T. H., 1988. Computer Simulation of Airflow and Particle Transport in Cleanrooms. *Journal of Environmental Science* 31(5):21–27.

Kuehn, T. H., et al., 1988. Comparison of Measured and Predicted Airflow Patterns in a Cleanroom. Proceedings of the 9th International Committee of Contamination Control Societies Conference, pp. 331–336, September 26, 1988, Los Angeles.

Ljungqvist, B., 1988. Air Movements: The Dispersion of Pollutants in Theory and Reality. Proceedings of 9th International Committee of Contamination Control Societies Conference, pp. 594–599, September 26, 1988, Los Angeles.

Melchert, H., & Hogh, J., 1985. Examples of Application of Laminar Flow Technology in the Pharmaceutical Industry. *Swiss Pharma* 7(11a):43–48.

Oh, H. D., Lim, H. K., & Lee, Y. L., 1990. Unsteady Propagation of Contamination Particles in the Turbulent Flow Field for the Dynamic Analysis of Clean Rooms. Proceedings of the 36th Institute of Environmental Science Annual Technical Meeting, pp. 283–289, April 1990, New Orleans.

Otake, N., & Watanabe, N., 1988. Development of HEPA Filter Applicable to High Temperature up to 400° C. Proceedings of the 9th International Committee of Contamination Control Societies Conference, pp. 180–182, September 26, 1988, Los Angeles.

Perez, G., 1986. Optimizing HVAC Systems through a Testing, Balancing and Adjusting Program for Environmental Systems. *Pharmaceutical Manufacturing* 3(2):23–26.

Sadjadi, R. S. M., & Liu, B. Y. H., 1990. Supply Plenum and Air Flow Uniformity in Cleanrooms. Proceedings of the 36th Institute of Environmental Science Annual Technical Meeting, pp. 157–162, April 1990, New Orleans.

Schicht, H. H., 1988. Engineering of Clean Room Systems: General Design Principles. *Swiss Contamination Control* 1(6):15–20.

Schneider, R. K., 1984. Energy Conservation in Clean Room Design. Proceedings of the 30th Institute of Environmental Science Annual Technical Meeting, pp. 112–116, May 1984, Orlando, FL.

Settles, G. S., & Via, G. G., 1988. A Portable Schlieren System for Clean-Room Airflow Analysis. *Journal of Environmental Science* 30(5):17–21.

Yamamoto, T., Donovan, R. P., & Ensor, D. S., 1988. Model Study for Optimization of Cleanroom Airflows. *Journal of Environmental Science* 31(6):24–29.

17

Cleanroom Work Support Component Areas

Satisfactory cleanroom operations require adequate support components. The same care in design required for the cleanroom proper must also be used in layout of all ancillary and support areas. All items that enter or leave and activities that are carried out in the cleanroom must pass through, are stored in, or are processed in one of these auxiliary support areas. Whenever possible, the support areas are located and maintained out of the cleanroom, but some items must be within the work area. The support areas include facilities for personnel activities, for process materials, for process equipment and tool storage and maintenance, for in-process products, and for cleanroom maintenance. The support areas often contain components that can degrade cleanliness in the work area. Operation of these components must be controlled so that contamination of products within the support area is minimized and transport of contamination from these components to the cleanroom is avoided.

Product component and subassembly storage areas are frequently required for incomplete process work. These areas may involve a production control and testing operation normally carried out either in the cleanroom proper or in an auxiliary area. The external storage and test areas may require environmental control in terms of temperature, relative humidity, or light level, as well as being contaminant-proofed to some extent. Many auxiliary support areas can be operated as a clean area that is two or three class levels more contaminated than the cleanroom proper. Because the cost of energy for air-cleaning and HVAC systems is so high, the cleanroom management philosophy should be to carry out as many activities as possible in the auxiliary area and to minimize the required floor space for the cleaner areas, which are more expensive to build and to operate.

The same contamination control requirements exist for process equipment and tools but with one additional consideration. Not only must the equipment be kept in a controlled clean environment but also any potential contaminants produced by the equipment must be prevented from contacting the product or being emitted to the cleanroom environment. Special materials, seals, and moving parts designs may be required for cleanroom processing devices. The equipment is usually designed and installed so that maintenance service can be handled from chase areas exterior to the cleanroom proper, particularly for large tools. Examples of some semiconductor manufacturing equipment operation and associated contamination sources are given in Table 17-1. Most equipment in this table is operated at pressures below the cleanroom operating level so that no contamination that may be present or generated within that tool is released to the cleanroom environment.

If manufacturing operations may involve changes in procedures as technology advances, then cleanroom layout may require modifications. Some cleanrooms may be in a constant state of flux insofar as design is concerned. Product modifications or improvements may require or allow different cleanroom operations from month to month. In this situation, temporary partition components may be used. It has been shown that low

TABLE 17-1 Some Semiconductor Processing Equipment Contaminant Sources

SEMICONDUCTOR PROCESSING EQUIPMENT CONTAMINATION	
Process	Contaminant Sources
CVD	Loaders, Quartz Oven Tube, Reactor Chamber, Vacuum System, Vacuum Controls, Gas Supplies
Diffusion	Loaders, Diffusion Tube, Gas Supplies
Epitaxy	Loaders, Reaction Chamber, Chemical Supplies
Flow Controls	Filters, Seals, Valves, Flow/Pressure Controllers
Ion Implantation CVD, Sputtering	Loaders, Vacuum Controls, Ion Source Materials, Beam-Induced Sputtering
Photolithography	Loaders, Resist Spinners, Chemicals, Pumps, Dryers Defective Photomasks or Reticles

Although not specifically stated, essentially all of the contaminant sources are in areas where gas, material, personnel, or equipment movement occurs in the tool operation.

Source: (Courtesy A. Gutacker, ARGOT, Inc.)

release of particles can be obtained by use of temporary wall partitions made of aluminum sheet taped and riveted to an aluminum tube frame (Kohler 1988). A number of panel configurations were designed and shown to be from 3 to 30 times cleaner in terms of particle emissions than similar panels made of fire-resistant plastic over wood frames.

Storage of components requires careful planning. The operator must be sure that ancillary cleanroom items (records, working and cleaning tools, etc.) are stored adjacent to the cleanroom or with exposure to the cleanroom environment, where they do not contaminate the product. An area adjacent to the cleanroom may be required to stage and prepare equipment for the cleanroom. Tool cleaning should be done outside the product handling area but may still require a clean facility. Cart storage and wipedown facilities may be required for both component transport and for cleaning or maintenance carts. Cleaning systems should include facilities for storage of soiled or contaminated cleaning tools. Cleanroom surfaces, including walls, floors, and furniture, require scheduled cleaning. This is a firm requirement, particularly for pharmaceutical areas.

Some special precautions are required for cleanroom equipment. Special wheels of polyurethane or similar material may be required for carts, along with sealed bearings. The carts should be balanced so that they do not produce excessive vibration that can be transmitted to production tools while the carts are being moved. Most conventional janitorial supplies must be examined for suitability. Vacuum cleaning should be carried out with centralized vacuum sources. If portable cleaners must be used, then these devices must be fitted with HEPA filter exhausts. Storage containers for work in process must be specially selected to minimize contamination. They should be designed so that the required routine cleaning of interior surfaces does not leave debris in the container.

Normal office equipment cannot be used in cleanrooms. Although it should be obvious, liquid paper use should not be allowed in any cleanroom. Whenever possible, data, notes, and messages should be handled by electronic means. Records to be stored should be transmitted out of the cleanroom area for magnetic storage. If hard copy is required in the cleanroom, special papers and marking devices are needed. It has been shown (Michaels, Fickel, and Donovan 1986) that specially treated papers designed for cleanroom use emit particles at a rate 2 orders of magnitude less than the rate from normal bond paper under typical handling conditions.

Personnel facilities include the change rooms where personnel store their street clothes, wash up, and don and remove cleanroom garments. Because personnel and their cleanroom garments enter the cleanroom and products are exposed to them, it is necessary that the garment storage areas do not add to the garments any contaminants that can be deposited in the cleanroom. These rooms must have lockers for outside clothing in one area, with lockers or racks for clean garments in another area. Booties, gloves, and hoods used in the cleanroom must be kept clean and stored so that soiled items are not put back in use. Tacky mats to clean outside footwear should be maintained and replaced on a regular schedule. Access into the cleanroom from the garment-donning area is usually through a dual-door air lock, frequently with an air shower facility in the lock area. If an air shower is used, then the door lock timing, airflow controls, and seals must be kept in good order. It has been shown that air showers with jet flows of 10 to 30 meters per second remove larger particles from clothing reasonably well (Hirasawa et al. 1984). Particles 5 μm and larger were removed with efficiencies ranging upward from 50%. Particles of 2μm and smaller were not removed well; because they are continuously generated by personnel activity in or out of the cleanroom, however, air shower operation is not of concern for these particles.

In some areas, storage lockers for hanging garments are fitted with HEPA-filtered air supplies so that no dust can settle on the garments during off-shift storage. The areas where the garments are donned must be laid out so that there is no possibility that clean garments can contact soiled floor, wall, or locker surfaces during the time that they are being donned. Clean foot coverings are stored in the clean side of the donning areas. A barrier can be provided so that the personnel sit with their feet on the "dirty" side to remove normal foot covers and swing their feet over the barrier to the "clean" side to don clean booties. Precautions such as these are important in keeping contaminants out of the clean areas. These requirements are especially important to minimize nonproductive time during shift breaks and changes. Lunch and rest breaks during a work period may require that personnel leave the work area, change to street-type clothing, store clean garments while eating or using the rest area facilities, and don the clean garments again to resume work. The design of the personnel support area must take into consideration the level of cleanliness required for a particular cleanroom product. A class 10,000 cleanroom should not require the same level of clothing types or storage facilities as a class 10 room. Gowning time should also be much less for the class 10,000 area personnel.

Personnel working in the cleanroom usually require hand tools for some operations. These items are basically simple assembly tools or inspection devices. The assembly items should be stored in areas out of the cleanroom work space. The storage areas should also include facilities for spare parts and for tool cleaning, as required.

Before any production process, the raw materials used must be received, inspected, stored, converted to product, and shipped from the facility. It has been pointed out that some 15% of cleanroom production costs are used for material transport to and during processing operations (Schmutzler 1988). Passive and active material locks, as well as mobile transporters with their own HEPA-filtered air supplies, are examples of devices needed to control contaminant acquisition during product transport. The same considerations apply to these materials, whether the products are semiconductors, pharmaceutical materials, or spacecraft. The raw materials must be received and placed in an area where their suitability for use can be verified. For the most part, this step is not part of the cleanroom process. The receiving and inspection areas can be removed from the cleanroom operations. However, after inspection and approval, the process materials should be cleaned as much as practical before they are used in the cleanroom. At least the initial cleaning operations should be carried on out of the cleanroom so that the raw materials do not contaminate the cleanroom any more than necessary. Packaging materials may have to be removed and disposed of, protective surface coatings may have to be removed, and fluids may require some processing (pressurization for some gases, filtration, etc.) before use. The specific operation depends on the process material, the product, and its needs. The difference in requirements for handling a shipment of ultrapure gas for a chemical vapor deposition process in a semiconductor fabrication area and a shipment of metal sheets that must be formed, cleaned, assembled, and cleaned again for an aerospace device is obvious.

Once received and inspected, the raw materials are stored until ready for use. At that time, batches are transported to the clean manufacturing area for processing or assembly. Therefore, the transport device is part of the cleanroom support facility that must not contaminate the product or the cleanroom. The transport device or system is one of the cleanroom support components that must be controlled so that cleanroom integrity is not degraded. This requirement applies whether the material is moved by overhead crane, piped in, wheeled in on a cart, or hand carried in a clean container.

Compressed gases are used widely as both processing and dilution materials. Large quantities of ultraclean inert gases are used in semiconductor processing. Either tanks of clean gas are delivered to the consumer's compressed gas storage and delivery system by a supplier, or on-site preparation and purification is carried out. For more reactive processing gases, used in smaller quantities, high-pressure cylinders are delivered for connection to the process line. Features in the design of pressurized gas transport and control systems are good examples of the need for great care in system design (Davidson and Ruane 1987; Jensen

1987; Henon and Overton 1988). Both the materials of construction and the structural design of the gas flow control components are important in controlling contamination for ultrapure gas systems (Bourscheid and Bertholdt 1990). The materials of construction should have inherent corrosion resistance, should be suitable for electropolishing operations, should not contain hazardous material inclusions, and should be formable to produce systems designs that are acceptable for minimizing contaminant inclusion during use. Construction of systems for transport of high-purity gases includes specification for 316L stainless steel for as many components as possible. In addition, surface processing by electropolishing to remove surface inhomogeneities that may entrap particles is common. Packless, bellows valves are usually chosen. Components are joined by automatic orbital welding equipment, using tungsten inert gas welding processes.

Similarly, the flow systems for large-quantity process liquids, such as deionized water, must also be carefully controlled. A deionized water system includes filters, ion exchange systems, reverse osmosis units, carbon filters, ultraviolet radiation components, and a variety of flow control systems. Materials must be chosen for resistance to the local water supply ionic contaminants as well as for normal use with the cleaned water (Thaman and Greiner 1988). The flow system must be designed so that contaminant removal is positive during operation and so that contaminant retention or generation during operation is minimal.

Much of the work on small products is done in the cleanroom in clean benches. These devices are enclosed work surfaces that are supplied with a flow of clean air moving at sufficient velocity to resist ingestion of contaminants from the surrounding environment or as a result of the operators' activities. Careful design is required for these clean air devices to ensure that adequate airflow is present within the clean bench and that sufficient air velocity at the bench face is present to prevent contaminant entry to the work space. If any wet processing is to be carried out, then liquid-handling systems must be provided with assurance of clean liquid supply as well as adequate waste material drainage. Work surfaces must be designed so that no contamination traps can be present in the airflow. The operator must be reminded that overnight storage of process materials cannot be allowed in the clean bench without suitable protection for both the material and the bench surfaces.

The process tools used in many cleanrooms carry out essentially the same operations as are carried out in most manufacturing operations. Some are more sophisticated than standard manufacturing tools, but the similarity still exists. For example, semiconductor manufacturing operations use a variety of precision optical systems both for device fabrication and for inspection. Similar tools are used in many photographic processing operations, but tolerances are much closer in the cleanroom process; for this reason, interferences from contaminants are a greater problem in the cleanroom than in the normal manufacturing process. The pharmaceutical manufacturing operation uses tools and devices very similar to those in many other fluid-handling processes, such as petroleum product preparation. In this situation, contamination is a relatively minor problem, whereas for pharmaceutical products freedom from contamination, particularly from microorganisms, is a positive requirement. Bottle-washing systems used to ensure cleanliness of pharmaceutical product containers are not located in the cleanroom. The sterilizing drying tunnel air exhaust is vented out of the filling line area. The process tools used in cleanroom production should therefore operate in such a manner that their normal operation does not produce contaminants that affect either the product or the surrounding cleanroom environment. In some cases, normal operation of process tools produces contaminants that must be removed before further processing can take place. Both particulate and gaseous contamination sources are present in most of these devices. Some reduction of contamination is possible by modifying components and operating procedures.

Areas of concern with these devices include the cleanliness of the consumable materials they use. Semiconductor processing systems use a variety of fluids in wafer treatment. The photoresists, solvents, flush liquids, and process gases used in these tools must be clean. The tools cannot contaminate these fluids as a result of wear particle generation or of outgassing from tool surfaces. The operating tools cannot be allowed to emit into the cleanroom environment contaminants that can degrade operations in neighboring areas. Many tools transport products being worked on and materials used to carry out that work on moving belts, hydraulic or pneumatic lifts, or air cushion pads. Moving parts cannot lose any fluid power material or lubricant to the internal product handling area or to the surrounding cleanroom.

All manufacturing tolls require routine maintenance or may have to be set up for specific tasks. If either of these operations involves access to the interior of the tool or requires significant component handling (similar to cutting tool change in a lathe or milling machine), then these operations must be carried out so that contamination is not produced by manipulation of the components. For large tools, it may be necessary to locate the tool at the wall of the cleanroom with access to the tool interior through a service chase. Where this is not possible, major work on the process tool may have to be carried out during off-shift periods.

Very few cleanroom products are produced in a single shift. In the very early days of cleanroom manufacturing, the production personnel simply covered the incomplete work at the end of a shift. Requirements for maintaining cleanliness have become more strict. Most components now require storage in clean containers during the time when actual processing is not taking place. Even though many product-cleaning operations are carried out, it is known that the cleaning process may not be totally effective. For that reason, clean, sealable containers are required for storing and transporting work items. In some semiconductor facilities, robot transfer tools are used to minimize exposure to contamination from personnel. It has been shown (Workman 1988) that transport of wafers in a SMIF container to a generic vacuum process tool results in a particle ($\geq$0.4 μm) gain of some 0.2 particles per wafer, whereas uncontained transport results in gains ranging up to a factor of 20. The in-process storage containers must be cleaned and stored in a secure location. Neither the container-cleaning operation nor the storage area should be located in the cleanroom. Either a suitable clean storage area close to the cleanroom is required, or each container must be inspected before use. Inspection of containers for cleanliness may be carried out in the cleanroom if these containers are not too large and if space is available in the cleanroom at the time.

References

Bourscheid, G., & Bertholdt, H., 1990. How Production Technologies Influence the Surface Quality of Ultraclean Gas-Supply Equipment: Material and Construction. *Microcontamination* 8(6):43–46, 103–106.

Davidson, J. M., & Ruane, T. P., 1987. Gas-Handling Hardware: Considerations for Ensuring Gas Purity. *Microcontamination* 5(3):35–39.

Henon, B. K., & Overton, J. S., 1988. Design Planning for Class I UHP Stainless-Steel Process-Gas Piping System. *Microcontamination* 6(2):43–47.

Hirasawa, K., et al., 1984. Particle Removing Efficiency of Air Showers on Clean Room. Proceedings of the 7th International Committee of Contamination Control Societies Conference, September 1984, Paris.

Jensen, D., 1987. Reducing Microcontamination Generation and Entrapment within High-Purity Gas Distribution Systems. *Microcontamination* 5(5):52–65.

Kohler, M. S., 1988. A Temporary Partitioning System for Clean Room Rearrangements. Proceedings of the 35th Institute of Environmental Science Annual Technical Meeting, pp. 355–359, May 1988, Anaheim, CA.

Michaels, L., Fickel, L. M., & Donovan, R. P., 1986. An Experimental Study of Particle Emissions from Cleanroom Papers. *Journal of Environmental Science* 29(6):38–40.

Schmutzler, W., 1988. Material Transport into Clean Rooms and Sterile Areas. Proceedings of the 9th International Committee of Contamination Control Societies Conference, pp. 360–364, September 26, 1988, Los Angeles.

Thaman, M. G., & Greiner, M. E., 1988. Implementation of a High Throughput Facility-Wide Pure and Deionized Water System. *Microelectronics Manufacturing and Testing* 11(5):36–38.

Workman, W. L., 1988. Contamination Control for a Submicron CMOS VLSI/ULSI Factory. Proceedings of the 9th International Committee of Contamination Control Societies Conference, pp. 104–109, September 26, 1988, Los Angeles.

18

Cleaning, Maintenance, and Janitorial Needs

The contamination control facility, including all items that may come into contact with products, requires regular maintenance and cleaning. The products produced in the cleanroom facility are typically sensitive to material that might be acceptable in other production areas but cannot be allowed in a cleanroom. The extent and frequency of both maintenance and cleaning depend on the cleanroom product or process, as well as its design level. Cleanroom maintenance includes regular cleaning of the room and of components, as well as mechanical component maintenance.

The purpose of cleaning is to make sure that no accumulation of contamination can build up to the point where it may be released to harm a product. Even though the cleanroom air supply is continuously treated to remove essentially all contaminants from the air entering the cleanroom, there will always be some contamination that enters the room, deposits on surfaces, and accumulates. Eventually the accumulation builds up to the point where it can be released as an agglomerated or composite contaminant. In addition, the internally generated contaminants accumulate in the cleanroom even more rapidly, requiring scheduled cleanroom maintenance and cleaning. Training of both janitorial personnel for cleanroom work and of operating personnel is required to make sure that contaminants generated by personnel activity are minimized and that contaminants generated by normal tool operation are removed expeditiously.

All components in a cleanroom facility require cleaning. Some components require regular maintenance operations, which may require later application of special cleaning procedures, especially if maintenance requires access to tool or component interiors. The critical elements that require cleaning include the room enclosure surfaces, the processing equipment in the room, the tools in use, the devices used to bring material and products into and out of the room, personnel clothing, and the furniture in the room.

Cleanroom walls are usually protected from dust accumulation by the lack of thermal gradients that could cause thermal precipitation of particles from the surrounding air. In addition, the surrounding air is filtered very well in a cleanroom. Even so, after use, some material deposits on wall surfaces as a result of small-particle diffusion or impaction of larger particles. It must be removed before it builds up to the point where it may be entrained into a critical airflow. In addition to removal of nonviable particles, maintenance requirements for pharmaceutical manufacturing areas include regular sterilization of room surfaces from which material may be released to products.

Floors must be cleaned regularly. Some debris is brought in on personnel shoe surfaces or on the wheels of carts moving equipment or material into the room. Some debris is produced by normal wear of even the best floor-covering materials. Special care is needed when cleaning perforated floor elements in vertical downflow areas. The cleaning equipment must be chosen so that the procedure does not degrade or damage cleaning equipment on sharp edges of perforations. Ceiling surfaces almost never need cleaning in unidirectional flow rooms, but mixed-flow room ceilings may collect some debris, especially near inlet air diffusers and overhead lights, where temperature differences can occur. Thermophoretic forces result in deposition of small particles on the colder surface.

Cleaning solutions used in cleanrooms fall into two major areas of application: Some are used mainly to remove particulate or film contaminants from surfaces; some are used for disinfecting the surfaces to which they are applied. In addition, surface-active materials aid in reducing electrostatic charges upon surfaces. Selection of the best cleaning material is often difficult. Solvents must be used with caution; some materials can dissolve easily in organic solvents, many solvents are flammable or toxic, and all organic solvents may cause problems in environmentally acceptable disposal. Even very clean deionized water has limits in use as a cleaning material; some surfaces can be corroded, and some surfaces are not wetted well unless a surfactant is added to the water. Ionic surfactants frequently contain metallic ions and are usually quite reactive to many surfaces. Nonionic surfactants are often used, especially in electronic fabrication areas, because these materials do not contain potentially harmful metal ions and are least reactive.

When room floor or wall cleaning is required, either vacuuming or wet wiping are two methods suitable for use in cleanrooms (Whyte and Davidson 1987). Dry sweeping should never be used because of the particle production. Vacuuming with HEPA-filtered exhaust can be used in the cleanroom, but it is grossly inefficient for particles smaller than approximately 50 μm. If vacuuming is required, either a central vacuum supply or vacuum cleaners with HEPA-filtered exhaust should be used. For these reasons, wet wiping is often used on cleanroom surfaces. For some pharmaceutical manufacturing areas, regular sterilization is required. In some areas, this is accomplished by releasing a wet vapor sterilant mixture such as moist ethylene oxide vapor into the room *during off-shift periods*. Care is required to ensure that no equipment that may be susceptible to the moist atmosphere is exposed. Because of hazard, personnel are not to be in the room when this process is carried out. Most pharmaceutical cleanroom sterilization processes involve application of wet sterilizing material to room surfaces for this reason.

Processing equipment in the cleanroom also requires cleaning. Many operations produce harmful debris that can reduce yield. Semiconductor production requires wafer movement to and from the tool, along with orientation within the tool. The wafers may be taken from and replaced within the cassette. In all these situations, if the tool is generating wear particles that are allowed to accumulate, then wafer yield is degraded. Pharmaceutical products are handled in machinery that involves similar activity. Photographic film is produced on machinery with a large number of moving mechanical parts, each capable of producing large numbers of particles. These parts must be designed so that emission of particles is minimized and the components can be cleaned as required. Many aerospace products are fabricated and moved about within the manufacturing area with the aid of hydraulic and mechanical devices that must be kept clean for reliable product manufacturing. All of the processing equipment and the instruments used to validate production equipment performance must be kept clean to avoid deposition of contaminants on products. When maintenance is required, the maintenance staff must make sure that the maintenance operation has not produced excess contamination and that the device is cleaned after maintenance has been completed. The cleaning procedure must take into consideration the tool material, its operation, and its materials of construction. Gross contaminant particles can be removed from exterior surfaces by vacuuming; final cleaning is usually carried out by careful wiping. Organic solvents may be used for components that may be corroded by exposure to some polar cleaning solutions. When organic solvents are used, they must be selected so that toxicity is minimal and any vapor emission that may cause environmental concern is contained safely or at least well controlled.

In-process materials and small tools are often hand-carried into the cleanroom. The item has usually been previously cleaned in an ancillary area and placed in a cleaned container or package before bringing it into the room. Larger items are brought into the cleanroom on a cart or a hand truck, or for very large items an overhead crane can be used. Hand-carried containers are cleaned by wet methods, just as incomplete products are cleaned. The choice of cleaning material depends on the container and the nature of the product to be carried in that container. It is mainly necessary to ensure that no reactivity occurs with the cleaning material. Larger material transport devices are cleaned mainly by wet wiping. If an overhead crane is required, then the pulleys and the hydraulic operators always produce contaminant particles. Fortunately, these tend to be large enough so that gravity is the most important force affecting them. In one installation, the movable contaminating crane elements were contained within a flexible protective enclosure. The enclosure was similar to an inverted umbrella, with access at the upper side and the lower portion sealed so that no particles could fall upon sensitive components suspended below the crane.

Cleaning solutions can be selected from a wide variety of materials. A filtered surfactant solution can be used for a variety of cleaning processes. Nonionic surfactants are normally used in cleanrooms because these materials do not contain metallic ions that may cause problems, particularly in electronic product manufacturing areas. After cleaning, some surfactant residue may remain on surfaces when the cleaning solution film dries. The residue must not cause flakes to form as a source of particles. It is necessary that the material selected be nontoxic, nonflammable, and noncorrosive. It must evaporate fast enough to leave a dry surface in reasonable time, but not so fast as to cause excessive surface cooling. There should be no particulate or film residue that can harm products produced in the area. Many cleaning solutions, including the quaternary ammonium compounds, are also effective as disinfectants for many bacteria and fungi.

A standard procedure for room enclosure surface cleaning should be established and followed by janitorial personnel. The cleaning operation should begin with the ceiling. Where the ceiling is composed of filters, use vacuum cleaning of solid surfaces only with minimum contact to the filters. Be sure to inactivate any overhead ionizers before cleaning. Overhead light fixtures can be wet wiped with a wiper moistened with deionized water. Care is required to ensure that the wiper does not touch any electrical connections. Weekly cleaning of the ceilings may be needed for class 10 areas.

Cleaning of walls, windows, and doors should begin in the most critical areas and proceed to areas of lesser concern. The cleaning operation may use a surfactant solution followed by a rinse with deionized water. Wiping in critical areas should be done with disposable, nonwoven wipers. Wiping in other areas can be done with cleanable, reusable wipers. Surfaces, such as doors, that are contacted most frequently may need daily cleaning in class 10 areas and less frequent cleaning in areas with larger class designations. Floor cleaning should begin with a vacuuming operation to remove "large" debris. Next, wet mop with surfactant solution and follow with a deionized water mopping. Daily cleaning may be required in class 10 areas. Work surfaces can be cleaned by using a moistened cleanroom wiper, starting at the back of the work surface, wiping across the surface, and using a fresh area of the wiper for each pass across the surface. The choice of cleaning solution depends on the process or product being handled at that work area. The cleaning frequency depends upon the criticality of the operation. Cleaning operations on work surfaces and tools several times per shift is not necessarily excessive for very critical products.

Responsibility for cleaning may be with a janitorial staff as well as with the cleanroom supervisor to make sure that the area is adequately clean. Responsibility for ensuring that the cleaning operations have been effective is rarely formalized in most cleanroom facilities. Even so, the cleanliness of the room should be verified, just as the performance of any instrument or tool used in the cleanroom must be validated. A suggested verification procedure for walls, floors, and furniture surfaces would be to specify a sequence of test locations of the specific area to be wiped with a precleaned or sterile lint-free wiper after any scheduled cleaning operations. Test wiper material would be chosen so that it is soluble in a selected solvent whose particle content or soluble chemical content could be measured before and after the wiper is dissolved in a specified volume. If more than an allowable increase in contaminant is seen in the solvent after the wiper is dissolved, then additional cleaning is required. Responsibility for testing should be shared between the cleanroom manager and the janitorial supervisor.

Because any personnel activity can generate contamination, major room-cleaning activity should be carried out when no work is being performed in the room. The maintenance and/or cleaning frequency is usually based on experience and knowledge of the product, the area, and its requirements. Sufficient time after janitorial activity should be allowed for the room airflow to clear out contaminants generated by the cleaning operation. Remember, during the time that the cleaning operation is being carried out, a very expensive facility is not producing. For this reason, cleaning should be done only when it is necessary; however, it should not be ignored when it is needed.

A bilevel cleaning operation may be required when new equipment is brought into the area. Even now, many equipment manufacturers are not aware of the cleanliness needs for cleanroom operations; some equipment may need rigorous cleaning before use, with routine wipedowns being adequate for later cleaning. Before entry into the cleanroom, all parts, tools, equipment, and material must be cleaned. The cleaning method depends on the item to be cleaned and the area in which it will be used. For example, small hand tools should be cleaned ultrasonically (if possible) before entry to the cleanroom, at scheduled intervals, and whenever routine observation indicates a need for cleaning. A cleaning and inspection protocol document is helpful in controlling this contamination source.

As previously mentioned, the areas to be cleaned in the room may include walls and floors, upper surfaces of overhead lamps, clean benches, storage cabinets and equipment, work surfaces, and chairs. Following regular cleaning or maintenance, a record of the action and results should be provided, along with verification of the cleanliness level achieved.

Reference

Whyte, W., & Davidson, N., 1987. Cleaning a Cleanroom. *Microcontamination* 5(11):49–54.

19

Component and Equipment Testing

Component and equipment are tested to make sure that these items carry out their design function but do not generate unacceptable quantities of contaminants while doing so. Most production tools generate some quantity of contaminants in operation. Modification of the tool so that contaminant generation is completely eliminated may be prohibitively costly. However, significant reduction in generation of harmful materials in tool operation can be achieved with economically acceptable improvements in components and/or operations (Silverman and Gruver 1990). Such steps as incorporation of improved filters in a chemical process liquid line or requiring adherence to inspection schedules and immediate remedial action where indicated can reduce contamination from tools and processes to a great extent.

The accepted quantity of contaminants is based on evaluation of the cost penalty of the tool's contaminant generation as opposed to the cost of modifying or replacing the tool. Manufacturing tools should not generate excessive quantities of contaminants directly to products or to the environment, where they may affect other operations. Therefore, we must have a good idea of what the contamination limits on product environment must be for acceptable yield. These limits include the particulate, gaseous, ESD, and any other contaminant types that can affect the product. If these values are known, then the contamination level produced by the equipment must be determined. First, request information from the manufacturer as to what contamination levels were measured for the product. This data should include both the internal contamination that may affect the product in process and the external contamination emitted by the device that may affect the cleanroom. If that information is not known, use historical data from similar devices and tools as a starting point; then verify the performance of the equipment with a suitable sampler or analyzer. A rigorous test protocol for gas valves used for flow control of high-purity gases can serve as an example of a qualification procedure that may become necessary for some situations (Cohen 1990). Stainless-steel valves were tested for emission of metal particles and of previously deposited particles as small as 0.01 μm using both an optical particle counter and a condensation nucleus counter. Helium testing for leakage rates in the range of 1.6×10^{-11} atmospheric cc per second was carried out for leakage into and from the valve, as well as across the valve seat. Finally, sample valves were tested by disassembly and sectioning to expose the internal surface in order to define surface roughness levels.

Experience with processing tools indicates the types of contaminants that are generated. For most semiconductor manufacturing tools, particulate contaminant generation during gas flow changes is a common occurrence. Particles are released from many surfaces as a result of dimensional changes caused by sudden gas pressure variation from very rapid valve operations; particles are generated by abrasion of moving parts in the tool or by wear of wafer surfaces; particles present in process liquids are deposited on the wafers and are not removed satisfactorily by cleaning operations. With older tools, plasticizer vapor release from some polymer surfaces may occur; with other devices, emission of previously adsorbed vapors can occur from otherwise refractory surfaces. Mechanical devices of all types can release particulate materials from frictional processes. The use of rotating elements or moving parts also introduces the possibility of lubricant droplet or vapor release, especially at higher temperature. In the pharmaceutical industry, cleaned glass containers are sterilized and dried by passing them through a drying tunnel in which filtered high-temperature air is flowing. If the filter has been damaged or if any of the tunnel parts are worn, then particulate contaminants can be deposited within the containers. Photographic film is handled and moved by mechanically driven sprockets during the coating and rolling parts of the production cycle. Wear particles are continuously generated by the mechanical operations of the handling system. Great care is required to keep these elements operating well with minimum contamination generation.

For electronics manufacturing, it has been known that personnel are a potent source of contaminants and that component orientation and location during manufacture must be extremely precise. These two points indicated that robotic handling systems could be extremely useful in increasing product yield. Robotic handling devices can also generate contaminants in their operation; it is therefore necessary to make sure that these devices do not generate excessive contaminants in operation. A test program development was carried out (Bailey and Rogers 1985) and data developed that showed that the major contamination from robotic devices would be generated by either mechanism wear or lubricant emission. However, care in design and material selection can reduce contaminant emission to a very low level.

A wide range of testing has been carried out for several specific application areas. Robots have been evaluated for certifying the memory disks used in computer rotating memory systems. Tests have shown that robotic test-operating devices have improved yield significantly over testing with personnel manipulating the disks (Hill and Eassa 1986). Semiconductor manufacturing involves a great deal of handling of wafers during the process. Because these devices are highly susceptible to contamination effects, the use of robotic systems for moving wafers to, from, and within process tools has received much attention. It has been found that robot handlers for wafers and cassettes generate as much as 2 orders of magnitude fewer particles than do human operators (Ferris and McConnell 1985; Muka and Mayer 1985). Careful design for wafer-handling robots can produce a device that is capable of meeting FS209D class 1 levels (Hoshizaki and Morgan 1986). This process includes selection of mechanisms that can provide smooth wafer motion with minimum particle production, as well as use of working and surface materials that do not emit particles in use. Further, the exterior configuration of the robot is designed to minimize airflow disturbances that might cause eddy formation that would entrap particulate contaminants near the robot. In a similar manner, aerodynamic equipment design principles can be used to minimize eddy formation in the airflow around almost any manufacturing equipment (Kern 1989). A unidirectional flow of clean air over critical products is thus assured with minimal possibility of reverse flow fields near the product.

High vacuum pumping systems are widely used in the semiconductor processing areas. Problems arising from passage of pumping fluids into the processing components have resulted in the development of new pump designs and pumping technologies (Hablanian 1989) that do not use oils or other vaporizable materials to produce vacuum. The problems of potential backstreaming have made it necessary to avoid these materials where extreme cleanliness is required. Ion gettering pumps, turbomolecular pumps, cryopumps, and other oil-free pumps are primary vacuum sources in the semiconductor areas. In addition to the use of pumps that produce very low levels of contamination, the pumping operations have been examined. Rapid pumping operations have resulted in entrainment of particles from interior tool surfaces into the gas stream that may pass over critical products. A combination of reduced pumping speeds and close control of cleanliness of critical surfaces has aided in reducing contamination during either evacuation or pressure recovery in tools.

In the pharmaceutical industry, process equipment must be used in aseptic operations. All foreign material, whether inert particulate or microbial contaminant, must be controlled. A major area of concern is in the packaging line for product. The product containers must first be cleaned and sterilized; then they are filled and sealed. Either glass or plastic containers must be washed first, sometimes with an initial rinse of 2% H_2O_2, and then flushed with clean water. After washing, the containers must be dried with clean air. Continuous-flow convection tunnels expose the glass containers to filtered, heated airflow for a time sufficient to ensure sterility and then move the heated, sterile containers to a filtered air-cooling zone if the product can be degraded by heat. Liquid product filling machines are constructed so that liquid lines can be flushed and sterilized easily and with components that do not trap materials that may remain in place long enough for bacterial growth to occur later. Worst-case control capability is also required for the possibility of container misfeed or breakage resulting in spills.

Powder filling and tablet presses both are notorious for potential dust generation. A pharmaceutical product material cannot be allowed to deposit or mix with other products. If cross-product contamination occurs, then the product may not be usable. Therefore, in addition to the normal powder flow and handling lines within these machines, a local air-filtering and control system is required to avoid emission of product into the local cleanroom filter system. Pharmaceutical product packaging machines differ from most other cleanroom product machines and tools in that they must be capable of sterile operation. This requirement is satisfied by selection of mechanical designs that allow effective cleaning and by using materials that are not affected by gas or thermal sterilization (Koblis 1985).

For nondestructive determination of particulate contamination from tools, we can observe particles added to the work environment or deposited on a witness plate near the equipment or in areas where product work flow exists. We should obtain samples at obvious locations in or near the device, that is, sample close to moving parts such as gear drives, bearings, spinning elements, and liquid lines and preferably where process products will be located. Two sample-acquisition methods have been used to characterize particle emissions from semiconductor process equipment. Where particles are generated by a flow distribution system or during tube furnace operation, an isokinetic sample probe is used to monitor flow surrounding the suspect contaminant generator. Changes in particle concentrations in that airflow can be stated easily (Cheung and Roberge 1987).

Unfortunately, only relative changes in concentration can be stated, because the actual concentration that is measured is also dependent on the airflow patterns and rates surrounding the suspect source. As production requirements become more affected by contamination in the work environment, better monitoring procedures are necessary. A primary requirement in this area is application of real-time contaminant monitoring to critical assembly operations (Monkowski and Freeman 1990). The large investment in facilities, materials, and labor for modern semiconductor production absolutely demands well-planned, statistically reliable real-time monitoring of particles, gases, and ionic and molecular contaminants in the production area in order to optimize production and yield factors.

Measurement of the size-dependent concentration of very fine particles generated in semiconductor processing tools operating at atmospheric pressure has been carried out by sampling with a combination of diffusion battery stages and condensation nucleus counters in parallel with a high-sensitivity optical particle counter (Viner et al. 1989). In this way, the capability of the optical particle counter for size definition is extended below the present lower limit of 0.05 μm to approximately 0.01 μm.

A similar approach places a small-particle observing system within a process tool (Weisenberger 1988). The system can be used for measurements within tools where operations take place in enclosures at atmospheric pressure or in a vacuum (Borden and Gregg 1989). For vacuum operations, most problems occur when the product is in the small loadlock chamber for pressure reduction. Observation of particle levels during pressure change in the loadlock indicate the need for care in pressure change. With this system, particle generation trends can be shown. The particle emission rate cannot be stated definitively because the particle-measuring device depends only on diffusion and normal gas movement to pass particles through the observing system with undefined efficiency. For this reason, an alternate method has been used that defines particle emission levels in a more quantitative manner (Rogers and Bailey 1987; Donovan, Locke, and Ensor 1987). The tool to be tested is enclosed in a sealed chamber with sufficient mixing that the particles emitted from the device are uniformly distributed within the chamber. Particle loss rates to the chamber walls are known by previous measurement. In operation, the device to be tested is loaded into the chamber, and all particles within the chamber are removed by recirculating the air with HEPA or ULPA filtration. When the particle load is zero, the device is activated and particle concentration monitored. This method is extremely sensitive for devices with low particle emissions.

For information on release of gaseous contaminants, first note the composition of the equipment elements and decide if there is potential for gaseous emission of specific contaminant materials from suspect elements in order to select analytical methods. Then draw samples for analysis at locations where heat may be generated that may accelerate gas, vapor, or fume release from device surfaces. Electric field or charge measurements are made where susceptible products may be exposed to high ion levels.

For many process tools and work environments, identification of contaminant sources can be accomplished easily if the contamination generation rate is high enough to be easily detected. For some semiconductor process tools, contamination levels must be so low that statistically valid samples are difficult to define. For example, some 150-mm and 200-mm wafer processing limits define acceptable levels of particulate contamination from process tools at less than 10 0.3 μm particles per wafer pass (PWP) added to the wafer surface by passage through the tool. Tullis (1985) proposed use of the PWP measurement as a means of quantifying tool performance. He proposed use of PWP measurement on unpatterned wafers, along with statistical definitions of confidence limits for the collected data and consideration of the effects of exposure time on the PWP results. Appreciable care is required in experiment design because the number of particles deposited in each operation may be very low and the particle measurement device may also be a contaminant source. If unacceptable levels of contaminant generation are seen from components or equipment, then remedial measures must be taken, such as addition of protective systems, modification of present equipment, additional cleaning steps, or replacing the problem tool with another one.

References

Borden, P. G., & Gregg, J., 1989. Measurement and Control of Particle Levels inside Vacuum Processing Equipment. Proceedings of the 35th Institute of Environmental Science Annual Technical Meeting, pp. 325–327, April 1989, Anaheim, CA.

Bailey, L. G., & Rogers, G. G., 1985. Reducing Contamination in Automated Systems by Design. *Microcontamination* 3(11):80–84.

Cheung, S. D., & Roberge, R. P., 1987. An Inside Look at Measuring Particles in Process Equipment. *Microcontamination* 5(5):45–51.

Cohen, R. M., 1990. Qualifying High-Purity Gas Valves: One Company's Experiences. *Microcontamination* 8(7):41–44, 88.

Donovan, R. P., Locke, B. R., & Ensor, D. S., 1987. Measuring Particle Emissions from Cleanroom Equipment. *Microcontamination* 5(10):36–43.

Ferris, D., & McConnell, W., 1985. Assessment of Robot Cleanliness for Wafer Handling and Cleanroom Applications. *Microcontamination* 3(5):55–59.

Hablanian, M., 1989. New Pumping Technologies for the Creation of a Clean Vacuum Environment. *Solid State Technology* 32(10):83–86.

Hill, J. E., & Eassa, K. D., 1986. The Use of Robots for Memory Disk Certification in the Cleanroom. *Microcontamination* 4(3):47–52.

Hoshizaki, J., & Morgan, D., 1986. Design and Assessment of a Robot for VLSI Wafer Fabrication. *Microcontamination* 4(3):41–45.

Kern, F. W., 1989. Integration of Numerical Simulation and Flow Visualization in the Design of Equipment for the Manufacture of Submicron Semiconductor Devices. *Journal of Environmental Science* 33(3):19–24.

Koblis, R. J., 1985. Key Considerations in Selecting Machinery for an Aseptic Packaging Line. *Pharmaceutical Manufacturing* 2(5):35–40.

Monkowski, J. R., & Freeman, D. W., 1990. Real-time Contamination Monitoring of Process Equipment: The Challenge of the 90s. Proceedings of the 36th Institute of Environmental Science Annual Technical Meeting, pp. 393–395, April 1990, New Orleans.

Muka, R. S., & Mayer, S. D., 1985. Particulate Contamination during Robotic Wafer Transport. *Microcontamination* 3(11):76–79.

Rogers, G. G., & Bailey, L. G., 1987. A Closed Chamber Method of Measuring Particle Emissions from Process Equipment. *Microcontamination* 5(2):43–48.

Silverman, R., & Gruver, R., 1990. Particle Reduction by Tool and Process Improvements. Proceedings of the 36th Institute of Environmental Science Annual Technical Meeting, pp. 350–354, April 1990, New Orleans.

Tullis, B. J., 1985. A Method of Measuring and Specifying Particle Contamination by Processing Equipment. *Microcontamination* 3(11):67–74.

Viner, A. S., et al., 1989. Measurement of Ultrafine Particles from Processing Equipment. Proceedings of the 35th Institute of Environmental Science Annual Technical Meeting, pp. 414–420, April 1989, Anaheim, CA.

Weisenberger, W., 1988. Particle Control in High-Current Ion Implanters. *Semiconductor International* 11(6):188–192.

20

Personnel Selection and Training

Normal manufacturing produces products with a yield that approaches 100%. The products that are made can operate with some manufacturing debris and contamination deposited within the product, but these materials do not seriously affect the quality of the product. The requirements of the product types produced in most cleanrooms are significantly different. The cleanroom products are usually high-technology objects that are affected by even the smallest particulate contaminants that deposit upon the product. Scores of such products require cleanroom fabrication. They vary in dimensions from 80-μm laser fusion targets to aerospace devices many meters long. Many cleanroom products are produced with yields that are much less than those in normal fabrication processes. In some semiconductor areas, acceptable yield is less than 25%. The cost of repairing a semiconductor chip with short circuits or open circuits is very high. In many situations, a contaminated product cannot be repaired or recovered for use. The critical products cannot be contaminated during the fabrication process.

For these reason, some significant differences exist in the definition of acceptable activity patterns for the cleanroom worker and for workers in other manufacturing environments. The worker in most manufacturing areas can clean up at the end of a work shift, but the cleanroom worker must keep the work area and the products therein clean at all times. Products that are manufactured in standard production areas are seldom damaged if a worker touches a surface with bare hands or if the worker breathes upon the product. The amount of time spent in preparation for work simply to ensure that the worker is clothed correctly is much greater for the cleanroom worker than for the normal production operator.

The ideal production worker in any area is a person who produces the maximum quantity of acceptable-quality product in minimum time. The definition of *acceptable* is quite different for the cleanroom product than for the standard production area product. Most production workers are trained in those disciplines that are known to be important for a specific product line or for a particular production operation. The production takes place in an area where personnel and material access and activity are controlled mainly for personnel safety. Otherwise, the worker can enter or leave the work area whenever necessary. In addition to scheduled break periods, departure for comfort reasons and return to work occur with no special activity upon return. Preparation for departure and reentry is minimal. Special garb is required only for safety reasons. Tools and equipment are stored where convenient and are cleaned normally when gross manufacturing debris may interfere with work operations. In process work, components are stored and handled mainly on the basis of convenience and available space. Movement of materials and process equipment into and out of the production area uses any equipment or carrier that is convenient. Contamination of in-process work material is of minor concern for most products. Cursory cleaning to remove visible debris is acceptable during production. After the product is finished, final cleaning can take place.

In cleanroom operations a primary concern is to produce a product that is free of contamination; the contamination generated by personnel is a major problem to be controlled. This contamination consists of the material that may be emitted from the worker's clothing or skin, as well as chemicals that may be exhaled or otherwise released. In addition, contamination present upon working surfaces that may be released to the atmosphere as a result of worker activity must also be controlled. The cleanroom worker must develop work habits and procedures that minimize contaminant deposition on products and materials. Quite often, these procedures are completely different from those that would be required in standard production areas. Because many cleanroom products are built or assembled by nearly automatic machinery, worker speed is not necessarily a critical requirement. The most important requirement is that the worker control the processing tools to maximize product yield; that is, the cleanroom worker should understand the process tool operation, the effects of changes in environmental and operating parameters on yield for the particular product, and the effects the worker's presence can have on the product. As an example, one of the requirements for working in a cleanroom is development of work habits that include careful, slow movement when near exposed products. In this way, air turbulence from rapid worker movement that might cause possible particle deposition on the product is avoided.

The ideal cleanroom technologist would be a person with more capabilities than can be expected from any individual. He or she would have training in civil engineering to help in laying out the ideal facility. The technician would know chemistry and chemical engineering well enough to set up the best cleaning systems for material surfaces or process fluids, select optimum construction and process materials, and understand the product line operations. The ideal technician would also know mechanical engineering well enough to understand all of the mechanized production processes, product component structures, and HVAC systems. The technician would know electrical engineering well enough to understand monitoring instrument operations, enough statistics to understand sample size needs and data validity, and enough biology to control any microbial contamination. The perfect technician would also have to be personally observant, thorough, well organized, credible, inquisitive, and communicative. In the real world, no workers are available with all these qualifications. The successful technologist at any level is the person who has had sufficient training to know when the task at hand is being performed in the best possible manner; when the "best procedure" is not known, that technologist knows where to get the best answers quickly to those particular problems from experts in those fields. Many sources exist for information and answers to cleanroom problems. The cleanroom manager and the technicians should be aware of as many of them as possible.

The real-world cleanroom organization usually has access to skilled personnel with a variety of technical backgrounds. For example, computers and control devices are essentially electronic products. However, semiconductor manufacturing is essentially microscale chemistry and chemical and chemical engineering disciplines that are needed for good based on materials science, physics, and mechanical engineering principles. The challenges to chemical engineers in electronic production cleanroom areas have been presented in detail by Larrabee (1985). He described many of the operations in semiconductor manufacturing and the chemical and chemical engineering disciplines that are needed for good products.

Among the considerations in real-life personnel selection for cleanroom duty are the educational and training levels expected from present employees and from local labor sources. Not all people can work well in a cleanroom environment. Cleanroom areas have been reported as high-stress work areas. Some cleanrooms are very noisy because of the large airflows in the room and HVAC systems that are not optimized in this respect, along with illumination that is not the most restful, but special skills, work habits, and attitudes are needed. The selected personnel must be trained in cleanroom procedures and in the prework cleaning and clothing requirements for the cleanroom. Some people can accommodate easily to the cleaning, cosmetics limitation, and other personal activity limitations that may be required for some cleanroom operations, but many workers cannot maintain the discipline required for the cleanroom.

Consideration of local union and governmental Fair Employment Practices Commission activity is another important part of the personnel selection process. Where a new cleanroom activity is implemented, whether it is part of an existing plant or part of a completely new facility, it may be wise to consider discussing some of the special requirements with union personnel. The cleanroom can be a high-stress environment because of the type of work and the light and noise levels present in many cleanrooms. Personnel transfers and reassignments may cause problems if potential problems are not worked out first.

Training and motivation of personnel are as important in the cleanroom as in any operation. Motivation tools such as the use of special titles and salary incentives can be used here. In addition, because the cleanroom products are typically high-technology new devices, distribution of information on operation and application of the special high-technology products in today's world can help to generate a feeling of pride that leads to better product manufacturing. Experience has shown that normal work training procedures, such as on-the-job instruction, classroom discussion, short courses, video presentations, and literature dissemination, are very effective.

Notwithstanding all of the procedures for individual technologist training, disciplines, and motivation, a central authority for cleanroom operations is still required. Depending on all the operations and supervisors to be thoroughly trained and consistently performing correctly sooner or later results in a major or minor catastrophe in the cleanroom. An administrative system is required to control all of the different personnel operating in the cleanroom. Fabrication technologists, assemblers, inspectors, supervisors, maintenance and cleaning personnel, visitors, and other people can be in the area at any time.

One method that has been used quite successfully to control the disparate groups is to implement a cleanroom manager system. The manager is responsible for all personnel and activities in the cleanroom, including personnel discipline, area cleanliness, product cleanliness, and safety. Personnel responsible for contamination control and for reliability, safety, and monitoring report to that manager. The contamination control group would be responsible for cleanliness of all items entering the cleanroom, for cleanliness testing procedures, and for cleanliness of the area. The reliability, safety, and monitoring group would be responsible for defining all acceptable manufacturing and handling operations; for personnel, product, and equipment safety; and for routine monitoring. (In some cleanrooms, the "group" may be a single person, but that person still has responsibility for the assigned task and reports activity in the task area to the cleanroom manager.) The manager would oversee all these areas and ensure that supervisory staff are aware of and responsible for personnel procedures. He would also set up working procedures so that overlap of authority and responsibility does not cause any problems.

Much cleanroom activity at this time is centered about the semiconductor industry. The product values there appear to compose a major fraction of the dollar value of cleanroom product output at this time. However, there are requirements for special personnel knowledge in other industries that should be known in all cleanroom operations. Training techniques and subject areas from the pharmaceutical industry emphasize control of microbiological contamination; information transfer from the aerospace industry brings knowledge of better ways of handling large clean objects. In one Swiss pharmaceutical organization, new personnel basic training includes three half-day sessions that emphasize personnel hygiene needs for good pharmaceutical product manufacture (Zuger 1984). In-depth training periods from 2 to 10 hours long cover such areas as basic microbiology, methods of detecting microbes, personal hygiene procedures, cleaning, disinfection and sterilization methods, and methods of disposal for potentially septic wastes. A training program at the University of Kentucky uses a lecture plus demonstration approach to provide information on microbial, pyrogenic, and particulate contamination in the pharmaceutical industry (DeLuca 1983). The lectures cover personnel and facilities effects on contamination, anticipated bioburdens in the manufacturing process and monitoring needs, processing and sterilization methods in use along with sterility testing, and effects of pyrogens and endotoxins on product sterility and acceptability. Regulations and standards that must be met in the pharmaceutical industry are reviewed. Finally, hands-on work in the laboratory is required.

A series of training programs has been developed for the aerospace industry. Again, lectures and demonstrations have been used widely, and hands-on training has been found invaluable. In one area, a series of five courses has been described (Manguray 1988) for training in contamination control of large cleanroom products. The first course is a general 1- to 2-hour lecture for groups of up to 30 students to introduce them to the cleanroom concept. All further classes are restricted to no more than five students. The first of these is a 6- to 8-hour demonstration of precision cleaning techniques using video and classroom demonstration. The second series is 24 hours of combined further demonstration and hands-on training aimed at cleanroom technicians. The third series is 30 hours of demonstration and hands-on training for cleanliness verification techniques aimed at the quality assurance inspectors. The fourth series is a 2-hour session on cleanroom maintenance procedures including both demonstrations and hands-on work for the cleaning staff. In all courses, teamwork and information transfer concepts are emphasized. It is extremely important that cleanroom personnel with specialized background education or training be taught to recognize that contamination control is not dependent on a single specialized phenomenon or process (Moller 1989). It can be affected by several factors that can interact in a variety of manners. The optimum contamination control program is based on consideration of all the processes, with special emphasis on those that are most important for a particular problem at a particular time. Even though most training programs emphasize the problems of a particular industry, information from other industrial areas that are sensitive to contamination should be pointed out to operators in any industry. Some problems may be new to one industry but commonplace in another.

Work procedures for personnel in cleanrooms are based upon special behavior patterns required to minimize any contamination produced by personnel in the cleanroom. All personnel must be reminded that procedures and behavior in the cleanroom also apply in preparatory areas, such as gowning areas and clothing and tool storage areas. Garment change procedures, for example, can involve shoe cleaning and removal and storage of street clothes and cleaning in one part of the change room. This procedure may be followed by donning cleanroom garb, except for booties, at a border area. For very clean areas, temporary special gloves and hair covers may be used to protect the garment to be worn in the cleanroom (Miller 1988). The booties are put on while the worker is seated at a boundary that separates the clean and dirty sides of the gowning area so that the clean booties are not exposed to the "dirty" side of the boundary. Face masks or covers are usually the last item to be added before entering the cleanroom, particularly for operations in cleanrooms where complete face covers for breathing air supply different from that in the cleanroom are used.

The procedures just summarized do not apply in all cleanroom areas. For personnel working in a class 100,000 area, the discipline is minimal: prohibition of eating, smoking, or drinking in the area and control of traffic in the area. Cleanroom garment requirements are minimal, consisting mainly of head covers, smocks, and street shoe covers. Any potentially dirt-generating operation must be followed by immediate cleaning of products and of the contaminated work area. For personnel working in class 1,000 to class 10,000 areas, the disciplines are more restrictive. A complete cleanroom uniform is required, with gloves and breathing zone cover.

Work location and operations are defined to avoid interference with clean airflow to the product. Rapid body movements are frowned upon. Cosmetics are forbidden. For personnel working in class 10 to class 100 areas, even greater control is required. Complete change of clothing is recommended, and face masks must be worn in the cleanroom. Special gloves and foot coverings are required. Strict gowning protocols must be followed. Parts of the cleanroom where products may be exposed should be avoided by personnel unless no contact with the product that may cause contamination is positively assured. For cleanroom areas operating at class 1 or better, personnel should not be present in the area. Clean robotic devices should be used for production whenever possible.

As part of the preparatory disciplines, cosmetics removal is usually required; use of specified medication must be reported if potential emissions of trace quantities may be a problem with a specific product. The content of some typical cosmetic materials is given in Table 20-1 (Phillips 1983); consider the effects of emissions of these materials in some manufacturing areas for semiconductors. The metal ion content of most cosmetic materials matches many forbidden contaminant materials in several manufacturing operations. Personnel usually understand the reasons for control of cosmetics in the cleanroom, but potential emission of trace quantities of medication may require a more detailed explanation of physiological actions and the extreme sensitivity of some products to complex

TABLE 20-1 Cosmetics Components

COSMETICS COMPONENTS		
Cosmetic	Particles ≥ 0.5 μm per Application	Elements
Lipstick	1.1 E9	Bi
Blush	6.0 E8	Mg, Si, Fe, Ti
Powder	2.7 E8	Si, Mg
Eye Shadow	8.2 E7	Bi, Si, Mg
Mascara	3.0 E9	Fe

Most cosmetic materials are ionic, particulate in nature, and a few micrometers in diameter. Especially in semiconductor manufacturing areas, personnel should not wear cosmetics in the working areas.

Source: (From Phillips, Q. T., et al., 1983. Cosmetics in Clean Rooms. *Journal of Environmental Sciences* 26(6):27–31.)

organic molecules. This subject should be brought up during training sessions.

Reference to control of human metabolic product emissions can be a politically dangerous subject, particularly in a strongly unionized area. Special emphasis may be required for extraordinary personal hygiene procedures. Contaminating emissions can be produced by both healthy and ill individuals with contamination effects that can be hazardous to products in both cases. Hazard to products must be spelled out to personnel, and methods of minimizing these hazards must be defined (Luna 1986).

The extent to which disciplinary activity on the part of management is possible varies with location, industry, and company. For example, if a cleanroom employee has received excessive exposure to summer sunlight, greater than normal emission of skin fragments can be expected until the sunburn is healed. Even though the cleanroom garments should contain skin fragment emission, the question arises as to whether effective control of emissions is possible from sunburned personnel who may be releasing skin fragments at a rate much greater than normal. Because most control devices used in this world are not 100% efficient, some skin particle penetration through cleanroom garments may be expected. After a person has been in the sun and received a tan, some of the natural oils in the skin are dried out, and surface skin layers can flake or peel faster than normal. The presence of a sunburned employee in the cleanroom can result in product yield loss. Several control methods are possible: application of emollient cream to reduce skin flake emission (if the product will not be harmed by the cream), temporary transfer to another work location, or temporary layoff (if union agreement has been established). Repeated infractions should not be allowed. An effective training program is needed to control repeated infractions.

Locker-room procedure is designed to keep stored cleanroom garments from unnecessary exposure to contaminants. Some cleanroom protocols require personnel to pass through an air shower just before entry into the cleanroom. Air lock controls maintain airflow exposure for a period long enough to remove most fibers and large dust particles from the cleanroom garment exteriors. Correct procedures seldom carried out without good training include even the simple need to maintain a correct attitude in an air shower. Because the high-velocity air jets are emitted from fixed locations, it is usually necessary that the person in the air shower rotate and move about so that all parts of the cleanroom garment are exposed to the air jets in the air shower.

Normal working procedure includes a lunch break and a midmorning or midafternoon break period. Before leaving for the break, in-process materials must be safeguarded by covering or storing them correctly, and tools must be placed in a defined location. When leaving the cleanroom (and returning, as well), personnel must move in an orderly manner so that airflow turbulence generated by personnel motion does not transport contaminants into clean areas. Clean garments should not be worn during the break period out of clean areas. The cleanroom garments should be removed in a suitable clean area and stored so that they are not contaminated. In this way, some reuse is possible. The operator or another designated person should inspect the garment for cleanliness before it is brought back into the cleanroom. In extremely clean areas, garments may not be used again. During the break, any actions that may result in subsequent contaminant emission by personnel must be controlled. These actions may include eating, violent exercise, and smoking.

Work procedures should follow the precepts laid down during training; for example, no sudden motions that can cause uncontrolled airflow should occur, avoid sneezing or coughing into clean spaces, do not scratch or touch the skin with clean gloves and then handle clean tools or products, do not lean over a work surface unless absolutely necessary, use clean wipes, and change gloves when needed. The cleanroom supervisor should ask experienced operators to point out any areas where they think different procedures, materials, or tools could help the manufacturing process. An experienced operator can spot potential problems and point out solutions where theoretical analysis may not provide the best answers.

Some personnel actions that can cause problems in cleanrooms are:

1. Not using shoe cleaner or air shower as directed
2. Wearing clean garments outside clean area
3. Opening garments in cleanroom or removing items from beneath clean garment
4. Wearing garments in need of repair or closures
5. Smoking, eating, chewing gum or tobacco in cleanroom
6. Wearing cosmetics or using external medication
7. Working when ill with skin problem, cold, or allergy
8. Scratching or rubbing skin with clean gloves
9. Fast motion, running, exaggerated arm movements
10. Not cleaning items entering cleanroom area
11. Not reporting actual or potential CC problems
12. Bringing nonapproved items into cleanroom
13. Not cleaning work station at job or shift end

Commercial sources for cleanroom training provide in-plant and lecture hall programs and short courses presented at academic centers where cleanroom work has been carried out. As of 1990, the best-known university courses include those at the University of Arizona, the University of Minnesota, and Rochester Institute of Technology. The approaches differ at each one; the first is more empirical, the second more theoretical, and the third emphasizes optical system areas. In addition, private organizations present short courses in this area. A recent paper summarizes an educational and training approach that is quite effective (Gutacker 1988).

The importance of selecting a training program to satisfy specific approaches is emphasized. The program should target a specific audience group. The same program cannot apply for both the cleanroom designer and the cleanroom operators. Next, the communication path from the instructor to the audience must be effective and straightforward. The program must motivate the audience to recognize their importance in the contamination control program. The training must reflect management concern in the overall process. Finally, the information must be technically correct and reflect the latest state of the art in both the technology and the training methods. Good-quality audiovisual productions are strongly recommended.

References

DeLuca, P. P., 1983. Microcontamination Control: A Summary of an Approach to Training. *Journal of Parenteral Science and Technology* 37(6):218–224.

Gutacker, A. R., 1988. Personnel Training in Contamination Control Technology. Proceedings of the 9th International Committee of Contamination Control Societies Conference, pp. 247–253, September 26, 1988, Los Angeles.

Larrabee, G. B., 1985. A Challenge to Chemical Engineers: Microelectronics. *Chemical Engineering* 92(12):51–59.

Luna, C. J., 1986. Introducing People into the Clean Room. *Pharmaceutical Engineering* 6(1):15–19.

Manguray, J., 1988. A Hands-on Approach to Clean Room Training. Proceedings of the 9th International Committee of Contamination Control Societies Conference, pp. 262–266, September 26, 1988, Los Angeles.

Miller, D. P., 1988. The Architectural Ergonomics of Sandia's RHIC-II Gowning Facility. Proceedings of the 9th International Committee of Contamination Control Societies Conference, pp. 98–103, September 26, 1988, Los Angeles.

Moller, A. K., 1989. Ten Years of Training in CC-Technology in the Nordic Countries. *Journal of R^3–Nordic* 18(3):33–36.

Phillips, Q. T., et al., 1983. Cosmetics in Clean Rooms. *Journal of Environmental Sciences* 26(6):27–31.

Zuger, A., 1984. Education and Training for CleanRoom Personnel. *Swiss Pharma* 6(11a):35–42.

21

Clothing, Gloves, and Wipers

Cleanroom clothing, gloves, and wipers are used by all personnel in cleanrooms. These items are made from both natural and manmade fibers into fabrics with a wide range of designs, weaves, and structures. Cleanroom clothing and gloves are the primary barrier to keep contaminants generated by personnel from being emitted into the cleanroom and depositing upon products. Wipers are used to clean surfaces that may have collected dirt. The importance of all of these items is evident. They must be clean to begin any operations, and they must not generate or pass contaminants when in use. They must be fabricated so that they do not contaminate the cleanroom even if they are damaged in use. Some aspects of use and application are covered here for the materials, design, fabrication, and handling of these products.

The cleanroom garment is used to protect the cleanroom product from any contaminants that are generated or transported by the cleanroom operator. In addition, where personnel may be working with hazardous materials, some special garments may also be required to protect personnel from exposure or direct contact with hazardous solids, liquids, or vapors. Special gloves and safety glasses, for example, are required for working with many acids.

The need for adequate garments has been shown in studies carried out to measure both airborne and settled particle generation levels with personnel wearing different garment types. It was shown that use of smocks results in nearly twice the airborne and settled particle levels seen in a similar area where personnel are wearing head covers, coveralls, and boots (Weber and Wieckowski 1982). The additional cover on personnel emissions is indeed effective. Care is required in selecting fabrics for cleanroom garments to minimize dispersion of particles from personnel working in the cleanroom. A wide variation in retention of particles emitted from the body has been shown to exist for various clothing materials and designs (Whyte and Bailey 1989). Passage of particles ≥0.5 μm varied from 25% to more than 90% through some materials.

Based on measurements such as these, recommendations have been made for garment requirements in various cleanroom conditions (Dixon 1989). The type of garment required varies with the cleanroom and product specification. The garments may be as simple as a buttoned or snap-closure smock worn over street clothing, with a head cover and booties worn over cleaned street shoes; they may involve a complete clothing change with gloves, face cover, a separate air supply, and special cleanroom foot covering. Many areas operating at levels over class 10,000 can be serviced by personnel wearing smocks, booties, and head covers. Gloves are used if hazardous materials are present or if sensitive products must be handled. In a class 10,000 area, the recommended minimum garment system is defined as a head and facial hair cover, a coverall or zippered frock, footwear, and gloves. For a class 1,000 area, the same items are recommended except that frocks should never be used; a coverall is required and a breathing zone cover is recommended. For class 100 areas, a hood with face cover, secure coveralls, boots and gloves are required. For class 10 areas, the same items are required with the face cover replaced by a complete breathing enclosure. In class 1 areas, personnel should be excluded insofar as possible.

An effective cleanroom garment has a number of important functions. It must contain any contamination produced by the operator's body or carried on any body or clothing surfaces under the garment. It must not generate any contaminant material as a result of wear, gradual fabric or seam degradation or as a result of contact with products. It must be designed so that the operator can work effectively while wearing the garment. This means that enough freedom of movement is allowed so that the operator can reach to any necessary lateral or vertical work locations without degrading garment protective integrity. The garment should allow sufficient water vapor to pass through its weave so that the operator can be comfortable at the cleanroom temperature and does not perspire excessively while working. The weave should not allow particle passage. Even those body movements that produce particles should not cause them to be pumped out of the fabric pores or from seams.

Materials used for garments include a number of polyester fabrics including some containing electrically conductive fibers, spun bonded nylons, polypropylene composites, fluorocarbon fiber laminate, and nonwoven Tyvek. The major parameters to be considered in garment construction are stitching, seam construction, fasteners and zipper type, cuff construction for both arms and legs, and neckline closures. The garment fronts and cuffs are closed securely with snaps or buttons for light-duty areas or with zipped seals for better-controlled areas. Stitching and seaming are done so that all cut fabric edges are enclosed. The Institute of Environmental Sciences has published a useful recommended practice on parameters and specifications for garment construction (IES 1987). These procedures should result in clothing and gloves with minimum contamination generation and transmission of contaminants generated by the wearer, but they may result in clothing and gloves that are not comfortable to wear because moisture and heat transfer through the garment is also reduced. Materials and designs should be chosen to minimize discomfort and still maintain good contamination control.

Table 21-1 (Salvo 1985) shows properties of some fabrics used in cleanroom garments. The characteristics of concern for cleanroom use are emphasized. A recent British definition emphasizes essentially the same fabric properties presented in this table (Redlin and Neale 1987). Pore size is important because it controls both contaminant particle and air passage (in both directions) through the fabric. Because particles generated by personnel should be contained within the garment, the pore size should be

TABLE 21-1 Cleanroom Fabric Characteristics

Characteristic	100% Polyester		Carbon-impregnated 100% Polyester	Expanded PTFE Laminate	Tyvek
Weave	Taffeta	Herringbone	Taffeta	Taffeta	Nonwoven
Pore size after wear	30-100 um	15-30 um	94% exclusion at 1 um	<0.5 um	<0.5 um
Acid resistance	Poor	Poor	Poor	Good in short duration	Good in short duration
Vapor transmission	Good	Good	Good	Good	Good
Relative cost per unit	Low	Low	Medium	High	Low, but must be replaced often
Antistatic properties	Poor; topical required	Poor; topical required	Excellent; + carbon thread spun in	Poor; topical required	Poor; topical required
Miscellaneous information	Washable, durable fabric	Washable, durable fabric	Carbon thread can break down with use or acid exposures; washable	2-ply style durability is not established, 3-ply more durable; washable/dry cleanable	Disposable; washable to limited extent if thicker grade is used

New fabrics are developed continually for cleanroom use.

Source: (From Salvo, N. M., 1985. Practical Considerations in Contamination Source Control: How to Make the Best of a Retrofit Cleanroom. *Microcontamination* 3(8):43–48.)

small. However, small pore size means low airflow and possible personnel discomfort. Acid resistance is of concern in areas where corrosive vapors may be present. In this situation, personnel safety may be compromised, and the integrity of the garment is reduced to the point where it will begin to emit fiber fragments very quickly. The garments must be either resistant to the acids, or a rigid garment replacement schedule must be kept. All fabrics transmit inert vapors easily. This property is of minor concern. Cost is always important in cleanroom operation, but even the most expensive garments are a necessary portion of cleanroom costs. Antistatic properties are important primarily in semiconductor manufacturing areas. Use of garments with poor antistatic properties in those areas requires additional charge dissipation measures. Both ion generation for the air supply and use of grounding straps for personnel help to control electrostatic discharge if such garments are used. Some garments have been made with conductive carbon-impregnated fibers to reduce electrostatic charge buildup. In addition to this material, some more recent garment developments use stainless-steel fibers that reduce friable fiber fragment emission.

Cleanroom gloves are used to handle delicate components that may be affected by organic effluents and particles emitted from bare skin. The gloves must also be thin and flexible so that the sense of touch is not degraded and small objects can be handled easily. Many protective glove fabrication processes include application of talc, silica, or cornstarch to the packaged gloves. Needless to say, these materials must be removed before the gloves can be used in any area sensitive to particulate contamination. Many disposable gloves are made of natural rubber, vinyl, or polyurethane. Some care is necessary in using these materials. If it is necessary to use rubber gloves, then gloves that are free of sulfur must be selected. Vinyl gloves may contain up to 40% by weight of organic plasticizer, which can be a problem material in areas where organic film contamination must be avoided. Some ionic extractable materials were also detected in both the rubber and polyurethane gloves (Bordoloi 1989).

Where garments are worn in cleanrooms associated with medical products and processes, the control of microbial dispersion becomes important. The areas of concern include both pharmaceutical manufacturing areas and hospital operating and recovery rooms. Tests have shown that garments with moderate air and particle permeability (pore diameter of approximately 20 μm) reduce microbial dispersion appreciably (Whyte and Bailey 1985). However, closures at neck, arms, and legs are required to control pumping out bacterially contaminated air during body motions. No specific relationship has been established between airborne particle concentration levels and those of airborne bacteria (Takahashi et al. 1988) or of surgical sepsis and other postoperative complications (Scheinberg et al. 1983). However, it is usually accepted that similar concentration trends can be expected for both inert particles and microorganisms in the air.

Garments are categorized as either disposable or washable and reusable. Both types are delivered in packages that must be cleaned before the garments are packaged. Frequently the "disposable" garments are laundered and used many times before disposal. The laundry should be in a cleanroom area to avoid deposition of contaminants on the garment during the laundering process. A cleanroom laundry requires special equipment to maintain the cleaned garments in a condition satisfactory for use in the cleanroom. Washing and drying equipment should be constructed from stainless steel and designed to omit rough edges that may tear fibers or cause excessive garment wear. For those fabrics that require solvent cleaning, the solvent is filtered and frequently distilled to remove particulate and soluble contaminants from previous use. Care is required in this operation because some solvents may be toxic or harmful to the environment. Vapor recovery is required here and during the drying process. For those garments that are washed, detergents must be selected that do not leave ionic residues that may be harmful to the products manufactured in the cleanroom where the garments will be used. The rinse water must be filtered and passed through suitable ion exchange columns before use. In some facilities it is believed necessary to verify adequate removal of detergent from the last rinse solution. It has been suggested that cleanroom garment laundries avoid the use of electronics grade high-purity water (Whitsel 1989) because very pure water may dissolve small amounts of the fabric and may adversely affect its antistatic properties. Following the washing operation, the final drying operation uses heated, filtered air to avoid deposition of ambient air pollutants on the garments. The cleaned garments are inspected to ensure that no damage has occurred to the fabric, the seams, or the closures. Finally, the garment is packed in a specially cleaned container. The last two steps take place in a cleanroom. Once unpacked in the gowning area, clean garments should be stored in a clean area. Filtered airflow has been suggested for garment storage areas.

In one special application (Miller 1988), the outer cleanroom garments are antistatic-fabric polyester jumpsuits. A headgear system is used with a full face cover that contains a battery-powered blower exhaust system to remove exhaled air through a small HEPA filter. A custom-designed laundry is integrated into the gowning facility at this installation. In addition to the usual laundering operations, this laundry charges battery packs for the blower supply, tests and replaces filters, and periodically hand-cleans the suit hardware. The obvious additional expense for such an installation may be justified where very clean areas are involved. One series of tests in Japan has shown that dust generated by personnel can be reduced by a factor of 5 over that emitted from a conventional cleanroom garment by using a garment with internal air circulation vented through a HEPA filter before discharge (Suzuki et al. 1988).

Several testing methods are available to verify important properties of garments and other fabrics. Fabric and garment testing has been carried out for many years in the industrial hygiene area to determine electrical charge and particle transmission data. Many of the techniques are also applicable for cleanroom garments (Wilson 1963; Bohne and Cohen 1985). A summary of fabric features of importance for cleanroom use and some test methods to evaluate these features were provided by Siekmann (1983). Initial cleanliness and particle retention after garment use should be tested by measurement of particles removed by shaking or abrasion in either wet or dry conditions. Material shedding rates should be tested on a flex testing system. Particle filtration capability should be tested to determine the garment's capability to control personnel emissions. Air permeability should be tested to indicate a comfort level for personnel. Electrostatic charge decay rates should be measured to determine both airborne particle attraction and release, as well as potential damage considerations in semiconductor fabrication. Chemical resistance to corrosive liquids (if known) should be determined for safety reasons.

One specific garment cleanliness testing method is given in ASTM F-51 (ASTM 1973). This method is used to determine the particle emission rates from garments by drawing air through a garment under specific conditions, collecting particles on a membrane filter, and counting the particles $\geq 5\ \mu m$ in diameter, using bright field microscopy. Another test method that may be more representative of conditions in the cleanroom has been described (Lamb, Kepka, and Miller 1990), where determination of particulate emission as a result of controlled fabric abrasion is accomplished by rubbing the fabric against itself to determine the particle generation rate. Other methods use shaking or tumbling a garment and sampling the surrounding air with particle counters or with filters. Garment testing by agitation or shaking under dry conditions has been questioned for some time. The testing methods in use do not describe the particles actually upon the garment and are a poor approximation of the stresses that can be imposed upon a garment in use. These methods were shown to be suitable mainly for definition of those particles smaller than 5 μm actually released from the garment because the trajectory and sampling efficiency for larger particles is so poorly defined under the test conditions (Hill and Wieckowski 1983).

Another testing method for particulate cleanliness has determined the quantity of particles removed from the fabric by flushing with liquid. This method usually produces data that show many more particles removed from the fabric than are shown by dry test methods. However, the question is always raised as to the retention mechanisms for the particles on wet and dry fabrics and whether the particles would ever be emitted from the dry material in actual use. Whether the test matches the actual conditions of use is a moot point. If contaminant particles are present on or in a garment, that garment should not be brought into a critical cleanroom area.

Evaluation of face coverings has shown that not only will the mask reduce the emission of particles and droplets during exhalation but also velocity of air emitted to the cleanroom during sneezing or coughing is significantly reduced to the point where disruption of standard uniform-flow cleanroom airflow is controlled, at least to some extent (Sullivan and Trimple 1986).

Once adequate design, construction, and freedom from contaminants are assured, cleanroom garments must be donned by personnel who will assuredly generate contaminants that must be contained by the garments. The process of donning the clean garments must be carried out in such a way that contamination generated by personnel is not deposited on the exterior of the garment and subsequently released into the cleanroom or onto product surfaces. It is generally accepted that garments must be donned in a head-to-toe gowning procedure and removed in the opposite direction. Gowning should take place in a clean area just before the personnel enter the cleanroom. Any clothing worn under the cleanroom garb should be made of material that produces minimum lint. One suggested (Seemayer 1988) nonsterile cleanroom gowning protocol has been described as follows:

1. Remove outer garments and place in assigned lockers.
2. Remove loose jewelry and all cosmetics.
3. Wash hands and apply moisturizer.
4. Place cleanroom garment package on boundary bench between clean and nonclean area of gowning area, or leave on hanger in this location.
5. Cover hair with net.
6. Put on garment-handling gloves carefully.
7. Put on hood and face mask components.
8. Remove garment from package or hook. Step into the garment carefully, making sure that it does not touch any surface, especially in the nonclean area. Fasten the closures carefully.
9. Pull shoe covers on over the legs of the garment. Do not touch the shoes with hands.
10. Remove garment-donning gloves and put on cleanroom gloves, being careful to avoid skin contact with glove exterior.
11. Enter the cleanroom, over the final adhesive mat and through the air shower, if required.

Gloves are used to prevent perspiration and skin fragments from depositing upon the product as well as to protect the operator from any hazardous process chemicals. For cleanroom use in general, the gloves should be long enough and designed so that they can be drawn up over the garment sleeve end. Latex gloves are frequently used in electronic areas because they meet the first requirement and allow good contact feedback. For other products, sturdier and more chemically resistant materials may be required. Polyethylene or vinyl gloves are used in some applications. The IES Recommended Practice RP-005, *Cleanroom Gloves and Finger Cots,* describes testing and use for these items. Gloves, unlike most cleanroom garments, are seldom reused. The process of donning and removing the gloves tends to contaminate the gloves beyond any hope of cleaning either satisfactorily or economically.

Wipers are needed throughout the cleanroom operation in many product areas. They are used to remove visible debris or liquids from sensitive surfaces. Wipers are not usually reused or allowed to contact more than one surface. Even for single-use cleaning, it is wise to use a fresh wiper for a final wiping operation. In some pharmaceutical manufacturing areas, however, specially cleaned wipers are available for extended use. They are made of nonlinting, knitted polyester that is modified for increased adsorption capability; they are available with or without antistatic properties. Wiper material must be selected so that contaminant emission does not occur, and they must be flexible and absorbent. They are used to clean materials, tools, and structural surfaces. Depending on type of use, more than one type of wiper may be stocked. Because many fibers and fabrics today are made with some plasticizer or filler, the possibility of chemical vapor or particulate emission from wipers must be considered and tested. As with other cleanroom products, unused wiper packaging and storage, as well as used wiper disposal, must be planned for in the cleanroom layout.

Many fabrics used in cleanroom wipers are made of polyester, cellulose blends, polyamide, polyurethane, polypropylene, or nylon fibers. They should be selected for minimum extractable organic and inorganic materials. In addition to organic materials such as hydrocarbons, fatty acids, and alcohols, inorganic compounds such as sodium, potassium, sulfates, chlorides, and phosphates may be present in more than ppm levels. Several testing methods have been described for wipers. One series of tests was used to characterize extractable materials from a range of wiper fabrics when the wipers were soaked in solvents frequently used in a cleanroom (Hovatter and Hendrickson 1984). Organic and inorganic extractable materials, including both cations and anions, were measured along with particle removal from the fabric. Water and chlorinated fluorocarbon solvent absorption were also measured for a number of fabrics. In another test, emission of small particles is measured by flexing the wiper for a fixed time period and determining emission of submicrometer particles released from the wiper. Preliminary findings indicated that synthetic fabrics released particles at a small fraction of the rate observed from cotton fabric wipers (Paley 1985). Needless to say, any cloth that releases particulate material should not be used in a cleanroom.

Other similar cleanroom commodities, such as cleanroom safety sleeves, aprons, face shields, hoods, and grounding straps, require the same considerations. Even though these items are not all concerned with keeping personnel contaminant emission from the cleanroom and from the product surfaces, collection and subsequent emission of contaminants by these items must be minimized. Regular and careful examination of the components should be part of the normal cleanroom maintenance procedures to ensure that these items are in good repair and that they are not obviously soiled.

References

American Society for Testing and Materials, *Standard Method for Sizing and Counting Particulate Contaminant in and on Clean Room Garments,* F51-73. Philadelphia: American Society for Testing and Materials.

Bohne, J. E., & Cohen, B. S., 1985. Aerosol Resuspension from Fabric: Implications for Personal Monitoring in the Beryllium Industry. *American Industrial Hygiene Association Journal* 46(2):73–79.

Bordoloi, B. K., 1989. Characterization of Cleanroom Gloves. Proceedings of the 35th Institute of Environmental Science Annual Technical Meeting, pp. 428–432, April 1989, Anaheim, CA.

Dixon, A. M., 1989. Garments: A Clean Approach. *Medical Device and Diagnostic Industry* 11(2):44ff.

Hill, S. L., & Wieckowski, J. M., 1983. Comparison Testing of Clean Room Garments. Proceedings of the 29th Institute of Environmental Science Annual Technical Meeting, pp. 305–307, May 1983, Los Angeles.

Hovatter, W., & Hendrickson, E. A., 1984. Cleanroom Wipers: A Comparative Study. *Microcontamination* 2(6):47–55.

Institute of Environmental Science, 1987. *Garments Required in Clean Rooms and Controlled Environmental Areas,* Report RP-CC-003-87T. Mt. Prospect, IL: Institute of Environmental Science.

Lamb, G. E. R., Kepka, S., & Miller, B., 1990. Particle Release from Fabrics during Wear. *Aerosol Science and Technology* 13(1):1–7.

Miller, D. P., 1988. The Architectural Ergonomics of Sandia's RHIC-II Gowning Facility. Proceedings of the 9th International Committee of Contamination Control Societies Conference, pp. 98–103, September 26, 1988, Los Angeles.

Paley, E., 1985. Flex Test for Particulate Analysis of Dry Wipers. *Microcontamination* 3(5):82–86.

Redlin, W. J., & Neale, R. M., 1987. Garments for Cleanroom Operators: The Demands of the 1990's. *Journal of the Society of Environmental Engineers* 26(1):17–19.

Salvo, N. M., 1985. Practical Considerations in Contamination Source Control: How to Make the Best of a Retrofit Cleanroom. *Microcontamination* 3(8): 43–48.

Scheinberg, S. P., et al., 1983. Reduction of Fabric-Related Airborne Particles in the OR. *P & MC Hospitals* 2(4):71–77.

Seemayer, W., 1988. Dress for Cleanliness Using Head-to-Toe Gowning Procedures. *CleanRooms* 2(4):10–11.

Siekmann, R. O., 1983. Features and Testing of Clean Room Apparel. *Journal of Environmental Science* 26(1):36–40.

Sullivan, G., & Trimple, J., 1986. Evaluation of Face Coverings. *Microcontamination* 4(5):64–70.

Suzuki, Y., et al., 1988. Effects of Suction Type Clean Room Garment. Proceedings of the 9th International Committee of Contamination Control Societies Conference, pp. 475–478, September 26, 1988, Los Angeles.

Takahashi, Y., et al., 1988. Airborne Particle Dispersion from Cotton Surgical Gowns. Proceedings of the 9th International Committee of Contamination Control Societies Conference, pp. 420–423, September 26, 1988, Los Angeles.

Weber, C. W., & Wieckowski, J. M., 1982. The Effects of Variations in Garment Protection on Clean Room Cleanliness Levels. *Journal of Environmental Science* 25(6):13–16.

Whitsel, B. H., 1989. Controlling Water Quality Critical for Processing Cleanroom Apparel. *Microcontamination* 7(2):18–21.

Whyte, W., & Bailey, P. V., 1985. Reduction of Microbial Dispersion by Clothing. *Journal of Parenteral Science and Technology* 39(1):51–60.

Whyte, W., & Bailey, P. V., 1989. Particle Dispersion in Relation to Clothing. *Journal of Environmental Science* 32(2):43–49.

Wilson, D., 1963. The Electrical Resistance of Textile Materials as a Measure of their Anti-static Properties. *Journal of the Textile Institute* 54:T97–T105.

22

Gas Sampling Considerations

Gas sampling is carried out to measure the quality of a gas. Gas samples are sometimes acquired by in situ observation within the main gas body by using remote or visual observation for specific properties. A more frequent method of sampling consists of removing a portion of the gas from the main body and transporting it to a location where the pertinent properties of the gas can be defined. These properties may be physical properties, such as temperature or pressure; they may be the chemical properties of gas itself or of it's impurity content. In cleanroom applications, the particulate content of the gas is of major importance. Because particles are not uniformly distributed in the gas, sampling problems in defining these materials are atypical as compared to those for sample handling to define other properties of a gas. Sampling for particle content is discussed in some detail here.

The gas to be sampled may be the ambient cleanroom air. It may be the gas flow entering or emerging from an air-handling device such as a filter, a gas scrubber, or a clean bench; the gas may be flowing within an air duct, or it may be in a high- or low-pressure system. Although we are concerned mainly about air, the same considerations apply to most inert gases used in the cleanroom, such as nitrogen, helium, or hydrogen. The differences in sample handling and acquisition procedures for gases other than air are based on the specific gravity of the gas or the needs for special care owing to toxicity or flammability.

It must always be remembered that the definition of *sample* is "a portion or piece taken or shown as representative of a whole." The sampling equipment and the sampling operation must not modify the environment from which the sample is acquired or change any condition of the sampled gas during the sampling process. The sampling process may remove a portion of the sampled gas from the entire gas system, but it can neither add artifacts nor remove materials from the sampled gas, and it cannot change any physical or energy level conditions in that gas.

Before any discussion of sampling is given, two terms commonly used in gas sampling should be defined. The most common term is *isokinetic sampling;* the second term, *isoaxial flow,* is not used as widely, but it is very important in most gas sampling operations and describes a condition that probably occurs more frequently. *Isokinetic sampling* means that the gas sample is collected in such a way that the gas sample velocity is not changed from the sampled gas system as a result of the sample acquisition process. This requirement can be satisfied by ensuring that the sample acquisition system changes neither the original mean direction of the moving gas molecules nor their speed during that sampling process. The sample inlet tube velocity is the same as that of the gas stream being collected. *Isoaxial flow* means that the sample inlet faces into the prevailing gas flow so that the sampled gas flow does not change direction, even if sample inlet tube gas speed is not equal to that of the sampled gas flow. Thus it can be seen that isokinetic sampling always includes isoaxial flow, but isoaxial flow does not necessarily mean that isokinetic sampling occurs.

The fundamental considerations in gas sampling to define particle content are concerned with losses in the sample-handling system. Sampling probe inlet efficiency and transfer losses owing to deposition in the transfer line following the sampler probe are the major considerations. Anisokinetic isoaxial sampling inlet error increases with an increasing Stokes number and a velocity difference between the free stream and the inlet probe velocity. The extent of error is shown in Figure 22-1 (Liu et al. 1989). Error increases with particle size, along with differential velocity. Deposition losses in the transfer tube occur as a result of several factors. Some turbulent or diffusive deposition (depending upon particle size and flow conditions in the tube) occurs for tubes in any configuration. Settling occurs for particles large enough to be affected by gravity during their residence time in the tubing. Losses occur at bends in the tube as a result of inertial impaction. Figure 22-2 (Liu et al. 1985) shows particle penetration through sampling tubes, taking into account diffusion, inertial impaction, gravitational sedimentation, and turbulent diffusion. Particle losses in the size range 0.1 μm to 1.0 μm or so are insignificant at either the lowest flow rate (maximum residence time) or the highest flow rate (maximum turbulence), which is shown here. Particle losses increase for larger particles because of inertial effects and for smaller particles because of their greater diffusion rates.

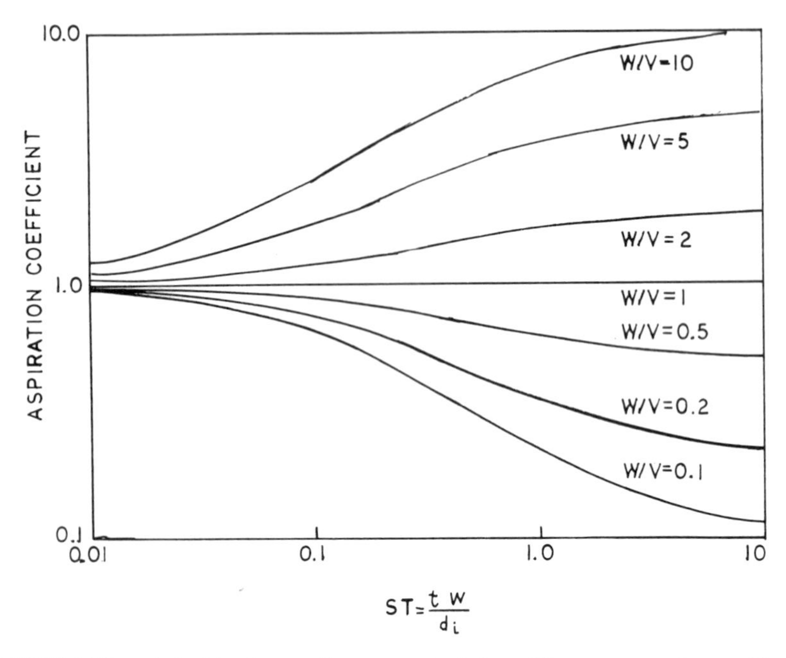

FIGURE 22-1. Anisokinetic sampling error calculations. The errors are shown with increasing Stokes number (*ST*), expressed in terms of the ratio of particle relaxation time to sampler inlet dimension at various ambient-to-sampler flow ratios. For Stokes numbers less than 0.03 or so, the errors are negligible. When the aspiration coefficient is 1.0, the sampling error is zero. (From Liu, B. Y. H., et al., 1989. A Numerical Study of Inertial Errors in Anisokinetic Sampling. *Journal of Aerosol Science* 20(3):367–380.)

Particle losses can also occur within transfer lines as a result of electrostatic forces. Image forces can result in losses of charged particles to conducting surfaces; dielectric tubing material can build up significant charges, resulting in coulombic attractive forces removing particles from the sample stream. The importance of these various loss effects is summarized by Pui, Ye, and Liu (1988).

When sampling takes place in a gas stream with varying gas flow rates, isokinetic sampling may not be possible without the use of sophisticated, complex sampling systems. Devices may be needed to vary the sample inlet system flow rate and/or orientation to match varying gas stream conditions. This situation is not usually seen. However, small offsets in sampling and sampled gas speeds do not generate large sampling errors, especially when small particles are of concern; therefore, isokinetic sampling errors may be small enough to be accepted in many situations. However, offsets in sample inlet and sampled gas flow directions (aniso-

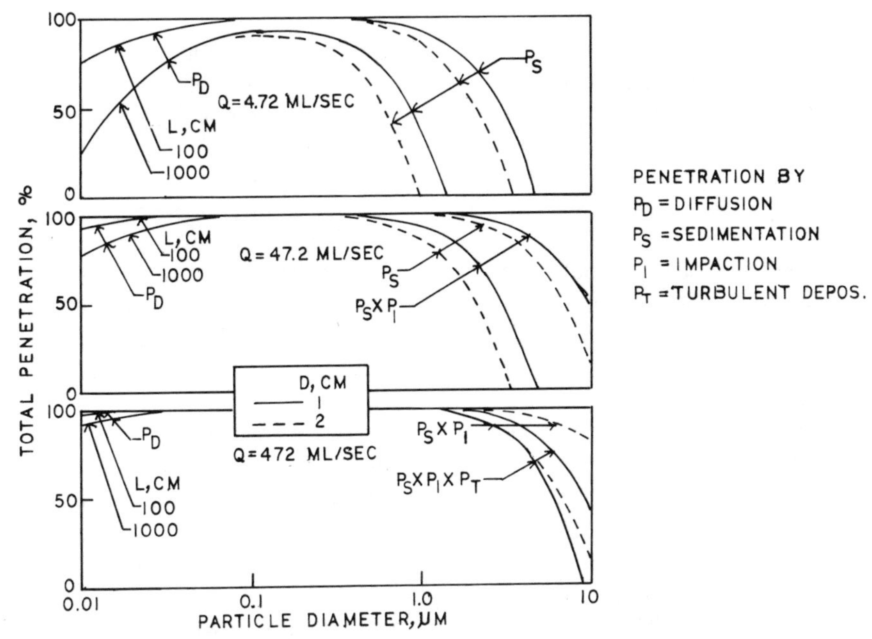

FIGURE 22-2. Overall particle penetration through transport tubes. Losses from all but electrostatic attraction are shown to be quite low for particles larger than approximately 0.05 μm and smaller than approximately 0.7 μm. (From Liu, B. Y. H., et al., 1985. Electrostatic Effects in Aerosol Sampling and Filtration. *Annals of Occupational Hygiene* 29:251–269.)

axial flow) can cause significant sampling errors even for small particles because of induced centrifugal force as the gas streamlines are forced to change direction at the sampler inlet. For this reason, isoaxial sampling conditions should be established whenever possible, even if isokinetic sampling cannot be attained.

Gas sampling is carried out in order to procure a portion of the gas system that must be characterized. Characterization is necessary for a variety of reasons. Classification of a cleanroom is based on the particle content of the air in a sample procured in a specifically defined manner. After the cleanroom class level has been verified, airborne particle content must be measured to ensure that conditions have not changed significantly. As previously stated, a monitoring program that includes a valid sampling protocol must be established. Many operations carried out in a cleanroom may generate particulate contamination that may affect subsequent operations. Air surrounding or emitted from such operations is sampled to define the particle emission rate or to verify performance of particle control systems. Performance of both new and in-place filters is characterized by sampling the filter challenge and effluent gas streams for particle content. In all these situations, the objectives of the sampling operation are similar: The sample must be representative of the population from which the sample is procured. The sample must be large enough that any measurement of its contents provides data with statistical validity. The sample must not be so large as to modify the nature of the population being sampled. The sampling method should be selected to match the requirements of the situation; for example, extreme care to procure an isokinetic sample may not be necessary if it is positively known that the sampled environment does not contain particles larger than 0.5 μm or so in size and if the point source location of the airstream is of little importance. If larger particles are of concern and the air being sampled is relatively clean, then a very long sampling period may be required to procure a sufficient quantity of such large particles to permit acceptable statistical validity. In one suggested means of reducing sampling time, the large-particle population is concentrated by use of a high-flow-rate virtual impactor (Marple, Liu, and Olson 1989). The impactor can be designed with a cut size in the range of 1 to 5 μm, and the flow containing the large-particle fraction can be collected on a membrane filter for detailed examination or can be isokinetically sampled with an optical particle counter that has been calibrated for large-particle measurement.

When sampling for gas composition analysis, the sampling system inlet efficiency is essentially 100%, although there may be some adsorption loss in transporting gases through long ducts with high surface-to-volume ratios. Inertial losses of all gases are essentially zero in entering the inlet probe, and diffusional losses in transport are essentially the same for all gases. A recent description of a real-time system for monitoring low-level impurities in very clean gases discusses the analytical methods and shows a gas sample acquisition system with a number of flow control devices, including several multiport valves, pressure regulators, and flow restrictions (Whitlock et al. 1988). Transport of gas molecules through such a convoluted system is very efficient. If such a system were to be used for measurement of particle content, then losses of 1 μm and larger particles would be unacceptable.

In sampling to analyze particles, particle losses always occur both at the inlet and along the walls of the sample transport system. Particle transport and deposition mechanisms depend on particle size as well as the forces exerted upon the particle. Sampling may cause a particle trajectory to be changed in order to collect it into a sample system inlet. Care is required to control that trajectory change to avoid any particle deposition on the sampler inlet surface. Figure 22-3 (ter Kuile 1979) shows airflow stream-

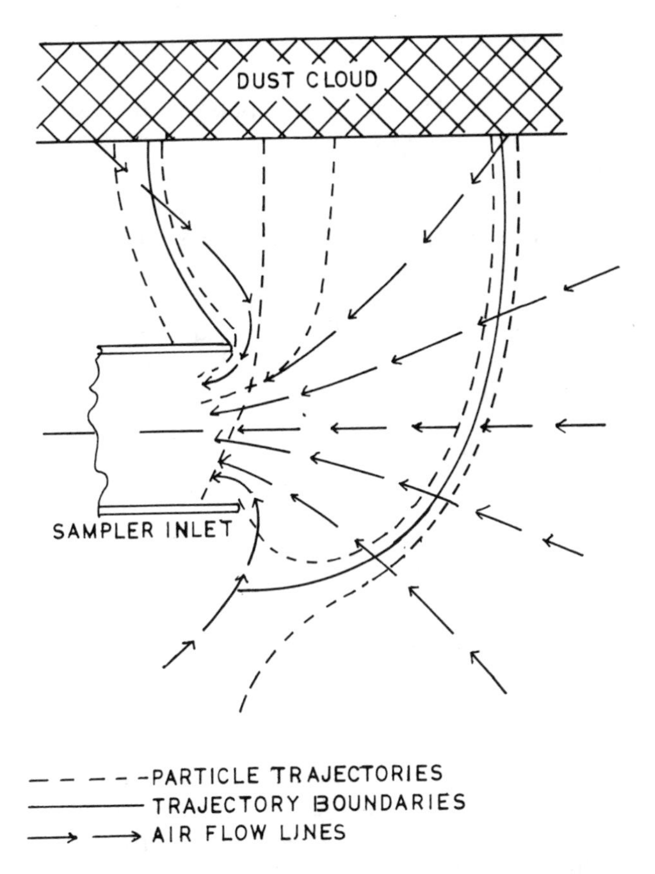

FIGURE 22-3. Particle trajectories to horizontal sampler inlet from a static aerosol cloud. These trajectories are calculated for 10-μm particles. It can be seen that gravity plays an important part in sampling these large particles. (From ter Kuile, W. M., 1979. Comparable Dust Sampling at the Workplace, Report F 1699, Project 6.4.01. Delft, The Netherlands: Research Institute for Industrial Hygiene, TNO.)

lines and some large-particle trajectories, as the particles move to a horizontal sample inlet from a surrounding quiescent cloud. The trajectories shown in this figure are calculated for 10-μm particles. It can be seen that the particle path deviations from the air streamlines increase as the radius of curvature of the streamline decreases. Smaller particles follow the air streamlines more closely. This figure shows streamlines and flow paths that approximate the particle paths and airflows in a nonuniform-flow cleanroom. Even though there are prevailing airflows from the inlet diffusers to the exhaust vents, the relative air velocity at almost any sampler inlet is so much greater than that in the overall sampled air body that the situation shown in the figure is close to the actual one in a nonuniform-flow area.

When sampling takes place in a uniform-flow air body, the insertion of the sampler head into the airstream causes a disturbance. A thin-walled sampling probe can still interrupt airflow to some extent at the probe boundary, even if overall isokinetic flow is attained at the probe inlet (Vincent 1987). Worse sampling efficiency occurs when anisoaxial sample probes are used at the same air speed than when isoaxial probes are used at air speeds different from that of the mainstream. It has been shown (Okazaki, Wiener, and Willeke 1987) that for anisoaxial sampling probe orientations varying from 30 to 90 degrees between the inlet and wind directions, the overall sampling efficiency is a function of a sampling parameter defined as the product of the particle Stokes number and a power of the ratio of the sampled air velocity to the probe inlet velocity. The power varies from -0.33 to 0.5 as the probe orientation angle to the prevailing airflow varies from 30 to 90 degrees.

A thin-walled probe, meeting the inlet design criteria of Belyaev and Levin (1974), is sharp-edged with the ratio of the exterior diameter to the interior diameter of the probe inlet face ≤ 1.05 and the angle of taper at the probe inlet ≤ 15 degrees. These dimensions minimize the rebound and reentrainment effects of dry solid particles at the edge of the probe inlet face. A thick-walled probe can be anything from a rounded tube end or a flat cut plastic tube end to an orifice in a plate. A study of the effects of the thickness of probe walls (Rader and Marple 1988) on sampling showed that even isokinetic sampling does not ensure representative sampling for thick-walled probes, but that thick-walled probes provide sampling efficiency closer to unity than that provided by a thin-walled probe at subisokinetic sampling conditions. At superisokinetic flow rate conditions, the effects of probe wall thickness were minor.

When samples are to be procured in a mixed-flow area, isokinetic sampling is very difficult because the airflow direction and speed may vary continuously. In this situation, sampler inlet dimensions and flow rates can be chosen to minimize sampling errors. Figure 22-4 (Agarwal and Liu 1980) describes the particle size and sampled airflow parameters that allow accurate sampling for a vertical, thin-walled sample probe with an upward-facing inlet. The particle sampling efficiency is shown as a function of Stokes number and the relative settling velocity. The Stokes number is the ratio of the stopping distance for the particle to the diameter of the sample probe inlet. The relative settling velocity is the ratio of particle settling velocity in quiescent air to the sampler probe inlet velocity. Although this figure suggests that "accurate" sampling is anything better than 90%

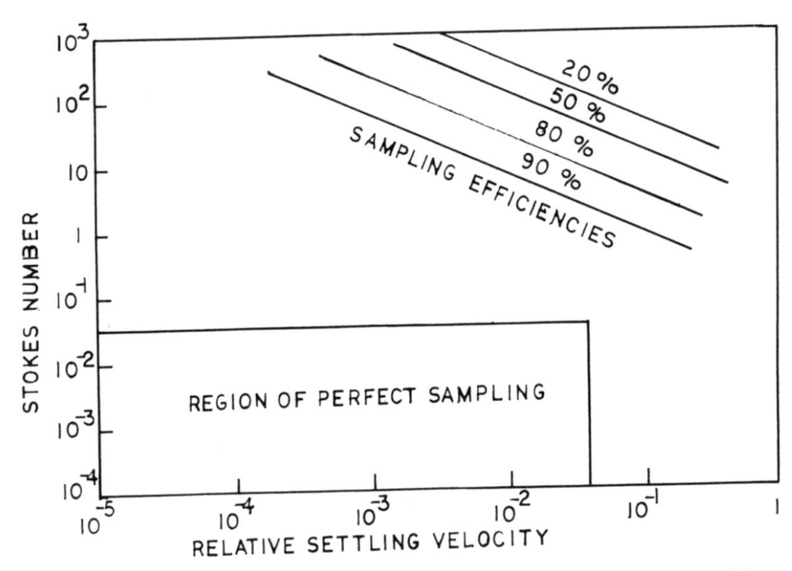

FIGURE 22-4. Vertical sampler inlet efficiency in turbulent airflow. Where neither isokinetic nor isoaxial sampling is possible, the major criterion for efficient sample collection is the Stokes number, expressed as the ratio of particle size to sample nozzle inlet size. Larger nozzle inlets are more efficient. (From Agarwal, J. K., & Liu, B. Y. H., 1980. A Criterion for Accurate Sampling in Still Air. *American Industrial Hygiene Association Journal* 41(3):191–197.)

efficiency, it is suggested that probe parameters should be used that provide a sampling efficiency of at least 95%.

For good sample inlet efficiency in well-defined airstreams, isokinetic sampling should be used. Figure 22-5 shows anisokinetic errors as sample inlet flow velocity diverges from the sampled air velocity. The larger the particle, the greater the anisokinetic sampling error. Remember that isokinetic means that both the inlet sample air speed and direction must be the same as those of the ambient air. Note the divergence in air streamline paths and particle paths for larger particles in the cases of anisokinetic sampling shown in Figure 22-5. For particles in the size range of 0.1 μm to 1 μm, anisokinetic errors are minor. However, isokinetic sampling is still recommended, especially if the sample source location must be pinpointed, as when locating leaks in filters or specific particle sources in any other device. Isokinetic sampling is also important when sampling ambient air in any uniform-flow area or in a duct.

Once the sample has entered the sample line, losses can occur in transport through the duct. The losses increase both for very large particles and

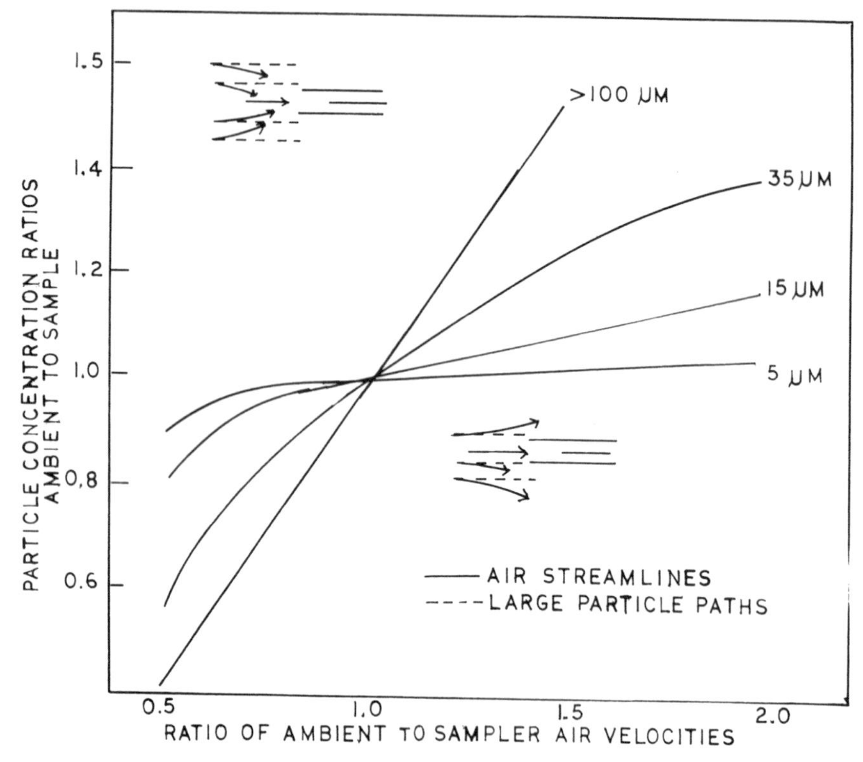

FIGURE 22-5. Anisokinetic sampling errors. These errors are shown for sharp, isoaxial nozzles with diameters as large as 20 millimeters.

for small ones. The large particles are lost because of gravitational settling and inertial effects; the small ones are lost because of their high diffusion rates. As shown in Figure 22-6, losses are highest for particles either greater than 1 to 2 μm or smaller than 0.1 μm in diameter. For these reasons, sample inlets should be sized to permit isokinetic sampling and should face into the airflow.

Sample transport lines should be as short as possible. If lines as long as 30 to 40 meters are required, then line sizes should be selected to permit a Reynolds number of 5,000 to 40,000 at the sample flow rate. In flow at Reynolds numbers below this range, residence time in the line may become so long that particle losses occur from large-particle sedimentation or small-particle diffusion. In flow at higher Reynolds numbers, turbulence may increase to the point where particle deposition on the line walls can become excessive. The lines should be as straight as possible, with a radius of curvature never less than 10 cm. Smooth, conductive line surfaces

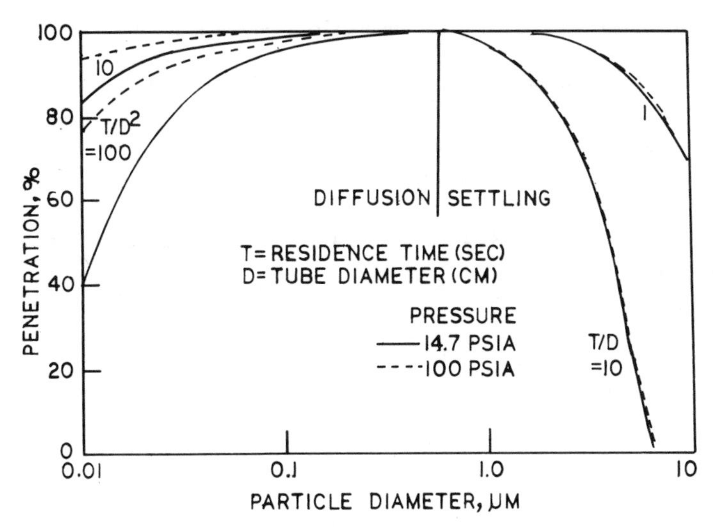

FIGURE 22-6. Particle losses in compressed gas sample line transport. Increased pressure has little effect on particle losses in transport lines. Loss in the particle size range 0.1 μm to 1 μm is not great. (From Pui, D. Y. H., et al., 1988. Sampling, Transport and Deposition of Particles in High Purity Gas Supply System. Proceedings of 9th International Committee of Contamination Control Societies Conference, pp. 287–293, September 26, 1988, Los Angeles.)

should be chosen, and the lines should be laid out with minimum changes in dimensions. Table 22-1 gives some recommendations for sample line materials in order of preference. The order is based on tube material properties. For minimum loss it is recommended that the tube material be electrically conductive, as smooth as possible, not reactive with the gases that will be in the tube, and capable of emitting no harmful materials. The last point applies both to particle loss caused by oxide-coating fragmentation from some metal tubes and to plasticizer loss from some polymer tubes. Particle emission in the former situation results in generation of erroneous particle data, whereas plasticizer loss is a problem only in areas where the organic vapors emitted by the plasticizer can be hazardous to the product in a particular cleanroom operation.

Testing was carried out on penetration of particles in the size range 0.024 μm to 0.420 μm through plasticized polyvinyl chloride (PVC), copper, standard PFA tubing and antistatic PFA tubing (Bergin 1987). It was shown that penetrations through the PVC and copper tubes were close to the theoretical values, assuming losses only by diffusion. Comparison of the two PFA materials showed that penetration through the antistatic PFA was 2 to 6 times that through standard PFA, depending on particle charge.

TABLE 22-1 **Particle Transport Line Material Recommendations**

PREFERRED AEROSOL TRANSPORT LINE MATERIALS
1. 316 Stainless Steel
2. Conductive Polymer
3. Polyester
4. Vinyl (if plasticizer does not interfere)
5. Polyethylene
6. Copper
7. Glass
8. Teflon
9. Aluminum

The recommendations are made for materials that will have minimum fragment emissions with time as well as minimum interference with particle transit through the lines.

Further testing (Emi et al. 1988) indicates that diffusion is the predominant deposition mechanism for particles with small charge in conductive tubes and that charged-particle deposition in nonconductive tubes is controlled by coulomb forces for both positively and negatively charged particles. Thus, the importance is shown of using conductive tubing and of neutralizing charged particles in the sample when sample transport is required.

Table 22-2 shows some flow parameters for lines of varying inlet diameters at a flow rate of 28.3 liters per minute. Although the line pressure drop and residence time are of minor importance in terms of particle losses in the line, they can produce serious practical problems. Many particle collection and measurement systems use vane pumps or axial blowers to draw sampled air. The sampled airflow rate may be seriously reduced if the line pressure drop is added to the pump load. Long residence time can result in collection or recording of sample data from air that has been in the line, rather from an air sample drawn from the open end of the line.

When the temperature or pressure at the point of sampling differs from that at the point of measurement, particle size change can occur as the condition of the sampled air changes. This phenomenon does not usually occur when the changes are within the tolerance limits for cleanroom ambient air. However, if the gas is sampled from a high-pressure line (3–10 bars) or from a device operating at a high temperature ($> 100°$ C), then the decrease in temperature to the point of measurement may increase the relative humidity in the gas sample to the point where particles smaller than ≈ 0.05 μm, which would not usually be counted, can act as condensation nuclei and grow to very large solution droplets. In this case, any

TABLE 22-2 Air Flow Parameters in Sample Transport Tubes

Inside Dia. (mm)	Reynolds Number	Pressure Loss psi/meter	Gas Velocity in/sec
4	9150	0.98	40.35
5	7360	0.34	25.9
6	6130	0.15	18.0
1/4 in	5780	0.11	16.0
7	5270	0.07	13.2
8 (5/16 IN)	4585	0.04	10.1
9	4070	0.02	8.0
3/8 in	3865	0.016	7.2
10	3670	0.013	6.5

The values are based upon a flow rate of 28.3 liters per minute at ambient pressure. The Reynolds number determines the turbulence during flow; pressure loss is important when a rotating vane pump is used to move air through the transport tube; gas velocity is important in calculating residence time and losses owing to diffusion to tube walls for small particles.

further change in gas temperature may cause the droplets to vary in size inversely with the temperature. This situation has been seen when exhaust air samples are collected while a batch of washed bottles is drying in a hot air sterilizing tunnel. A similar phenomenon has been observed when gas samples are being collected from high-pressure storage bottles (Kasper and Wen 1988). Evaporation of surface contamination material was followed by condensation of the vapors to form countable particles in the sampled gas.

In cleanroom areas where bacteria sampling may be required, the sample analysis may require data in terms of the number of colony-forming units that have been collected. For long-term sampling onto a nutrient surface or to a membrane that will later be cultured for viable bacteria determination, desiccation of the nutrient media or of bacteria on the collection surface has been a concern. When accurate data are required for very clean areas, then larger air volumes must be sampled. A study has been carried out showing few effects of desiccation of the bacteria normally seen in cleanroom air when these are on the surface of an agar plate (Whyte and Niven 1986); a water loss of up to 14% should not reduce indicated viable airborne bacteria concentrations by more than 8%. This rule does not apply to all bacteria, but it was reported that 16 Gram-negative and Gram-positive bacteria were not seriously influenced by this level of drying.

Specific sampling protocol documents are available for collecting air samples used to verify a clean area classification level, for characterizing HEPA and ULPA filter penetration, and for in-place filter leak testing in cleanroom operations. Valid procedures have also been described in the technical literature for sample collection from ducts, from low- and medium-pressure gas lines, and for general room monitoring. Unfortunately, although the procedures are based on good practices and a realization of the requirements for sample acquisition with minimum particle loss, many of these procedures have not been formally documented and accepted as referee methods. For example, recent discussions have been presented on defining test methods for measuring particles generated by components used in compressed gas systems (Koch and Pinson 1989; Wang, Wen, and Kasper 1989). Although the methods follow good engineering practice and are effective and relatively easy to apply, they are not widely accepted because they are not documented by an accepted standards preparation agency.

A system for reducing the pressure of compressed gas for measurement at ambient pressure is an example of a protocol developed with the aid of good modeling practice; it has been verified by extensive testing. With the aid of a simple orifice disk in the high-pressure line, gas system pressure was reduced to a point where conventional optical particle counters and condensation nucleus counters can be used effectively at sample points downstream of the orifice. Figure 22-7 (Schwartz and McDermott 1988) shows the calculated particle concentrations before and after the orifice is used for pressure reduction. No significant particle losses or generation was observed in this device; similar testing with bellows or diaphragm

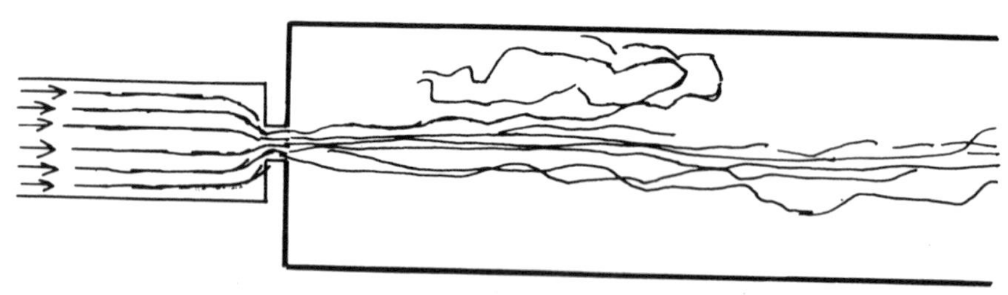

FIGURE 22-7. Particle transit through pressure reduction device. A sharp-edged orifice is used for pressure reduction. Even though some recirculation is seen downstream from the orifice, wall deposition is not significant, which contrasts with the losses when venturi expanders are used. (From Schwartz, A., & McDermott, W. T., 1988. Numerical Modelling of Submicron Particle Deposition in Pressurized Systems: Analysis of Fallout Particles for Quality Control Applications. Proceedings of the 9th International Committee of Contamination Control Societies Conference, pp. 78–86, September 26, 1988, Los Angeles.)

valves showed that both original particle capture and release of artifact particles were noted in these devices. The testing protocols satisfy requirements for adequate device operation and sample collection; until they have been further validated by others in the field and their limits defined on errors and range of applicability, however, they cannot be generally accepted.

An excellent presentation on general aerosol sampling has been prepared by Fissan and Schwientek (1987). This paper defines conditions for representative sampling for duct flow and provides equations to allow a first estimation of errors. Changes in the sample during transit in laminar and turbulent flow are also described. The procedures in this paper describe methods used for many years that produce results that can be justified on the basis of known and accepted physical principles.

When sampling in a cleanroom for validation purposes, the sample point location, sample size, and frequency selection are based on the requirements of FS209D. This document states the number of sample point locations for a unidirectional flow room to be "the lesser of (a) the area of the entrance plane perpendicular to the airflow (in square feet) divided by 25, or (b) the area of the entrance plane (in square feet) divided by the square root of the airborne particulate cleanliness classification." For nonunidirectional airflow rooms, the number of sample locations should be equal to the square feet of floor area of the clean zone divided by the square root of the airborne particulate cleanliness class designation. The sample size and frequency are specified so that the cleanroom classification can be defined with an upper confidence level of 95%.

It has been shown that sampling in very clean, large-area cleanrooms may require collection of thousands of samples to meet these requirements. Table 12-4 showed the minimum number of sample points for verifying cleanroom classification levels; for a 100,000-square-foot, class 100 cleanroom, 4,000 samples are required. If a particle counter with sensitivity of 0.2 μm at a flow rate of 1 cfm is used, then 6-second sample times are required for each sample. Assuming that each sample also requires 10 seconds to position the sample probe inlet and 10 seconds to produce hard-copy data, at least 33 hours of data collection time are required to classify the area. If two measurements are taken at each point, then the time is doubled. If more than one particle counter is used, then the cost of verification is increased accordingly. At this time, it appears that initial verification measurement time cannot be reduced without abrogating the requirements of FS209D. However, a realistic evaluation of the actual sampling conditions leads rapidly to the conclusion that the sampling program for verification of a cleanroom class involves a number of assumptions as to uniformity of conditions throughout the room and for each sample location. Optimization of sample point locations can be based on laying out the location of sample points so that each point is close to the center of a sampled area with an aspect ratio as close to 1 as possible (Cooper et al. 1990). Practically speaking, laying out sampled areas with a nearly square configuration appears to be an optimum choice for most cleanrooms. A statistical approach has been suggested to reduce the number of measurements required to ensure that the clean area is still in satisfactory conditions (Cooper 1988). The approach involves sequential sample collection with observation of the data trend; it has promise of reducing the required number of sample measurements appreciably.

References

Agarwal, J. K., & Liu, B. Y. H., 1980. A Criterion for Accurate Aerosol Sampling in Still Air. *American Industrial Hygiene Association Journal* 41(3):191–197.

Belyaev, S. P., & Levin, M., 1974. Techniques for Collection of Representative Aerosol Samples. *Journal of Aerosol Science* 5(4):325–338.

Bergin, M. H., 1987. Evaluation of Aerosol Particle Penetration through PFA Tubing and Antistatic PFA Tubing. *Microcontamination* 5(2):22–28.

Cooper, D. W., 1988. Sequential Sampling Statistics for Evaluating Low Concentrations. *Journal of Environmental Science* 31(5):33–36.

Cooper, D. W., et al., 1990. Selecting Nearly Optimal Sampling Locations throughout an Area: Application to Cleanrooms and Federal Standard 209. Proceedings of the 36th Institute of Environmental Science Annual Technical Meeting, pp. 257–263, May 1990, New Orleans.

Emi, H., et al., 1988. Deposition of Charged Submicron Particles in Circular Tubes Made of Various Materials: Analysis of Fallout Particles for Quality Control Applications. Proceedings of the 9th International Committee of Contamination Control Societies Conference, pp. 73–77, September 26, 1988, Los Angeles.

Fissan, H., & Schwientek, G., 1987. Sampling and Transport of Aerosols. *TSI Journal of Particle Instrumentation* 2(2):3–10.

Kasper, G., & Wen, H. Y., 1988. On-Line Identification of Particle Sources in Process Gases: Analysis of Fallout Particles for Quality Control Applications. Proceedings of the 9th International Committee of Contamination Control Societies Conference, pp. 485–489, September 26, 1988, Los Angeles.

Koch, U. H., & Pinson, J. H., 1989. Particle Measurement in Gas System Components: Defining a Practical Test Method. *Microcontamination* 7(3):19–21, 61–68.

Liu, B. Y. H., et al., 1985. Electrostatic Effects in Aerosol Sampling and Filtration. *Annals of Occupational Hygiene* 29:251–269.

Liu, B. Y. H., et al., 1989. A Numerical Study of Inertial Errors in Anisokinetic Sampling. *Journal of Aerosol Science* 20(3):367–380.

Marple, V. A., Liu, B. Y. H., & Olson, B. A., 1989. Evaluation of a Cleanroom Concentrating Aerosol Sampler. Proceedings of the 36th Institute of Environmental Science Annual Technical Meeting, pp. 360–363, April 1989, Anaheim, CA.

Okazaki, K., Wiener, R. W., & Willeke, K., 1987. Non-Isoaxial Aerosol Sampling: Mechanisms Controlling the Overall Sampling Efficiency. *Environmental Science and Technology* 21(2):183–187.

Pui, D. Y. H., Ye, Y., & Liu, B. Y. H., 1988. Sampling, Transport, and Deposition of Particles in High Purity Gas Supply System. Proceedings of the 9th International Committee of Contamination Control Societies Conference, pp. 287–293, September 26, 1988, Los Angeles.

Rader, D. J., & Marple, V. A., 1988. A Study of the Effects of Anisokinetic Sampling. *Aerosol Science and Technology* 8:283–299.

Schwartz, A., & McDermott, W. T., 1988. Numerical Modelling of Submicron Particle Deposition in Pressurized Systems: Analysis of Fallout Particles for Quality Control Applications. Proceedings of the 9th International Committee of Contamination Control Societies Conference, pp. 78–86, September 26, 1988, Los Angeles.

ter Kuile, W. M., 1979. *Comparable Dust Sampling at the Workplace,* Report F 1699, Project No. 6.4.01. Delft, Netherlands: Research Institute for Industrial Hygiene, TNO.

Vincent, J. H., 1987. Recent Advances in Aspiration Theory for Thin-Walled and Blunt Aerosol Sampling Probes. *Journal of Aerosol Science* 18(5):487–495.

Wang, H. C., Wen, H. Y., & Kasper, G., 1989. Factors Affecting Particle Content in High-Pressure Cylinder Gases. *Solid State Technology* 32(5):155–156.

Whitlock, W. H., et al., 1988. Continuous Monitoring of Gaseous Carbon Impurities in Electronic-Grade Bulk Gases. *Microcontamination* 6:43–45.

Whyte, W., & Niven, L., 1986. Airborne Bacteria Sampling: The Effect of Dehydration and Sampling Time. *Journal of Parenteral Science and Technology* 40(5):182–187.

23

Optical Airborne Particle Counter Operation

For most cleanroom airborne particle measurements, an optical single particle counter (OPC) is used. An OPC measures particle size in response to the amount of light level change caused by the presence of a particle as it passes through a sensing zone within the OPC. The light level change can be positive or negative, depending upon the nature of the OPC. Three types of measurements are made with OPCs. Particles suspended in gas or liquid or on surfaces are measured. Particles can be counted and sized in ambient air or in either compressed or low-pressure gases with these devices. Particles can be measured in a variety of liquids, including those that are extremely corrosive. Particles can also be measured on surfaces that are actual product surfaces or surrogates of the product or of the cleanroom environment. The last procedure is used to indicate deposition upon critical surfaces that cannot be removed from the product area for testing purposes.

Because these instruments are so widely used, their operating principles are described here in some detail. OPCs are used to define both the size and the number of particles that are present in or on a sample. This chapter is directed to operation of OPCs used for measuring particles in gases. Operation of OPCs used for measuring particles in liquids or on surfaces is discussed later. A separate chapter is devoted to a discussion of instrumental and operational errors and other problem areas that are common to all OPC systems.

Figure 23-1 shows the basic operation of the OPC for counting and sizing particles in gases. As single particles pass through the sensitive zone of the OPC, light is scattered from the particle. The light pulse produced as each particle passes through the sensitive zone is collected and transferred

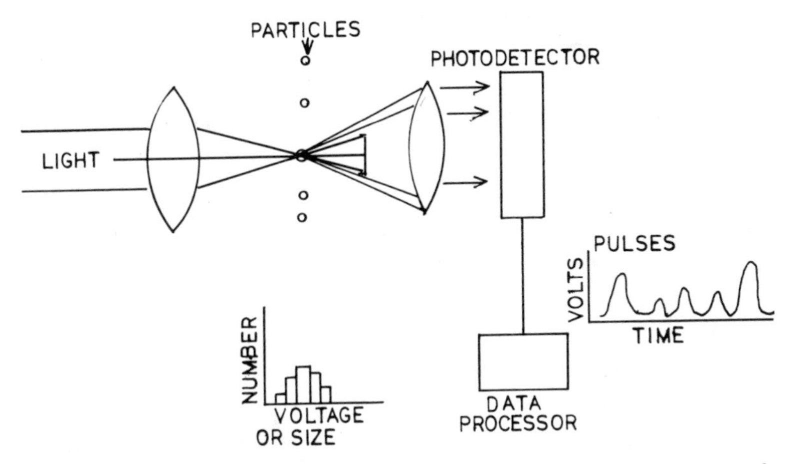

FIGURE 23-1. Optical particle counter operation basis. This illustration is shown for a coaxial forward-scattering optical system, but systems collecting light scattered at any angle operate in the same manner.

to a photodetector, which converts the light to an electrical signal. Because the scattered light intensity varies with particle size, an indication of particle size can be derived from the signal pulse amplitude level. Each pulse amplitude is measured, and the numbers of pulses within defined amplitude ranges are recorded. Because the sampled gas volumetric flow rate is known and the pulse population in each of the pulse amplitude ranges can be determined, the particle concentration in a number of size ranges within the gas can be defined.

OPC design is aimed at detection of the smallest possible particles, at differentiating clearly between particles of nearly similar dimensions, and at counting particles at the highest possible rate with minimum measurement error. The OPC usually contains sample aspiration components that aid in defining the quantity of sampled fluid. In addition to these basic operations, the electronic system within the OPC can verify and indicate operational integrity of the optical, electronic, and fluid flow components of the OPC. The electronic system can also be capable of storing data for further operations; it can carry out a variety of data processing and recording activities both during and after measurement.

The operation of the OPC is neither as simple as indicated by this quick summary nor as complicated as one might expect from the light-scattering optical theory, aerodynamic, and electronic designs for an operating OPC. Light scattering theory is complex, as shown in Figures 23-2 through 23-4. Figure 23-2 illustrates interaction of a parallel light beam with a spherical,

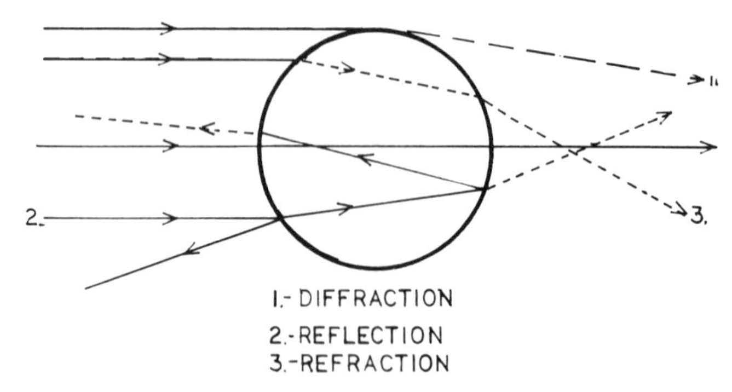

FIGURE 23-2. Geometric light scattering. The three major effects controlling light interaction with translucent or transparent particles are shown. The major parameters affecting the angle and amount of light emitted from the particles are the particle size, the refractive index of the particle relative to that of the surrounding medium, and the illumination wavelength.

partially transparent particle. The particle reflects some light at the interface between the particle and the suspending medium, refracts light passing into and out of the particle, diffracts light at the particle boundary, and may absorb some of the light passing through it. From the phenomena diagrammed in this figure, it can be seen that the scattered light amplitude varies with scattering angle as the several emitted beams interact. Figure

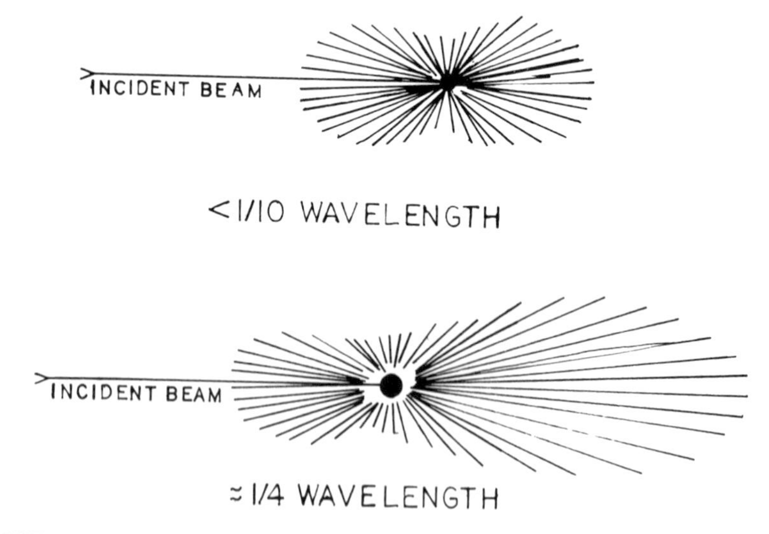

FIGURE 23-3. Small particle light scattering basis. The scattered flux variation with scattering angle is not great. The scattering angle function or the ratio of particle area to illumination wavelength is 10 or less.

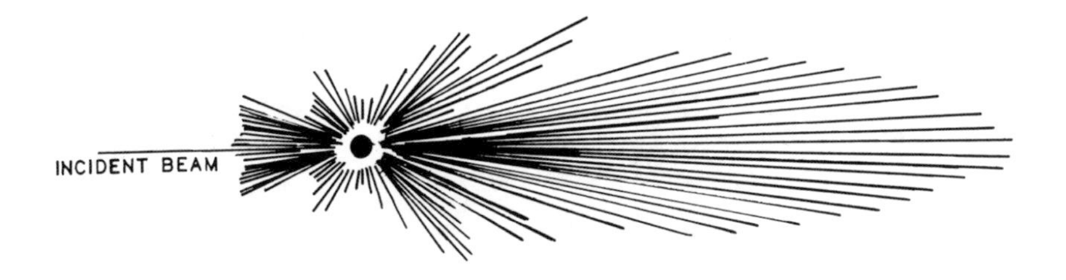

FIGURE 23-4. Large particle light scattering basis. As particle size increases, the variation of light flux with scattering angle becomes larger. A major portion of the scattered light is seen in the forward direction.

23-3 shows the relative flux of light scattered from small particles as a function of the scattering angle. For the most part, scattering intensity from such small particles does not vary appreciably with scattering angle. Figure 23-4 shows the same scattered light versus scattering angle relationship for larger particles. The variation of light with scattering angle is much greater at either polarization for larger particles. The result is that the scattered light pulse amplitude that is collected by a specific OPC optical system varies appreciably, depending upon the scattering angle for that OPC, the refractive index of the particle, its shape and size, and the wavelength of the light. Figure 23-5 shows how the scattering varies as a function of particle size and collecting angle at a single illumination wavelength. It can be seen that some OPCs with small, solid-angle forward scattering optical systems have polytonic responses for transparent particles. Optical systems with larger solid angle collection optics that reject collection of near foward scattered light have a less ambiguous response as particle size changes. Even so, physical construction problems associated with package size may be a major limit to design choices.

OPCs with a number of optical designs have been manufactured to allow good scattered light collection with acceptable package size and economy of manufacturing. The basic optical designs shown in Figure 23-6 were described by Hodkinson and Greenfield (1965). Right-angle scattering systems, forward-angle scattering systems, and wide, solid angle systems are shown in this illustration. One of the three optical designs shown here has been used in all of the OPCs that were ever built. Although the 1965 illustrations show point source illumination focused at the sensing volume, rather than coiled filament incandescent lamp or laser illumina-

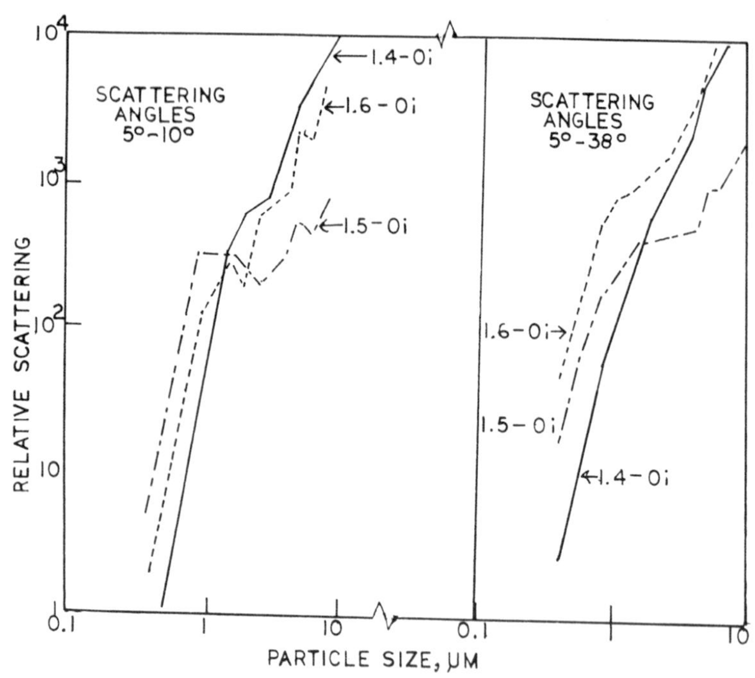

FIGURE 23-5. Scattered light flux versus size for some specific solid scattering angles. Small solid angle forward scattering response tends to be polytonic, especially for single wavelength optical systems.

tion, the same illumination and collection optics schemata are still in use. Figure 23-7 shows some 1980-vintage commercially available OPC optical systems. Note the similarity to the fundamental 1965 designs. Even in the 1990 period, the same basic designs are being used; incandescent filament illumination sources have been replaced by lasers and solid-state detectors are used to detect the scattered light, but the basic optical system has not changed. A recent optical design uses an active cavity helium neon (HeNe) laser for illumination by passing the aerosol stream through the operating cavity of the laser where the beam intensity level is in the range of 20 to 30 watts. Scattered light is collected at right angles to the illumination beam and airstream axes. This OPC, shown in Figure 23-8, is capable of counting and sizing single particles of 0.05 μm in an air sample stream at 2.8 liters per minute. Several very sophisticated optical and electronic design and component elements are used in this instrument, but the basic optical design is the angular scattering system described in Hodkinson and Greenfield's 1965 paper.

There are several factors that control both the performance and reliabil-

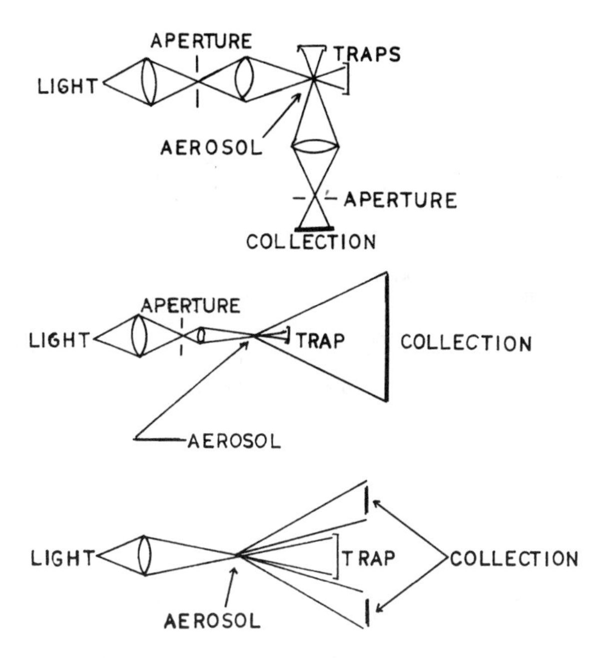

FIGURE 23-6. Optical designs for particle counters. The three designs are the fundamental optical systems used for all particle counters to this day. (From Hodkinson, J. R., & Greenfield, J. R., 1965. Response Calculations for Light Scattering Aerosol Counters. *Applied Optics* 4(11):1463–1474.)

ity of the OPC operation. These factors include the design and construction of the OPC, the nature of the system that is being measured, and the OPC optical system. OPC design and construction affect such phenomena as the uniformity of the illumination within the sensing zone and the velocity distribution of particles passing through that sensing zone. In combination with the type of electronic system used for data processing, these factors affect the particle size resolution capabilities of the OPC. The signal level from any particle can vary with location in the sensing zone, and the particle residence time and pulse rise time within that zone vary with fluid flow rate. The pulse shape and time factors affect electronic responses.

OPC performance is not controlled solely by the design and components of the OPC alone. The nature of the system being measured affects OPC performance in that the sample status controls OPC response to some extent. The OPC measures scattered light from a fixed volume in which there may be a mixture of fluid molecules, particles large enough to be identified, and a possible large quantity of particles somewhat smaller than

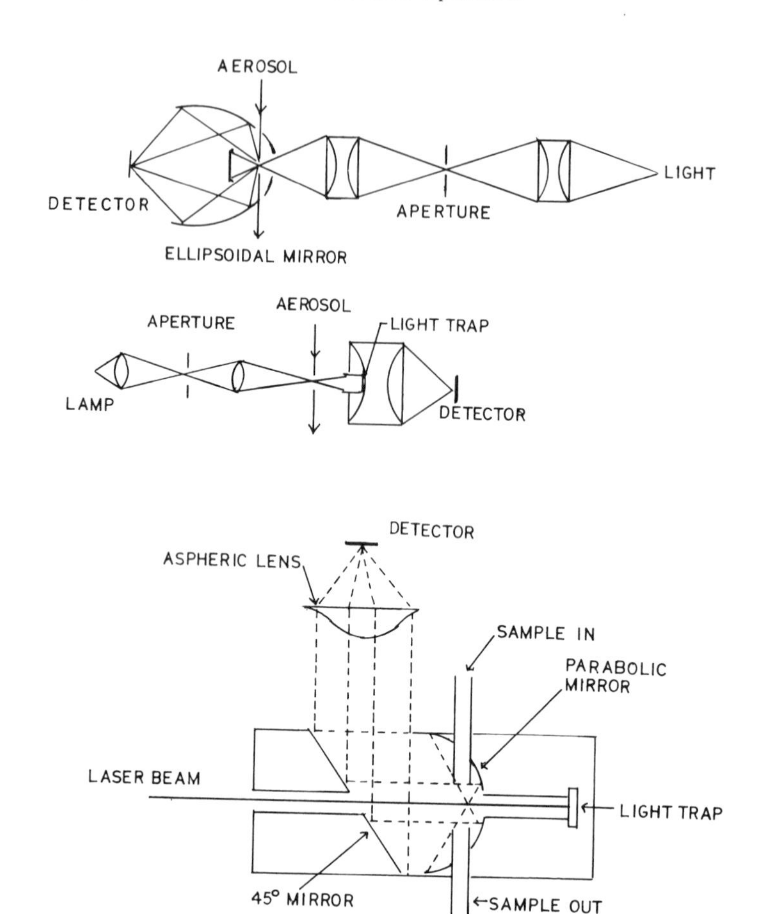

FIGURE 23-7. Commercially available OPC optical designs. Comparison with Figure 23-6 shows the similarities between the operating systems of recent commercial OPCs and the 1965 designs of Hodkinson and Greenfield.

the minimum size sensitivity level for that OPC; that is, molecular scattering from liquids or compressed gases usually results in higher steady-state background noise than that from ambient air measuring systems. As always, the environmental cleanliness affects response for all OPCs, particularly those that are used for measurement of clean fluids. If a high concentration of subcountable particles is present, they produce background noise signal that varies randomly as the background particle concentration varies in the sensitive zone. If particle concentration in the

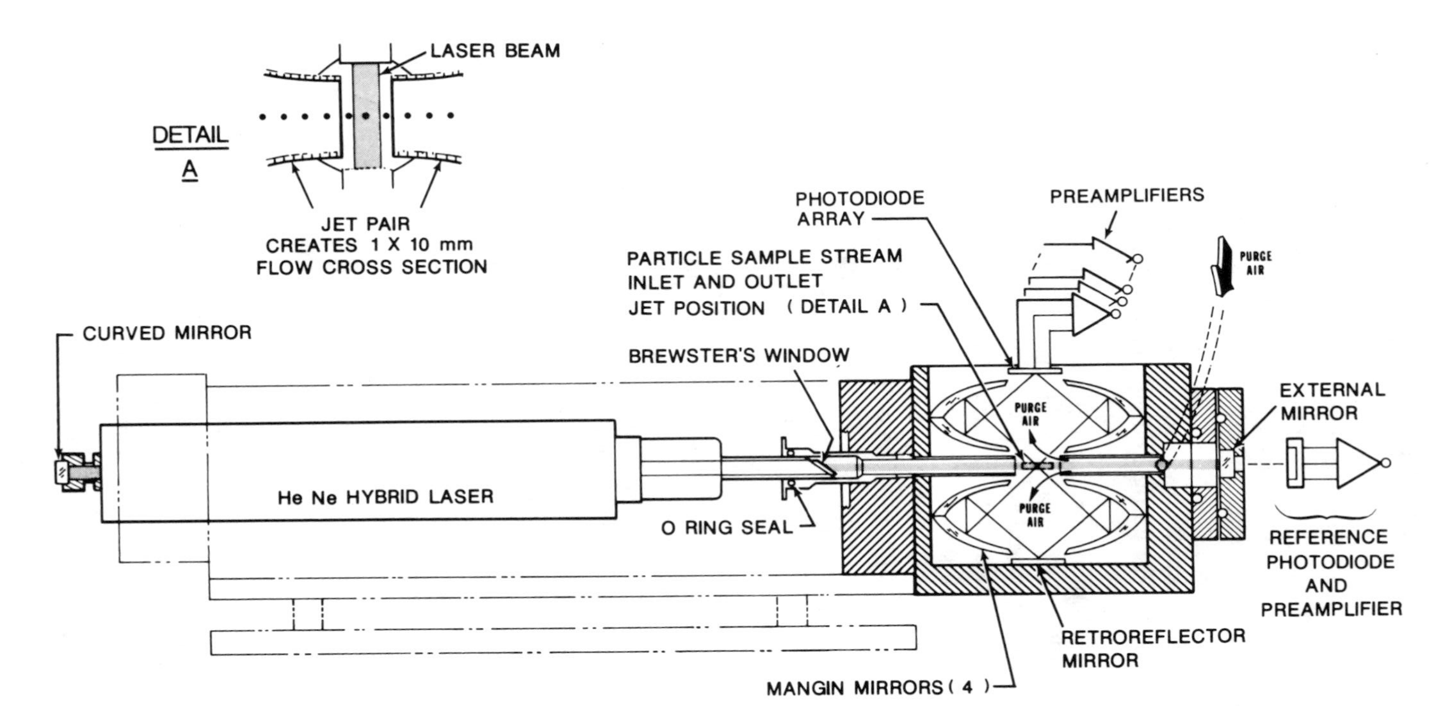

FIGURE 23-8. Optical system for 0.05-μm particle counter. The high sensitivity for this system is obtained by several means. An active cavity laser illumination system is used with light flux in the sensing zone of some 30 watts, and the detection system uses a photodiode array. Each photodiode element observes a portion of the sensing zone to reduce background air scattering noise; the electronic system logic sorts the data. (Courtesy Particle Measuring Systems, Inc., Boulder, CO.)

measured system is excessive, then two problems arise. Several particles may be present in the sensing volume at any time, or they may arrive so close together that their pulse repetition rate may be too fast for the OPC electronic system to handle. In either case, the OPC indicates fewer particles than are actually present and an incorrect size distribution can be reported. The array of particles in the sensing volume is interpreted as a single larger particle. The particle number is shown as less than the actual number.

The most important factors are still those associated with the optical design of the OPC. A number of fundamental optical parameters affect OPC response. These parameters are associated with the particle-fluid or particle-substrate system that is emitting the scattered light and with the optical system that is providing the light to be scattered and is collecting that scattered light. The pertinent parameters are (1) the illumination and collection optics configuration, which define the light scattering solid angle; (2) the light intensity and wavelength; (3) the particle size; and (4) the particle refractive index relative to that of the surrounding material. The solid angle scattering parameter controls the quantity of scattered light produced by particles of various sizes and the fraction of the scattered light that is collected by the OPC collection system. This quantity is a fraction of the light that illuminates the particle while it is in the sensitive zone. It is affected by the relationships between the illumination beam angle and the collecting optics solid angle. Note that these angles are based upon a real optical system. The illumination beam may be produced by an incandescent filament lamp, a gas laser, or a solid-state laser. Beam shaping is usually required to control the illuminated volume dimension and illumination beam direction. A lens system is normally used for this purpose. In any case, the beam is never generated from a true point source; it has finite dimensions, and it may be convergent or telecentric when it reaches the sensing zone of the OPC. The fraction of the total light scattered from a particle that is actually measured varies markedly with the scattering angle that is observed. Selection of the observed scattering angles also affects the monotonicity of the OPC response as well as the response to absorbing or transparent particles. The light-collecting elements may be mirrors or lenses. Mirrors are usually used when larger collection angles are desired. In either case, the light-collecting elements are also finite with dimensions so that the scattering angle always covers some range and is never a single zero-width angle. As an example, an optical system described as "a right angle scattering system" usually collects scattered light over an angular range of 90 degrees ± 45 degrees. A "wide angle scattering system" usually covers a scattering angle range from 15 degrees to as much as 135 degrees. A "forward angle scattering system'' usually collects light over a scattering angle range from 10 to 30 degrees or so. The angular values mentioned here are approximate because manufacturing tolerances may allow some variation from one instrument to another from the same producer. In addition, the finite dimensions of the optical components may allow the scattering angle to vary depending on the combination of optical layout and pneumatic control systems. The sensitive zone through which particles pass is usually large enough so that the scattering angle changes with the position of the particle in that zone.

The other parameters control the total amount of light scattered from a particle. As the illumination intensity is increased, smaller particles can be detected; as the wavelength is decreased, size sensitivity increases. Larger particles scatter more light, producing greater pulse amplitudes than small particles. As the refractive index varies, the relationship of the scattering intensity versus the particle size varies. The refractive index effects not only derive from the combined real and imaginary values for the particle but also are affected by the ratio of the index of the particle to that of the substrate medium. A complete discussion of light scattering theory would require many pages more than are justified here, but the points made summarize the changes that occur in scattered light level as system parameters are changed.

For OPCs with scattering solid angles that collect significant quantities of forward scattered light, the relationship between the scattered light amplitude and the particle size usually becomes polytonic, as shown in Figure 23-5. Polytonic response usually results in reduction of OPC sizing resolution. Counting accuracy may not be affected. This figure shows the change in response as scattering solid angle is varied for some particles with a small range of refractive indexes. The response for the system collecting mainly at the forward angle shows the more polytonic response. Further significance of this effect is discussed in more detail later.

The first optical particle counters were modeled after optical devices developed for detection of chemical warfare agents in the 1940s period. One of the first instruments capable of detecting and sizing airborne submicrometer particles was completed nearly four decades ago (Fisher et al 1955). This instrument used a coiled ribbon filament white light illumination and collected light scattered at a right angle to the illumination beam. The instrument was capable of detecting particles of 0.5 μm diameter and larger in a one tenth cubic foot per minute sample flow and sizing them in several discrete ranges. Data processing was carried out using a vacuum tube electronic system. A modification of this instrument was built for use in U.S. Air Force–supported laboratory scale cloud physics studies. The same USAF design instrument was also used in one of the first cleanroom monitoring studies (Lieberman and Stockham 1961). Along with general levels of particle concentration in the room, the effects of varying levels of personnel activity at the points of measurement were noted. When this device was used for the cleanroom monitoring study, it was so large that it could not be taken into the cleanroom. Samples were removed from the cleanroom through stainless-steel tubing passed through the cleanroom walls. The OPC was moved around the periphery of the cleanroom, and each sample tube was connected to the OPC inlet. Because only data for size ranges from 0.5 μm to 5 μm was of interest, loss of larger particles in the tubing was not a great concern.

Present-day high-sensitivity counters use an HeNe laser as the illumination source. The gas plasma is contained between one mirror and a Brewster's angle window. The sample passes between that window and the second laser mirror. The particles are contained within the active cavity of the laser, where beam intensity is several thousand times greater than the beam usually emitted from an HeNe laser. This configuration is shown in Figure 23-8, showing the optical system of a present-day 0.05-μm-sensitive particle counter (Knollenberg 1987; Lieberman 1988). As each particle passes through the laser beam, light is scattered at a level that is a power function of particle size and refractive index. Each scattered light pulse is collected over a total solid angle approximately 45 to 135 degrees by a dual-mirror system that then focuses the collected light to a sensitive photodetector. The photodetector consists of a 12-element array. The two end elements are used to define the bounds of the sensitive volume, and each of the central 10 elements observes a tenth of the total sensitive volume. The signal from a particle is measured only in comparison with the noise from that portion (10%) of the sensitive volume where the particle is present. This patented process allows a much better optical signal-to-noise ratio than a single-element detection system that observes all the volume that must be illuminated for passage of 28.3 liters per minute of airflow at an acceptable rate. The optical signal results from light scattered by the particle; part of the optical noise results from light scattered by the air molecules within the sensitive volume.

To minimize optical noise sources, the air sample stream is directed through the optical system without confining the stream in a transparent tube. In this way, the integrity of the illumination is maintained, and scattering from particles that could be deposited on a transparent tube wall is avoided. Air-handling system design for an OPC requires that the OPC be capable of aspirating an air sample at a fixed, known rate, that the aspirated air be directed through the optical sample viewing zone with minimum flow recirculating during passage through the optical chamber, that any particles in the sampled air be kept from deposit on the OPC interior, and that the sampled air be clean before it is vented from the OPC into the cleanroom. Ambient airflow rates from less than 0.28 to 28 liters per minute are in common use. Air-handling systems for flow rates up to approximately 15 liters per minute commonly use sheathed sample flow systems. The sheathed flow system draws an air sample into the OPC and passes the sample stream into the center of an annular stream of filtered air moving at the same velocity as the sample stream. Both streams are maintained in the laminar flow regime. Most 28-liters-per-minute sample flow systems do not use sheath flow; the constraints on air sampling pump performance and size may limit these systems to a purge air configuration. The sheath airflow systems minimize particle deposition within the OPC optical chamber more effectively than does a purge air system.

Figure 23-9 shows the basic operation of a sheath flow system. As the sample air is drawn into the optical chamber, it is surrounded with an annular sheath of filtered air to minimize jet divergence. In this way, the sample is directed through the center of the viewing zone where light levels are more uniform. Diffusion of particles out of the sample stream into the OPC, where they may deposit on sensitive optical elements, is thus minimized. Experience has shown that an effective sheath for a cylindrical sample stream should have a volumetric flow rate some 2 to 3 times that of the sample stream. The sample stream aspect ratio, as well as the total flow, must also be considered. A simple relationship for sheath flow vol-

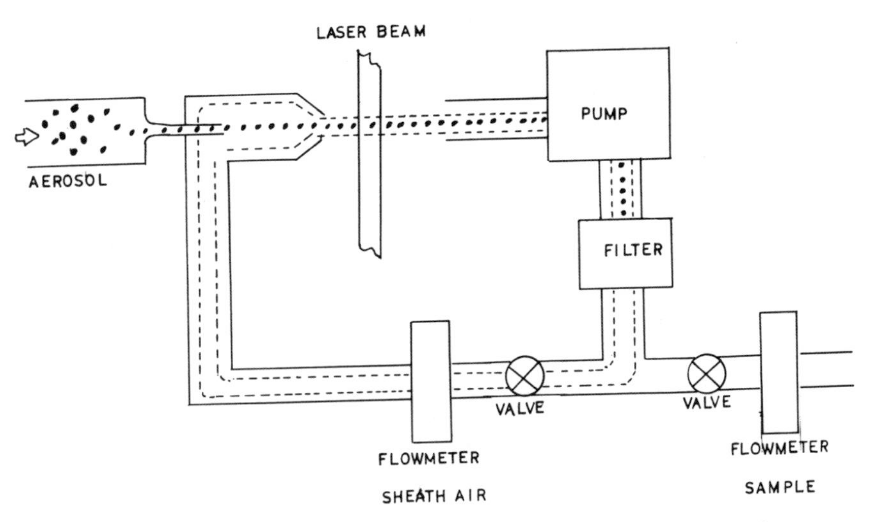

FIGURE 23-9. Sheath air low-flow rate sample flow control system. The annular sheath flow maintains the sample stream at a fixed location in the sensing zone where light level is relatively uniform and gas flow is constant.

ume for a rectangular sample stream cannot be stated. At any rate, the sheath flow system must be capable of moving air through the OPC and the high-efficiency sheath air filter at a high flow rate. The high flow rate, with its requirement for a large, high-power vacuum pump, is the reason why sheath air systems are not always used for higher flow OPCs. Figure 23-10 shows a flow system using only purge air to protect the optical elements. This system is effective in minimizing deposition of particles on sensitive optical elements as long as the purge airflow rate is sufficient to remove any particles emitted from the sample flow jet boundaries. In addition, it is still advisable that the sensitive optical elements be as far as possible from the sample stream. Some diffusive transport of small particles always occurs and eventually results in deposition on nearby surfaces.

When the sampled air is vented from the optical chamber, it is normally filtered through a high-efficiency air filter so that sampled particles and particles that may be generated by pump operation are not released into the cleanroom. Most OPC flow systems contain either a mass flow meter or a rotameter to define sample airflow volume. Both of these meter types can be affected by gas density and may require recalibration if volume flow definition at ambient temperature and pressure conditions is desired and either the gas composition or the ambient air pressure is changed. When an OPC flow meter has been calibrated at sea level, operation of that OPC at a

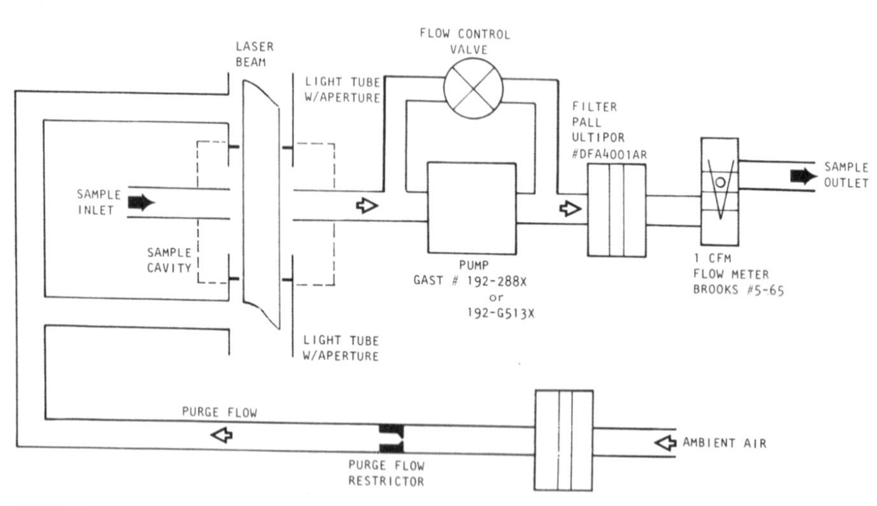

FIGURE 23-10. Purge air high-flow rate sample flow control system. Sheath flow cannot be obtained with a high sample flow system because the aspiration pump capacity is inadequate to provide the required flow rate. The pump is adequate for the sample flow, along with sufficient filtered purge air to maintain the optical components in a stream of clean air.

different altitude results in airflow at a different volumetric flow rate. FS209D states that the air volume used to define cleanroom classification must be measured at ambient pressure; therefore, the OPC flow metering system must be recalibrated for any altitude changes.

The electronic signal processing system receives light pulses upon the sensitive surface of a photodetector as each particle passes through the sensing volume. This device converts the light pulses into electrical pulses whose amplitude varies with particle size for any single OPC. Figures 23-3 through 23-5 contain some definition of the relationship between particle size and scattered light flux. As a generalization, scattering from particles smaller than approximately 0.1 μm varies with the sixth power of the radius, whereas scattering from particles larger than approximately 0.8 μm varies with the square of the particle radius. The shape and location of the power function transition is affected by the OPC optical system and the illumination wavelength (Cooke and Kerker 1975). The pulses can be sorted into as many as 16 amplitude bins by using a pulse comparator circuit, or into as many as 1,024 bins by using a pulse height analyzer. For many OPCs the sensor resolution does not justify sorting particles into more than six or eight bins. This is especially true for OPCs with polytonic response.

The pulse duration varies mainly with the air velocity through the sensing volume and with particle size. Most sensing volumes have dimensions in the range of 0.1 to 1 mm on a side. For particles whose size approaches the sensing volume dimension, the residence time is also affected by particle size. Most OPC pulses are generated with durations of 10 to 50 μsec. Average pulse rates of several thousand per second can occur with individual pulses appearing at random time intervals that may be as small as a tenth of the average interval. For an OPC dynamic size range of 20 to 1, the pulse amplitude measurement range requirement can vary from a minimum of 400 to 1 for large particles up to 300,000 to 1 for smaller particles. The larger range may require sophisticated electronic design to process pulse size and number data without losing or distorting pulse information. Once the pulse processing and recording system design is completed, the electronic system can carry out a variety of data recording and processing operations.

The data processing system of a modern-day particle counter is capable of many tasks. It can monitor status and control operation of the instrument. It sorts and counts the signals from the photodetector. It compares those signal levels with calibration records to provide particle size distribution information. It can record particle size distributions in cumulative and/or differential format. It can identify samples in terms of sample point location, sample collection time, and date. It can collect a series of measurements of preset sample durations and intervals and carry out simple statistical calculations, including those of FS209D. It can sound alarms when preset particle count or concentration levels are exceeded or when operating component failure is imminent. It can record, store, print, or transmit data in a variety of formats. It can be operated by front panel or remote controls.

Figure 23-11 shows a data format for a single-point particle counter, including alarm operations, selected data storage features, and preparation and transmission of information in desired format to a computer for further operations. These features are especially important when very clean areas are to be monitored or if a large facility's cleanliness is to be controlled. For monitoring in very cleanrooms, data recording within the cleanroom requires clean printing operations so that any data recording may have to include a data transfer step to a printer located outside the cleanroom. In any circumstances, impact printers should never be used in an area cleaner than class 10,000; thermal printers have been used even in class 100 areas, but paper that may release fibers during the printing process should not be used. When large facilities are to be monitored, long-term and large-area data trends can be recorded. Figure 23-12 shows some particle count data trend plots.

Most present OPCs observe particles drawn into a sensitive zone at a

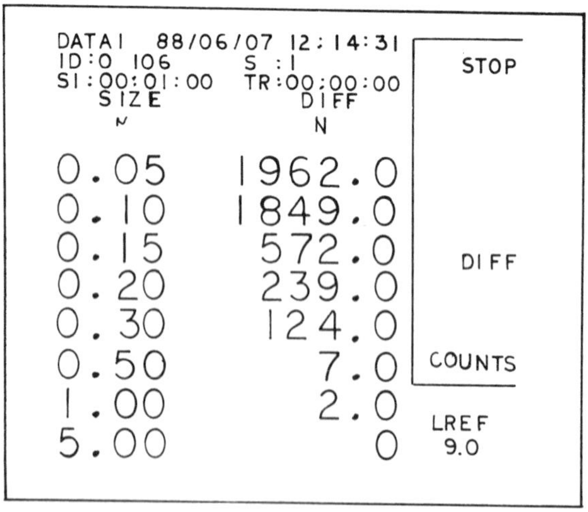

DATA1 88/06/07 12:14:31
ID:0 106 S :1
S1:00:01:00 TR:00:00:00
SIZE DIFF
N N

SIZE	DIFF N
0.05	1962.0
0.10	1849.0
0.15	572.0
0.20	239.0
0.30	124.0
0.50	7.0
1.00	2.0
5.00	0

STOP
DIFF
COUNTS
LREF
9.0

FIGURE 23-11. Single-point particle counter data format. This example of a data format shows particle count data in terms of size ranges and particles counted. Additional information is provided to define OPC operation.

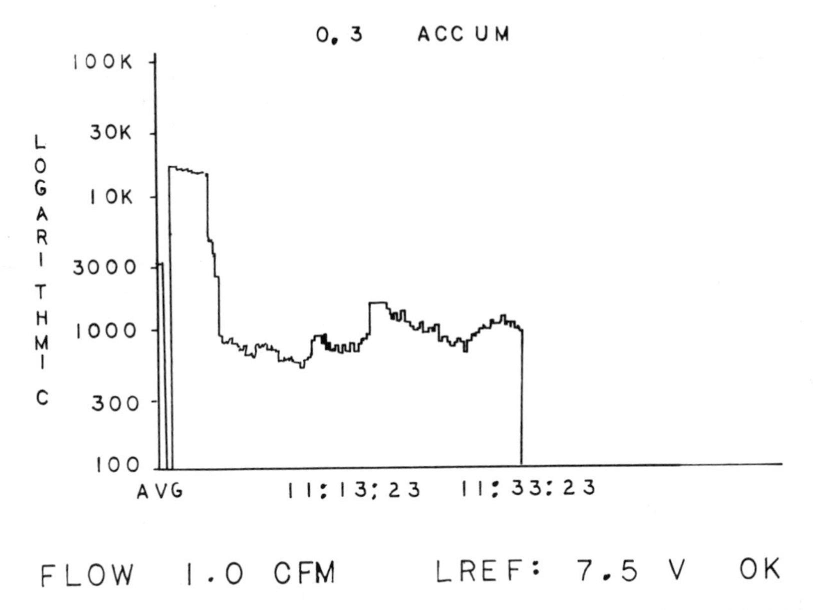

FIGURE 23-12. Particle count trend data plot format. This example of a graphical data display shows particle count trends over a period of time in an operational area. This presentation format may be the most valuable for normal, everyday control of cleanroom operations.

fixed air sample flow rate. These instruments are used to count and size airborne particles directly in a known sample volume. For detection of very small particles (ca. 0.01 μm and larger), the particles are used as condensation nuclei in a supersaturated vapor environment. The vapor condenses upon these very small particles and grows to liquid droplets in the 10- to 20-μm-diameter size range. The OPC then counts the number of droplets that correspond to the number of active nuclei drawn into that OPC, but data on original nucleus particle size distribution cannot be obtained. This OPC type is referred to as a condensation nucleus counter (CNC). The first CNC was described more than 100 years ago (Aitken 1888). In operation, an air sample was drawn into a chamber that contained a porous surface soaked in water. The air sample was compressed and the resulting heat removed by convective heat transfer from the chamber exterior surface to the cooler atmosphere. Then the relative humidity within the chamber rose nearly to saturation. After the temperature had returned to that of the ambient air, a valve was opened very quickly, and the compressed air was expanded adiabatically. The sudden decrease in temperature increased the relative humidity within the chamber to the point where supersaturation existed for a short period. The water vapor in the chamber then condensed on all of the particles within the chamber that were large enough to act as condensation nuclei. Droplets of some 10- to 20-μm diameter form on each nucleus. An illumination source and a means of determining the droplet population by photometric observation were incorporated in the chamber. This type of CNC is known as an expansion system because supersaturation occurs as a result of cooling owing to adiabatic expansion.

Another CNC type passes the air sample over a heated, liquid-soaked porous surface, where the airstream becomes saturated with the vapor from the liquid in the surface. Ethylene glycol and butyl alcohol have been used as vapor sources in these devices. The vapor-saturated airstream is then passed into a cooled condenser tube, and the vapor diffuses onto the small-particle nuclei in that airstream to form large liquid droplets as that vapor condenses on the particles. The droplets are passed through an OPC, where they are measured either by direct single-particle count or by photometric measurement using autoranging circuitry to switch to the latter mode of operation when the concentration becomes so high that coincidence errors are significant. This type of CNC is reported effective in detecting particles as small as 2.6 nm with 50% efficiency, although most CNCs are used for counting particles larger than approximately 0.01 μm (Stolzenburg and McMurry 1984). It has been noted that counting efficiency for the smallest particles varies with CNC design (Sinclair 1982; Bartz et al. 1985). The CNC can also be used in gases other than air; however, the detection efficiency may vary somewhat. It has been shown that the nucleus size for 50% efficiency for one particular CNC increases from 6.5 nm, when observing particles in nitrogen, to 10 nm for particles in carbon dioxide, whereas particles in helium are counted with greater efficiency (Niida et al. 1988).

Even though the CNC is not able to define particle size directly, and sample flow rate is less than that of direct-reading OPCs, these devices have been used in cleanroom applications. The CNC is now the only available real-time instrument that can detect particles smaller than 0.05 μm. For that reason, it can be used to measure airborne particles in cleanrooms where an overall particle count is desired for particles larger than approximately 0.01 μm (Ensor and Donovan 1985). It is used to characterize cleanliness of inert process gases where particles ≥0.02 μm are of concern. It can be used to detect leaks in ULPA filters; if a diffusion battery or a differential mobility separator is used in conjunction with the CNC, then particle size data can be derived for the very small particles, and fractional filter penetration data can be obtained for a range of particles smaller than 0.05 μm. A parallel array of diffusion batteries, each with its own CNC and each pair drawing an independent air sample, allows definition of particle size distribution in a size range starting at ≈ 0.01 μm (Ensor et al. 1989). Using multiple systems in the array allows measurement of very small particles in quite clean air in a reasonable time period.

Many semiconductor manufacturing occurs within tools operating at reduced pressures where standard OPCs cannot draw sample air adequately. A need has been stated for satisfactory real-time measurement of particles in these environments (Bowling, Larrabee, and Fisher 1989). Requirements for real-time in situ monitoring include measurement of particles within operating tools, sensitivity to particles at least as small as 0.1 μm, and acceptable counting accuracy. A particle flux monitoring OPC observes particles passing through an open sensing volume with no control of sample flow (Borden and Knodle 1988). That kind of OPC depends on diffusion, gravitational sedimentation, or other forces to cause particle passage through the sensing volume. This device establishes a rectangular illuminated volume between two mirrors by reflecting a laser beam back and forth to form an open net pattern that covers 10 to 20% of the rectangle. This monitor is reported to be sensitive to 0.5-μm particles. The flux monitor has been used within operating semiconductor process devices at both ambient and decreased pressures. Because no flow is required, this type of device can also be used to observe settling or depositing large particles (Borden et al. 1989). Large-particle (10–1000 μm) flux could be observed in real time and was found to compare well with data obtained from observation of deposits collected upon a witness plate located close to the flux monitor.

A similar approach has been used on a much larger scale (Hayakawa et al. 1984). A laser beam is passed through a cleanroom area, and forward scattered light at 10 degrees from particles in an optically defined unenclosed volume is collected by a telephoto lens and focused to a photodetector and a television camera (after magnification). This system has the advantages of operation with no interference to suspended particles in the cleanroom, but the sample volume is difficult to define very accurately.

Optical particle counters are used to determine the class level for the cleanroom and to monitor the continuing performance of the cleanroom, once it is in operation. Particle count data are procured at specific locations within the cleanroom to define the room classification; sample size is specified, and the data are processed so that the room classification can be stated within a 95% confidence limit. A cleanroom monitoring plan is implemented, and measurements taken to ensure that the cleanroom has not changed since it was last classified. For both cleanroom classification and monitoring, the measured particle size range and the air sample volume are selected so that an adequate quantity of particles are measured to provide statistically valid data. In some cases, monitoring data may be acquired, which simply indicates that a specific concentration is never exceeded or a particular particle size range is never seen.

In classifying a cleanroom, the sample points are first identified and located. Next the particle counter (or a sample line inlet nozzle) is located at a specific point. The desired particle size range is selected for measurement in accordance with the anticipated cleanroom class level. The particle counter sampling time is set so that the required sample volume is drawn into the counter. Data are recorded or transmitted to a central data processor. The particle counter is moved to the next sample point, and the process is repeated. For monitoring, a sample point(s) is selected that is representative of the cleanroom or of the process to be monitored. The particle counter makes repetitive or continuous measurements of the airborne particle concentration and size distribution at that location. The particle counter is required to procure accurate particle size and concentration information; particles may be very small and may be very sparse at the sample point; particle concentration may vary drastically with time, and the particle composition may change with operating conditions in the cleanroom. The requirements for cleanroom data acquisition include good OPC performance. The OPC must be capable of providing accurate particle size and concentration data and cannot introduce data artifacts in its operation. As an example of the factors driving particle measurement requirements, it is generally accepted that particles one tenth the dimension of components on the semiconductor wafer can cause failures. Considering that present-day semiconductors are being manufactured with geometry at the 1-μm level and that future developments at half that level are already being planned, the need for measurement of particles at and below the 0.05-μm level is obvious.

References

Aitken, J., 1888. On the Number of Dust Particles in the Atmosphere. *Proceedings of the Royal Society, Edinburgh,* p. 35.

Bartz, H., et al., 1985. Response Characteristics for Four Different Condensation Nucleus Counters to Particles in the 3–50 nm Diameter Range. *Journal of Aerosol Science* 16(5):443–456.

Borden, P., & Knodle, W., 1988. Monitoring Particles in Vacuum Equipment. Proceedings of the 9th International Committee of Contamination Control Societies Conference, pp. 204–207, September 26, 1988, Los Angeles.

Borden, P., et al., 1989. Real-Time Monitoring of Large Particle Fallout for Aerospace Applications. Proceedings of the 35th Institute of Environmental Science Annual Technical Meeting, pp. 394–396, April 1989, Anaheim, CA.

Bowling, R. A., Larrabee, G. B., & Fisher, W. G., 1989. Status and Needs of In-Situ Real-Time Process Particle Detection. *Journal of Environmental Science* 32(5):22–27.

Cooke, D. D., & Kerker, M., 1975. Response Calculations for Light Scattering Aerosol Particle Counters. *Applied Optics* 14(3):734–739.

Ensor, D. S., & Donovan, R. P., 1985. The Application of Condensation Nuclei Monitors to Clean Rooms. *Journal of Environmental Science* 28(2):34–36.

Ensor, D. S., et al., 1989. Measurement of Ultrafine Aerosol Particle Size Distributions at Low Concentrations by Parallel Arrays of a Diffusion Battery and a Condensation Nucleus Counter in Series. *Journal of Aerosol Science* 20(4): 471–475.

Fisher, M. A., et al., 1955. The Aerosoloscope: An Instrument for the Automatic Counting and Sizing of Aerosol Particles. Proceedings of the Third National Air Pollution Symposium, April 18–20, 1955, Pasadena, CA.

Hayakawa, I., et al., 1984. Image Processing on Remote Detected Particulates in Clean Room. Proceedings of the 30th Institute of Environmental Science Annual Technical Meeting, pp. 90–93, May 1984, Orlando, FL.

Hodkinson, J. R., & Greenfield, J. R., 1965. Response Calculations for Light Scattering Aerosol Counters. *Applied Optics* 4(1):1463–1474.

Knollenberg, R. G., 1987. Sizing Particles at High Sensitivity in High Molecular Scattering Environments. Proceedings of the 34th Institute of Environmental Science Annual Technical Meeting, May 1987, King of Prussia, PA.

Lieberman, A., 1988. A New 0.05 μm, 0.1 CFM Optical Particle Counter. Proceedings of the 20th DOE/NRC Nuclear Air Cleaning Conference, August 1988, Boston.

Lieberman, A., & Stockham, J., 1961. Automatic Techniques of Airborne Particle Counting. *Air Engineering* 2(1):37–39.

Niida, T., et al., 1988. Counting Efficiency of Condensation Nuclei Counters in N_2, Ar, CO_2 and He. *Journal of Aerosol Science* 19(7):1417–1420.

Sinclair, D., 1982. Particle Size Sensitivity of Condensation Nucleus Counters. *Atmospheric Environment* 16(5):955–958.

Stolzenburg, M. R., & McMurry, P. H., 1984. A Theoretical Model for an Ultrafine Aerosol Condensation Nucleus Counter. In *Aerosols*, ed. B. Y. H. Liu, et al., pp. 59–62. New York: Elsevier.

24

Optical Measurement of Deposited Particles

Contamination control in most cleanrooms is mainly directed to maintaining cleanliness of the cleanroom air, the processing materials used, and the products produced in that cleanroom. It is assumed that if the environment where the cleanroom product is manufactured is kept clean, then the product will not become contaminated. Unfortunately, keeping the manufacturing environment clean is only part of the task of avoiding product contamination. Any tool, process material, analytical device, and particularly the personnel used in the manufacturing process can contaminate the manufacturing process even if the entire operation is carried out in clean air. Handling materials or process components can contaminate them. Both chemical film and particle contaminants can be deposited on product surfaces with subsequent product quality degradation. For this reason, definition of the contamination deposited upon products during manufacturing is a necessary step in maintaining product quality.

As of 1990, the techniques of maintaining cleanroom work environment air to a level of fewer than 10 particles per cubic foot, $\geq 0.1\ \mu m$ in diameter, were known and implemented where needed. The procedures for handling sensitive products almost entirely by clean robot devices are known. Process tools and chemicals are a major source of contamination at this time. Control to the same cleanliness level as can be accomplished for cleanroom air has not yet been accomplished. The problem particles are mainly smaller than 1 micrometer in diameter; they are frequently generated within the process tool or are emitted to the work flow paths. These small problem particles as well as many larger ones seldom enter the airstreams, which are kept clean and will remove the particles from the environment. These problem particles can be deposited directly upon critical surfaces with minimum time in the airflow. For this reason, evaluation of tool and material cleanliness is becoming a measurement problem that can be solved mainly by observation of the particulate deposition on the product. Even the use of in situ monitoring systems must be finally verified by examining the product quality during and after fabrication. The surface particle analyzer appears to be the best way of handling the measurement problem for particles as small as 0.1 μm, whereas particles larger than 5 μm are measured with light microscope methods. Some discussions of the deposited particle problem and the solutions available at this time for the semiconductor industry have been presented recently (Batchelder 1988). Both manual and automated test methods are required to define the problem areas adequately.

Although this discussion is directed mainly to the measurement of particulate contamination deposition on surfaces, several sensitive measurement methods are used for characterization of organic films on surfaces. For example, luminescence spectroscopy can be used to detect many organic films at the μg per square centimeter level (Vo-Dinh 1983). The technique is based upon irradiation of a surface with ultraviolet energy and measurement of the light emitted at a range of wavelengths by a contaminant film as it undergoes a radiative transition between two different electronic energy states. Scanning laser acoustic microscopy has been used to examine interior connections of packaged integrated-circuit components (Adams 1985). A high-frequency (ca. 1 GHz) plane wave of ultrasound is passed through the component with absorption and attenuation of the sonic energy occurring at various interfaces within the component. Any air gap is more opaque to the sonic energy than most solid materials. The transmitted beam is detected by a laser scanner directed to a thin mirror facing the beam. The distortion pattern on the mirror is scanned and demodulated to image the interior of the device. Detection of defects as small as 10 to 20 μm appears reasonable.

Characterization of viable particle deposition rates on surfaces has long been important in ensuring the integrity of pharmaceutical and food products. Most measurements in those areas have been carried out by collection of airborne materials with possible viable content, followed by culturing the collected sample on a nutrient bed. In addition, surface deposition was measured by procuring "wipe samples" from areas where deposition of particulate material was expected. Sterile wipers were used to swab a suspect surface; the wiper-collected material was then cultured, and the bacterial count on the wiper surface was defined by noting the number of bacterial colonies grown on the surface of a nutrient culture medium.

In many cleanrooms, deposition of large, inert particles is measured by placing "witness plates" at or close to the location where deposition may be a concern. The witness plate is a substrate of known surface area whose particle collection characteristics are assumed to be similar to those of the product surface. Witness plates could be made of metal, glass, or plastic materials. Some witness plate surfaces are coated with an adhesive film to aid in retaining large particles that may have deposited on a vertical plate. In earlier cleanroom measurements, electrostatic field effects were seldom considered important because most particles that were collected and measured were 5 μm in diameter or larger. Electrical conductivity between the witness plate and the surface of concern was seldom sought. This situation has changed with improved measurement techniques for detection of submicrometer particles and the wider application of witness plate collection.

After exposure to the local environment for several hours or days, depending on the cleanliness of that environment, the witness plate was removed in a clean container to a microscope facility for definition of large-particle deposition. Any particles that were deposited upon and retained on the surface of the witness plate were counted and sized. The witness plate was usually in the form of a glass microscope slide or a polished metal sheet with dimensions ranging from 1 to 200 square centimeters. Data were reported in terms of the number of collected particles $> 5\ \mu$m per square foot per hour. Classes of cleanliness were based on the levels stated in MIL-STD-1246B (U.S. Department of Defense 1982). Attempts had been made previously to correlate fallout data with airborne particle populations. Some very general trend data have been noted (Hamburg and Shon 1984), but the lognormal particle size distribution functions applicable to stable airborne particle populations in cleanrooms did not fit settled particle populations well for particles more than 100 μm in diameter (Beeson and Weintz 1988). It was concluded that size distribution of these particles can be approximated with a Poisson distribution. Particles found on surfaces can be produced from more than one source. It would be surprising if any single size distribution function that can be ascribed to a single source material would fit this multiple-source material. Attempts to correlate deposited particle characteristics with those from any other single source appear pointless.

Observation of large particles on witness plates is usually carried out by optical microscopy, using a binocular microscope at 100× magnification with oblique illumination. Practically speaking, the reliability of this type of manual microscopic observation is minimal for many particles smaller than 5 μm. Even so, the usefulness of data relating to the particle deposition that may occur on product surfaces is great. For this reason, the need for more reliable definition of the large-particle deposition on witness plate surfaces has led to development of more automated means of measuring the particle deposition. For cleanroom areas with access to an automated image analyzer for observation of microscope images, the measurements can be carried out with much greater reliability and in much shorter time than when manual observation is used. An integrating particle fallout meter (Lindahl 1970) has been updated recently (Eesbeek et al. 1988). This photometric device characterizes particle collection on a 40- × 45-mm glass plate in terms of an obscuration value for deposited particles > 5 μm. This value is based on the ratio of area covered by the particles to the total plate area. Calibration of the device, based on polystyrene latex particle coverage, has been carried out, and testing shows reasonable reproducibility for this device. However, measurement of smaller particles, if present in low concentrations, requires more sensitive measurement techniques.

Present-day optical wafer scanning systems are available that can detect 0.1-μm particles on smooth silicon wafer surfaces. These devices can distinguish between wafer surface irregularities and deposited particles. In operation, the beam from a laser is focused by a system of lenses and/or mirrors to a beam that is some 50 to 100 μm wide. The beam is swept across the wafer while the wafer is moving below the beam. Scattered light from wafer defects or deposited particles is collected over a reasonably large solid angle and transformed to electrical signals by a suitable photodetector. The scattered light level can be correlated to the size of the particle or defect that caused the scattering. The high-intensity beam of specularly reflected light is not collected. In 1983, a system was described that was capable of detecting 1-μm particles using this general operating approach (Gise 1983). A review of the problems and some of the solutions for detection of particles on surfaces was presented by Lilienfeld (1986). That paper reviewed the application of light scattering to detection and measurement of particles on smooth surfaces. The optical properties of both specular and diffuse reflecting surfaces were summarized. The parameters influencing detectability of particles on surfaces and the principal design criteria of light scattering inspection systems were discussed. Both the laser illumination/scanning systems and the video inspection systems are discussed, along with a presentation of commercial systems available in 1986. The problems associated with the combined illumination of the particle by the primary beam and by specular reflection from the wafer surface are discussed.

Even though the primary application for many commercial surface observation systems is for scanning bare silicon wafers, the application of the laser scanning systems to observation of other critical devices has been increasing. These areas include inspection of rotating memory disk surfaces, both rigid and flexible. Witness plates with a polished metal surface have been inspected with laser scanning systems.

Some typical design and performance parameters of laser scanning systems, as reported by Lilienfeld (1986), are shown in Table 24-1. The same parameters that are important for characterizing performance of airborne particle counters are also important for surface analyzers. These parameters include particle size sensitivity and measurement accuracy, counting accuracy, dynamic range, resolution, repeatability, and stability. The means of verifying counter performance are similar to those used for other particle counters. An overall discussion of general particle counter

TABLE 24-1 Operating Parameters for Laser Surface Scanners

Laser Type:	2-10 mW HeNe, polarized
Laser Beam Diameter:	0.5-1.0 mm (1/e2 diameter)
Incidence Angle on Target Surface:	0°-87°
Focal Length of Beam:	up to 60 mm.
Scan Frequency:	100-1000 Hz
Illumination Spot Size:	10-1000 μm
Scattering Angle:	10°-120°
Detection Pulse Duration:	0.1-1 μsec
Surface Inspection Speed:	1-100 cm^2/sec
Minimum Detectable Particle Size:	0.1 μm
Position Resolution:	0.1-10 mm.

The values shown are those for one specific scanner only, but the parameters shown apply to many others as well.

Source: (From Lilienfeld, P., 1986. Optical Detection of Particle Contamination on Surfaces: A Review. *Aerosol Science and Technology* 5(2):145–165.)

operating requirements, calibration procedures, and problem areas is given later.

Some anomalous data occurs as a result of secondary illumination of and scattering from particles on the wafer (Knollenberg 1986). This occurs as a result of light reflecting from the wafer surface to the rear side of a deposited particle of irregular shape factor, in addition to illumination from the direct beam. Exploitation of the knowledge of surface effects has been used to design a surface analysis system with two orthogonally polarized lasers that may operate at different frequencies (Knollenberg 1988). Use of an optical system modeled after dark field microscopic systems allows measurement of particles on both transparent and reflective surfaces. Particles as small as 0.1 μm have been sized with this technique.

The procedure for calibrating instruments for particles deposited upon a surface are very similar to those used for other particle counters. Standard latex calibration particles are used. The monosize latex particles used for size calibration are deposited on the surface of a clean, bare wafer, and that wafer is examined by the surface analyzer. The amplitude of the signals produced by the particles on the surface of that wafer is used to define the size response and resolution characteristics of the surface analyzer system. One problem with the surface analyzers is the difficulty of maintaining cleanliness of calibration wafers during handling. Extreme precautions are required to minimize deposition of airborne particles during the calibration process and during storage and transport of calibration wafers. Counting accuracy for the analyzers is usually determined by using wafers with point features etched into the surface in specific patterns. These wafers can be used to define the counting accuracy, dynamic range, and spatial resolution capability of an analyzer (Berger and Tullis 1987).

At this time, most wafer inspection systems, based on laser scanning or video inspection are used for observation of unpatterned wafers. The response from patterns and deposited features would have to be programmed into the detection system microprocessor so that data from desired pattern surface structure would not be interpreted as contamination. Holographic and optical spatial frequency filtering systems have been used for inspection of patterned wafers (Billat 1988). A complete inspection (ca. 7,500 mm^2) of a 150-mm patterned wafer for particle defects of 0.5 μm and greater in size is carried out in 30 minutes. A spatial filter is generated in the Fourier transform plane of the optical system and is used to block the repetitive pattern information at the wafer fabrication stage of concern. A holographic recording is made of the test wafer, and defects in the repetitive areas are recorded as a hologram that is used in combination with the spatial filter to reconstruct a real image of the entire wafer surface. The image is scanned to define location and size of defects.

References

Adams, T. E., 1985. Acoustic Microscopy Improves Internal Reliability of IC Packaging. *Semiconductor International* 8(2):100–104.

Batchelder, J. S., 1988. Applications of Particulate Inspection of Wafers in Semiconductor Manufacturing. Proceedings of the 34th Institute of Environmental Sciences Annual Technical Meeting, pp. 359–362, April 1988, King of Prussia, PA.

Beeson, R. D., & Weintz, W. W., 1988. Cleanroom Facility Analysis of Fallout Particles for Quality Control Applications. Proceedings of the 9th International Committee of Contamination Control Societies Conference, pp. 628–631, September 26, 1988, Los Angeles.

Berger, J., & Tullis, B. J., 1987. Calibration of Surface-Particle Detectors. *Microcontamination* 5(7):24–29.

Billat, S. P., 1988. Holographic Wafer Inspection. Proceedings of the 34th Institute of Environmental Science Annual Technical Meeting, pp. 363–367, April 1988, King of Prussia, PA.

Eesbeek, M. V., et al., 1988. The Effects of Particle-Size and Particle-Colour on Integrating Surface-Particle Counters. Proceedings of the 9th International Committee of Contamination Control Societies Conference, pp. 622–627, September 26, 1988, Los Angeles.

Gise, P., 1983. Principles of Laser Scanning for Defect and Contamination Detection in Microfabrication. *Solid State Technology* 26(11):163–165.

Hamburg, O., & Shon, E. M., 1984. Particle Size Distribution on Surfaces in Clean Rooms. Proceedings of the 30th Institute of Environmental Sciences Annual Technical Meeting, April 1984, Orlando, FL.

Knollenberg, R. G., 1986. The Importance of Media Refractive Index in Evaluating Liquid and Surface Microcontamination Measurements. Proceedings of the 32d Institute of Environmental Sciences Annual Technical Meeting, April 1986, Atlanta.

Knollenberg, R. G., 1988. A Polarization Diversity Two-Color Surface Analysis System. Proceedings of the 34th Institute of Environmental Sciences Annual Technical Meeting, April 1988, King of Prussia, PA.

Lilienfeld, P., 1986. Optical Detection of Particle Contamination on Surfaces: A Review. *Aerosol Science and Technology* 5(2):145–165.

Lindahl, B., 1970. A New Method for Determination of Surface Contamination Caused by Particle Fall-Out. Proceedings of the European Symposium on Contamination Control, June 1970, Paris, France.

U.S. Department of Defense, 1982. *Product Cleanliness Levels and Contamination Control Program*, MIL-STD-1246B. Philadelphia: Naval Publications and Forms Center.

Vo-Dinh, T., 1983. Surface Detection of Contamination: Principles, Applications and Recent Developments. *Journal of Environmental Science* 26(1):40–43.

25

Optical Liquidborne Particle Counter Operation

Optical liquidborne particle counters are used to characterize the concentration and size distribution of solid materials that may be suspended in liquids used in cleanroom operations. The instruments are capable of counting and sizing single particles ranging in size from approximately 0.05 μm in diameter to the millimeter size range. Particles in suspensions containing up to 50,000 particles per milliliter of liquid can be sized and counted at rates to thousands per second. As is the case with any optical particle counter, operation of the liquidborne particle counter is such that data can be accumulated with both counting and sizing errors and no indication that errors may have occurred. Errors can arise either from faulty design or construction of an OPC or from incorrect operational procedures. Operation of these devices is discussed here so that many errors can be avoided. A brief description is also given of some methods other than optical response that are commonly used for counting and sizing particles in liquid suspensions.

Some of the methods of measuring particles in liquids used in cleanrooms are derived from technology used in other technical areas, similar to other measurements made in these rooms. Particle measurement in liquids has been carried out for many years before contamination control was of concern. Particle measurements were made to assure cleanliness of potable water and other beverages, to examine characteristics of body fluids such as blood and serums, to characterize atmospheric components, and to control mining operations. Several types of liquidborne particle counting instruments are used for measurement of particles in liquids (Lieberman 1984a). The instruments in use are based on a variety of procedures, including light and electron microscopic measurement of particles that have been collected from the liquid onto suitable substrate materials, electrical resistance measurement of liquids passing through a small orifice, acoustic echo sounding at high frequency, and several optical methods. Other methods used for larger particle characterization are not normally used in contamination control areas; these methods include observation of sedimentation rate of particles in fluids and retention on sieve surfaces. The optical methods often seen in contamination control include light absorption and scattering from arrays of particles in liquid suspensions and both light obscuration and light scattering from single particles in a liquid stream.

Microscopic measurements are used for off-line analysis of extremely clean liquids. The method is sometimes used when seeking data on concentration and morphology or composition of particles smaller than 1 μm in diameter. To produce accurate data by microscopic measurement when the particle count is very low requires a clean operating environment, a well-trained operator, and access to a very large amount of liquid. Acquisition of valid data through observing samples from an ultrapure deionized water system may require filtration of up to 100 liters of water through a membrane filter with a pore size of 0.1 μm or less, followed by scanning electron microscopy. Time to complete this procedure may total as much as a week or more (Gerard et al. 1988). The long time period results from the low flow rate of liquid through a membrane filter with that pore size and the need to accumulate sufficient particles on the filter to acquire enough data to be statistically significant.

Electrical resistance measurements are used in some pharmaceutical areas. Liquid is passed through a small orifice placed in a nonconducting barrier in a liquid. Electrodes are located at each side of the barrier, and an electrical current is passed through the orifice. The interruption of current caused by a particle passing through the orifice can be related to particle volume. The method has excellent sizing resolution because its response varies with the particle radius cubed. Its major drawback is the requirement that the liquid being measured must be conductive. In addition, the signal indicating a particle is derived from the current passing through the liquid; the conductivity of the long liquid path within the instrument makes it quite sensitive to external electrical fields that result in background noise signals.

Acoustic response systems are used to observe qualitative changes in heavily contaminated liquids. These devices are capable of nonintrusive measurement in flowing streams because an acoustic transducer can be mounted external to the liquid line. Acoustic energy can be transmitted through a suitable window in the flow line. Single particle sensitivity is approximately 5 μm, with response also varying with the speed of sound in the liquid being measured.

Most particle measurements in process or product liquids that may be sensitive to particulate contaminants measure either the scattered light flux or the light obscuration caused by an individual particle passing through a viewing volume. Light scattering instruments are used to measure particles 0.05 μm in diameter and larger; obscuration instruments are used to measure particles larger than approximately 2 μm. These instruments do not detect the image of each particle in the viewing volume; rather, they observe the total amount of light emitted from the viewing volume. The change in light level can result from passage of a single particle, an assembly of particles, a gas bubble, a filament of liquid other than the base liquid, or any change in the illumination level in that observed fluid volume.

The light level change is referenced to that from a "standard calibration particle" and is reported as having been produced by a particle with a diameter equivalent to that of the reference particle and with optical properties similar to that of the reference particle. This is similar to the response basis for an OPC used for airborne particles; however, the variability of refractive indices for the gases commonly measured is quite small compared to that for typical liquids. The response of all OPCs is affected by the difference in refractive index between the particle and the medium in which it is suspended. Therefore, the response of an OPC for liquidborne particles varies significantly for a particle of given size and composition, depending upon the liquid in which it is suspended.

Light scattering particle counters with submicrometer sizing capability use either gas or diode laser illumination. The direct beam from the laser is blocked after passing through the viewing volume, while the scattered light is collected by a suitable optical system that uses either mirrors or lenses for collection and direction to the photodetector. Figure 25-1 shows a commercial optical system with lenses for collection of scattered light, and Figure 25-2 shows a system using mirrors for collection. The basic operating principles are the same for both systems. Both systems shown use a HeNe laser for illumination. The lens system shown uses a smaller collection angle for operation than does the mirror system. Sensitivity of the illustrated mirror collection system is approximately 0.2 μm, whereas sensitivity for the lens system is approximately 0.4 μm. Observed sample flow rates from less than 1 milliliter per minute to several hundred milliliters per minute are used with these systems. Instruments with greater sensitivity are normally restricted to measurement in smaller sensing volumes and at smaller sample flow rates. The entire flow can be passed through the viewing volume, or a portion of the flow can be illuminated to provide an optically defined sample within the flow stream. The latter arrangement can be used for higher-sensitivity particle measurement. Figure 25-3 shows a lens system for light collection with particle size sensitiv-

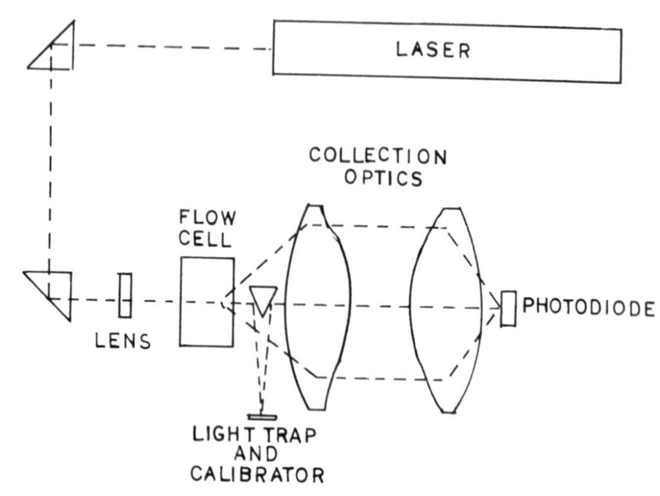

FIGURE 25-1. Submicrometer optical system using lenses for light collection. This 1979 system is capable of counting and sizing particles as small as 0.4 μm in liquids at a sampling rate of some 25 ml per minute.

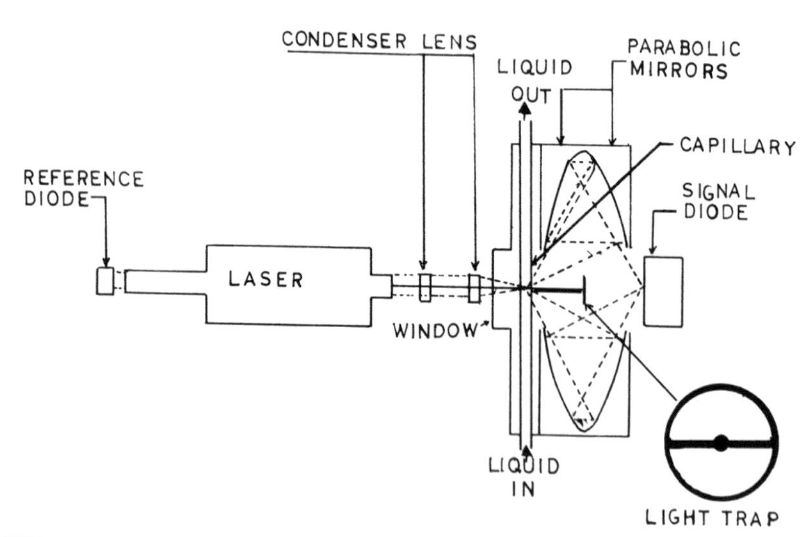

FIGURE 25-2. Submicrometer optical system using mirrors for light collection. This system collects light over a larger angle than is possible with a collecting lens. It can count and size particles as small as 0.2 μm in liquids at a sampling rate of some 20 ml per minute. (Courtesy Particle Measuring Systems, Inc., Boulder, CO.)

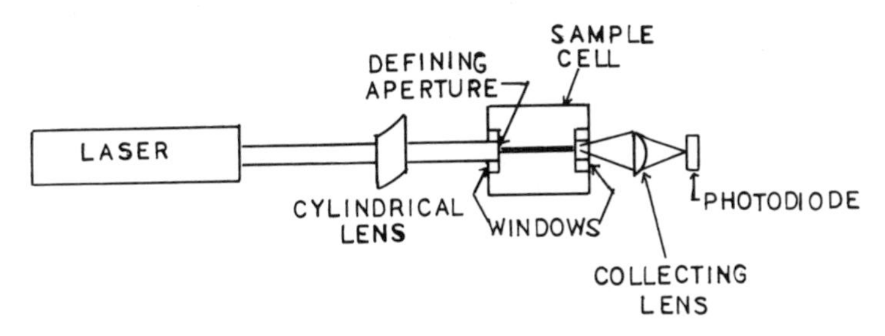

FIGURE 25-3. High-sensitivity optical system for liquidborne particles. This system uses a collecting lens and measures particles over a smaller size and over a lesser dynamic range than does the mirror collection optical system. (Courtesy Particle Measuring Systems, Inc., Boulder, CO.)

ity of 0.05 μm. The high sensitivity of this system is due to a small sensitive volume within the entire stream, a very high level of illumination beam intensity, and use of an optical system in which the windows containing the liquid are not imaged. Any debris that may deposit on the windows is out of the focus of the optical system in this way.

The particle size dynamic range for light scattering sensors capable of measuring particles as small as 0.05 μm is usually 20 : 1 or so; the range for sensors used for particles larger than 0.5 μm is 40 : 1 or so. The difference is due to the greater portion of the range where scattering varies with the sixth power of particle radius for the more sensitive instruments and the resulting requirement for electronic systems with a much larger dynamic range. A response curve for such an instrument is shown in Figure 25-4. Note the change in slope from the sixth to the second power at approximately 0.5-μm diameter. These instruments are used for counting and sizing particles in "clean" liquid streams, where particle concentration is no more than approximately 10,000 per milliliter. The volume/volume particle loading at this population concentration is in the ppb range. For these instruments, the sizing resolution is usually quite good.

Light obscuration or extinction counters use either laser or white light illumination. Either rectangular or circular flow passages, with dimensions ranging from 30 μm to a few millimeters in cross-section and lengths up to a few centimeters, are used. The entire flow stream is usually observed. Sample flow rates range from tens of milliliters per minute to liters per minute, with the smallest flow rates used for sensors observing the smallest particles. The usual limit of measurement for most extinction counters is approximately 2 μm. The extinction system defines the amount of light removed from a direct beam when a particle (or other interruption) is in the sensing volume. The transmitted light flux is decreased by a combination

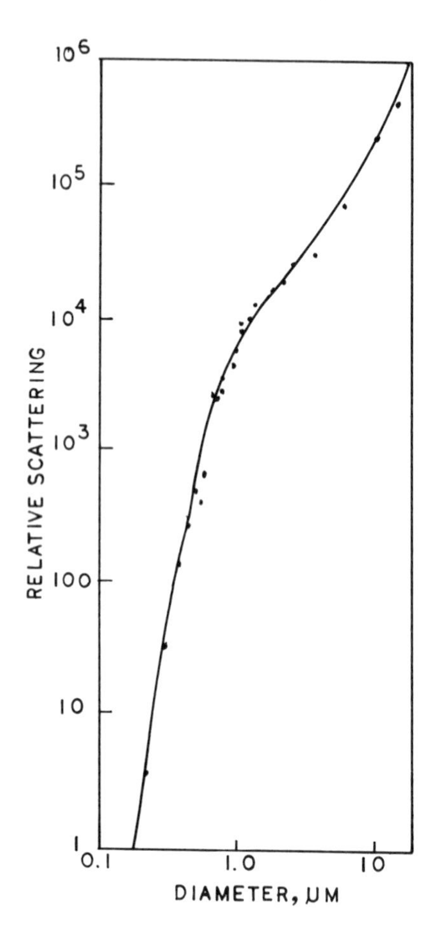

FIGURE 25-4. Response curve for submicrometer liquidborne particle counter. Note that the response curve slope shifts from sixth power to second power for particles larger than approximately 0.5 μm.

of light absorption by a particle with any opacity and light scattering out of the direct beam.

During the early development of light extinction liquidborne particle counters, it had been assumed that the response of the system was due only to removal of light by absorption. For that reason, many descriptions of these instruments still use the term *light blockage*. The system response was assumed to be directly related to the projected area of a particle that "blocked" that fraction of the transmitted light equal to the fractional portion of the sensing volume cross-section. In fact, the fraction of light that is removed by a particle from a direct beam is a function of the extinction coefficient of the particulate material; this function varies with the square of the particle radius. For that reason, the experimental response curves produced by these instruments agreed with the assumed geometric optical blockage mechanism. Figure 25-5 shows how the extinction coefficient varies for small spherical or irregular particles of differing refractive indices. Note that the variation decreases as particle size becomes larger than 1 to 2 μm. For that reason, data variability with composition and size of particles larger than that range is less than that for light scattering systems used for the same particle size range. Therefore, these devices are used if contaminant particles from a variety of sources may also have a range of shapes and compositions.

Figure 25-6 shows the basic operation of a light extinction system. Illumination can be provided by a gas or solid-state laser or by an incandescent lamp. The flow passage cross-section area (width and depth) is defined by a rectangular or circular tube with the illuminated sensing volume height usually defined by the beam height. Whether an incandescent filament or a laser illumination source is used, the optics of the systems are still similar. The main difference is in the means of defining the sensing volume dimensions. In most incandescent lamp systems, the flow passage width and the optically defined height are nearly the same; in the laser systems, the flow passage width and/or depth can be up to a factor of 10 greater than the optically defined height. The capability for a larger cross-section for flow can be very useful when working with viscous liquids or large particles. The gas laser systems have an additional advantage in that the telecentric beam allows more uniformity of illumination along the beam axis.

The terms *volumetric* and *in situ* refer to the sample conditions existent in the operation of the particular OPC. The volumetric OPC observes particles in a sample procured by mechanical or hydraulic means and then passed through the OPC. In a volumetric OPC the total sample flow passing through is observed for particle content. Samples are passed through the sensor by either aspiration or pressure. The samples can be derived from a batch container or from a side stream sample line. The sensor portion of the volumetric OPC can use either extinction or scattering for particle sizing. The entire sample stream in the sensor is illuminated. The maximum sensitivity for a volumetric sensor is typically 0.2 μm. A volumetric OPC with sensitivity for particles smaller than 10 μm or so can handle sample flow rates up to a few hundred milliliters per minute.

In an in situ OPC the optical system is used to observe a portion of a liquid system without disturbance, as through a set of windows. It may observe a sample flow rate up to 100 ml per minute or so within a liquid

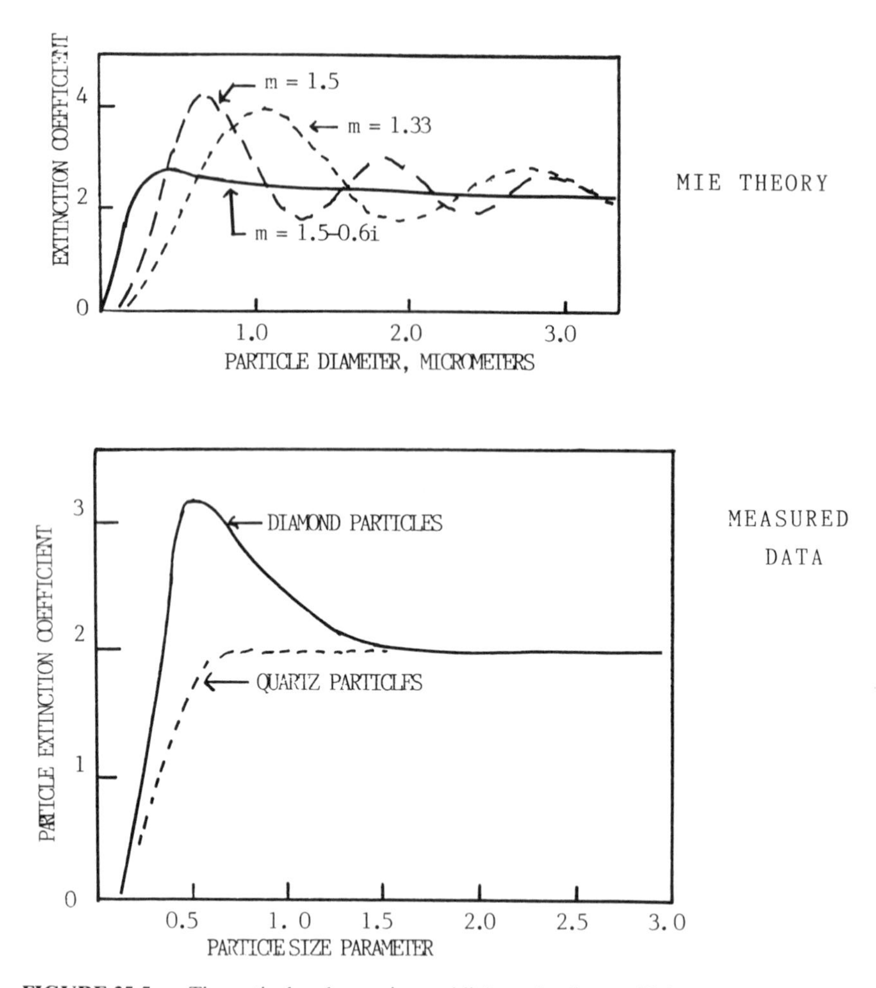

FIGURE 25-5. Theoretical and experimental light extinction coefficients. For particles much smaller than 2 μm, the response curve slope is not constant. For larger particles, the extinction coefficient is essentially 2.0, allowing square law response for extinction counters used for the larger particle size ranges. Experimental data follow theory closely. (From Hodkinson, J. R., 1965. The Optical Measurement of Aerosols. In *Aerosol Science*, ed. C. N. Davies. London: Academic Press.)

stream flowing at a much larger rate. The in situ OPC observes particles within an optically defined sample. Light scattering information is normally used to count and size particles. Larger samples can be observed if sensitivity can be decreased. Sensitivity to 0.05-μm particles (based on latex particle in water response) was available as of 1989. This response is based upon optical definition of a sample of approximately 0.5 ml per

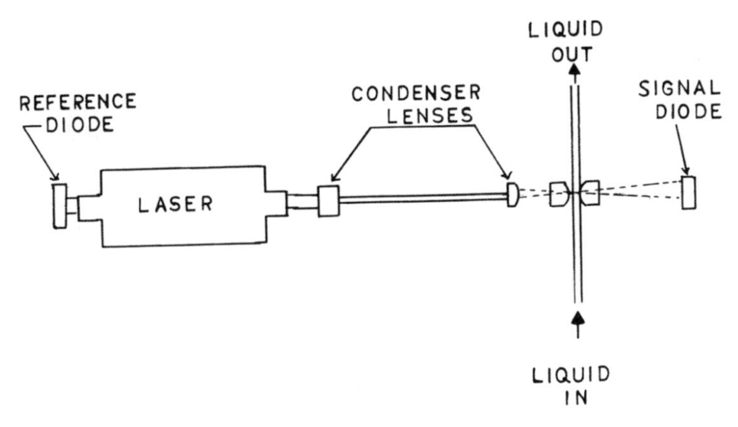

FIGURE 25-6. Basic operation of extinction optical system. Although a gas laser is shown as the illumination source here, many older extinction systems have used incandescent lamps as illumination, and many present systems use solid-state lasers. The laser source aids in providing the desired telecentric illumination pattern.

minute from a larger flow of some 300 ml per minute. This sensitivity is obtained by concentrating a laser beam into a small volume and by removing the windows from the focal plane of the optical system. The optical system for this device is shown in Figure 25-3.

Volumetric OPCs can be used to measure particles in batch samples passed through the sensor or in side stream sample flow through the sensor. When side stream sampling is used, it may be difficult to ensure isokinetic sampling with an orifice sized to withdraw no more than 100 ml per minute from a line in which liquid may be flowing at up to several hundred liters per minute. In addition, deposition of particles in sample lines occurs continuously with the possibility of particle emission if the lines are disturbed. In such systems, in situ OPCs would be preferable for better sampling. Knowledge of total liquid flow rate in the main line as well as the observed sample rate in situ is necessary to define particle concentration in the main line accurately.

As of 1990, filtration systems for deionized water used for semiconductor manufacturing require particle counting methods with sensitivity to particles smaller than 0.1 μm in diameter (Yang and Tolliver 1989). Production of semiconductors with features based on 1-μm geometry requires water with no more than 5 to 10 particles per milliliter in the 0.1-μm size range (Yabe et al. 1988). Measurement of particles under these conditions has been carried out in the past either with membrane filter collection followed by scanning electron microscope observation or with light scattering OPC measurement. Because of sample measurement time constraints, the only practical particle counting procedures involve light scattering OPCs with a sensitivity of at least 0.1 μm.

Particle counters are widely used for ensuring cleanliness of pharmaceutical liquids. Freedom from excessive particle contamination for both large-volume (> 100 ml) parenteral liquid containers and small-volume injections (< 100 ml) must be verified by particle measurement. The compendial method for the large-volume containers requires light microscope verification of particle content based on observation of membrane-filtered samples. The method for the small-volume injections requires the use of light obscuration OPCs with careful sample-handling methodology (Lieberman 1986). Light scattering OPCs are not used in this application. The pharmacopeial requirements for particle control in the United States at this time are based only on definition of particle content at 10 μm and 25 μm, even though particle content at the 2-μm size is controlled in other countries. OPC performance is verified in accordance with ASTM calibration methods (ASTM 1987).

Problems in OPC response to immiscible silicone oil lubricant droplets, as sometimes seen in parenteral liquids, can result in an indication of harmful particles in samples as a result of OPC response to the silicone oil droplet. There is a difference of opinion in the pharmaceutical industry as to whether silicone oil droplets are a hazard. This material has long been used as a lubricant for syringes, and no specific hazard has been associated with this use. However, questions have been raised as to the physiological effects of exposure to this material. The question will not be resolved or even discussed here.

Cleanliness of a variety of nonpolar liquids used in other industries must also be assured. These fluids are used widely in hydraulic fluid power applications, lubrication systems, and heat transfer devices. Particle count data are used (1) to verify theoretical model predictions of component wear and control device operation caused by presence of abrasive hard particles and (2) to rate the performance of filters used to maintain the cleanliness of these fluids. The smallest particle size measurement requirements are seldom below 2 μm, and the contaminant particle properties vary widely. Therefore, light obscuration OPCs are used widely with these liquids (Verdegan, Stinson, and Thibodeau 1988). These fluids are also frequently contaminated with large quantities of water. Problems in differentiating particles from emulsion water droplets of equivalent size suspended in the oil have always resulted in anomalous response from an OPC. A procedure was developed (Verdegan, Thibodeau, and Stinson 1990) wherein the water is incorporated into surfactant micelles to form a microemulsion of ca. 0.01-μm droplets. These droplets are too small to be detected as particles by the OPC. The surfactant of choice is ionic.

As with any particle-counting system, there are two error types of concern. Either the OPC is sizing particles incorrectly or it is counting them incorrectly. The two error types can be interrelated. A general discussion on OPC errors that is applicable to all OPCs is given in a later chapter. Some specific error problems peculiar to OPCs for liquids are pointed out here.

OPCs for particles in liquid are calibrated with latex spheres of refractive index 1.6 suspended in water with refractive index 1.33 or in oil of refractive index 1.47. Any particle passing through the OPC sensing volume is defined as having the same size as the latex sphere that would produce the same change in light level as was produced by the particle. Equal-sized particles in liquid of a refractive index different from that of the calibration liquid produce data indicating a different size for each liquid.

The OPC operational requirements for liquidborne OPCs are essentially the same as those for OPCs that are used for detection of airborne or surface-deposited particles. Because the basic requirements are discussed in some detail elsewhere, no further comments are made here. However, assurance of satisfactory liquidborne OPC performance includes some unique requirements that are discussed further. These requirements are associated primarily with sample-handling methodology and the effects of the chemical and physical property differences between gases and liquids.

Samples stored within containers for long periods of time may have heavy particles settling to the bottom of the container through the liquid boundary layer and becoming permanently attached to the container surface. These particles will never be analyzed. Some containers are extremely difficult to clean to the point where small particles are not continuously released into samples that are present in the containers. High liquid viscosities may result in extremely slow sample flow through small OPC orifices; as rise times are extended, pulse measurements are distorted. In addition, the gravitational settling rate for dense particles may be higher than the (usually) vertically upward flow rate through a sample transport tube to the OPC, resulting in none of the large particles entering the OPC sensitive volume. Liquids require greater energy than gases to pass them through the small cross-sectional area flow lines. Significant pressure drops are required, sometimes resulting in the pressure within the sensing volume being reduced to a point below the vapor pressure of a particular liquid. Bubbles are then released within the sensing volume. The bubbles are reported as particles. An experienced operator can usually distinguish between the particle size data produced by bubble generation and that from true particles, but the reliability of this observation is poor.

The wide range of application areas for liquidborne particle measure-

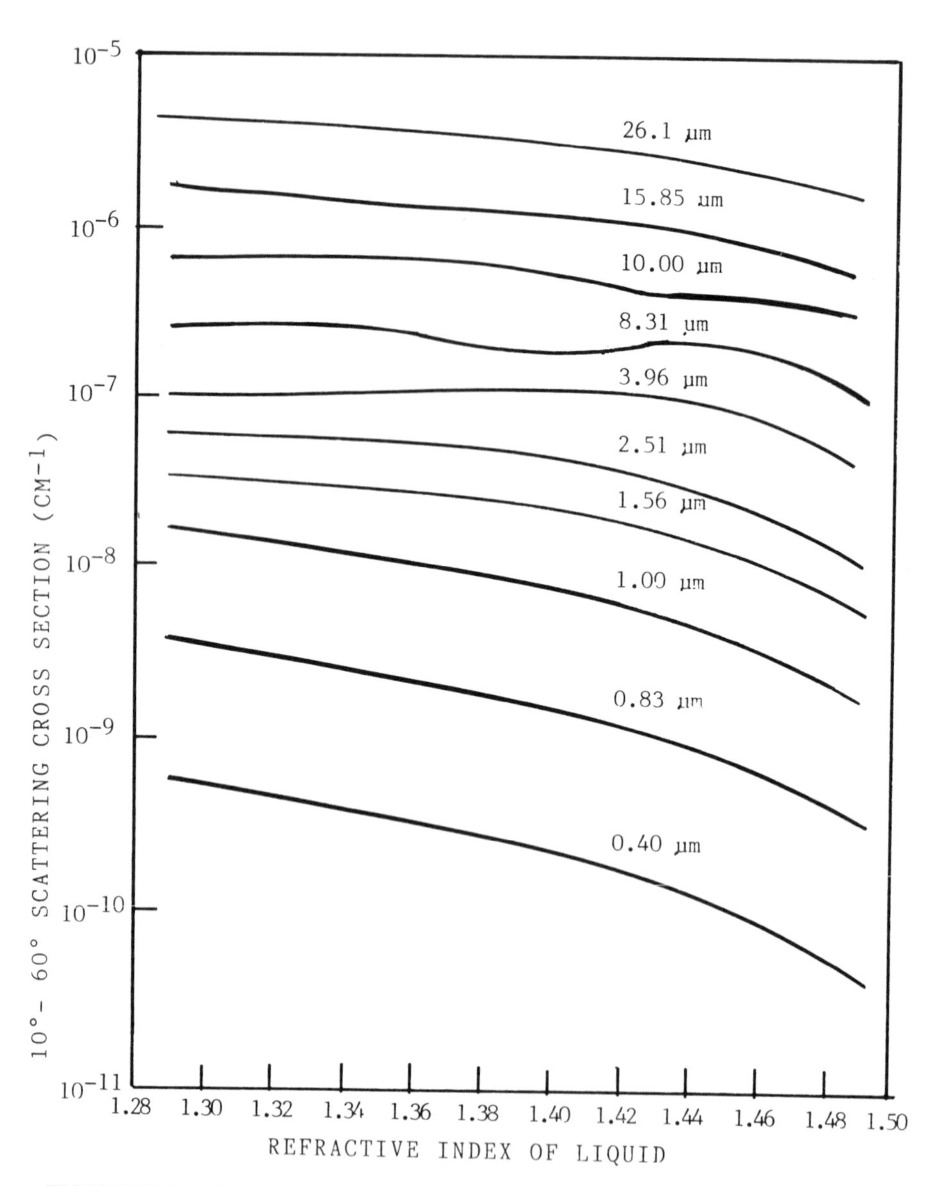

FIGURE 25-7. Computed scattering cross-sections for latex spheres in liquids with varying refractive index. As the differential refractive index decreases, scattered light flux from the particle decreases. Eventually, the scattered light level decreases so that small particle detection becomes very difficult. (From Knollenberg, R. G., 1989. The Importance of Media Refractive Index in Evaluating Liquid and Surface Microcontamination Measurements. *Journal of Environmental Science* 30(2):50–58.)

ment includes an equally wide range of liquid materials. Physical properties of the liquids vary significantly more than the properties of ambient gases in which particles may be present. A significant variation results from different refractive index values for different liquids. Response of OPCs depends strongly upon the ratio of the refractive index of the particle to that of the suspension fluid. As the ratio approaches unity, the particle-sizing sensitivity of the OPC degrades significantly, as shown in Figure 25-7 (Knollenberg 1989). Consideration of this problem has been recognized in at least one OPC calibration and measurement document, which specifies correction factors when measuring particles in different liquids (SEMI 1988).

A variety of problem areas have been identified and discussed. Some investigators have summarized problem areas and made some suggestions for remedial measures (Lieberman 1984b; Przybytek and Calabrese 1985). These suggestions are directed primarily at control of sample-handling problems and minimization of simplistic interpretation of OPC data presentations.

References

American Society for Testing and Materials, 1987. *Defining Size Calibration, Resolution and Counting Accuracy of Liquid-Borne Particle Counter Using Near-Monodisperse Spherical Particulate Material,* ASTM F658-87, Philadelphia: American Society for Testing and Materials.

Gerard, R., et al., 1988. Evaluation of the State-of-the-Art Particle Counting Methods for Ultrapure Water. Proceedings of the 9th International Committee of Contamination Control Societies Conference, September 1988, Los Angeles.

Knollenberg, R. G., 1989. The Importance of Media Refractive Index in Evaluating Liquid and Surface Microcontamination Measurements. *Journal of Environmental Science* 30(2):50–58.

Lieberman, A., 1984a. Fine Particle Characterization Methods in Liquid Suspensions. In *Particle Characterization in Technology,* ed. J. K. Beddow, pp. 187–232, New York: CRC Press.

Lieberman, A., 1984b. Problems Associated with Submicrometre Contaminant Measurement. In *Semiconductor Processing,* ASTM STP 850, ed. D. C. Gupta, pp. 172–183, Philadelphia: American Society for Testing and Materials.

Lieberman, A., 1986. Particle Counting in Pharmaceutical Liquids. Proceedings of the 32d Institute of Environmental Science Annual Technical Meeting, pp. 427–431, May 1986, Dallas, TX.

Przybytek, J. T., & Calabrese, K. L., 1985. Measuring "Low Level" Particle Counts in Solvents. *Microcontamination* 3(6):51–54.

Semiconductor Equipment and Materials International, 1988. *Specifications for Reagents; 3.9; Calibration and Measurement Method for Particles in Liquids,*

SEMI C1-88, Mountain View, CA: Semiconductor Equipment and Materials International.

Verdegan, B. M., Thibodeau, L., & Stinson, J. A., 1990. Using Monodispersed Latex Spheres Dispersed in Nonpolar Liquids to Calibrate Particle Counters. *Microcontamination* 8(2):35–39, 64–65.

Verdegan, B. M., Stinson, J. A., & Thibodeau, L., 1988. Accurate Methods for Particle Counting. Proceedings of the 43d National Conference on Fluid Power, October 11–13, 1988, Chicago.

Yabe, K., et al., 1988. High Sensitivity On-Line Monitoring of Ultrapure Water for ULSI Manufacturing. Proceedings of the 9th International Committee of Contamination Control Societies Conference, pp. 509–515, September 26, 1988, Los Angeles.

Yang, M., & Tolliver, D. L., 1989. Ultrapure Water Particle Monitoring for Advanced Semiconductor Manufacturing. *Journal of Environmental Science* 32(4):35–42.

26

Optical Particle Counter Operating Procedures, Calibration and Correlation Methods

Data from any OPC are derived on the basis of a number of OPC design and operating parameters. Most operators assume that the data produced are valid for any material and under any circumstances. However, it must be realized that the OPC is not a primary measurement instrument. Even though it reports only the size for each particle and the number of particles observed, it responds to several parameters such as particle size, the ratio of particle to fluid refractive indexes, and particle shape. OPC response is affected by the fluid flow field from which the sample is produced. Response also varies with OPC optical and electronic design. The particle size data produced depend on several system parameters. These effects occur even when the OPC is operating at its best and has been recently calibrated. Therefore, the OPC is calibrated to a known standard material using an accepted and standard method. Further, when more than one OPC is used in a large cleanroom area, it may be necessary to control operation of the several instruments to assure that each one provides similar data in similar environments. Minor differences in response to particle properties can result in a large change in indicated particle concentration. It is advisable to carry out effective correlation procedures whenever more than one OPC is to be used in a single installation in order to minimize this problem. These procedures may include selection of OPC designs so that similar systems are used for this purpose, as well as verifying calibration of each OPC.

Although the remainder of the discussion on calibration is concerned mainly with airborne OPCs, the content is also applicable to calibration problems for OPCs that are used for counting and sizing particles in liquids or on solid surfaces. Both the methods and the materials required for size calibration must be considered. The data produced for particle numbers are also affected by the operation of the OPC and the nature of the environment where the OPC is located. Pertinent environment properties include the fluid flow rates, pressure, and composition, as well as the particle characteristics of size distribution, concentration, and composition. Sample-acquisition procedures must be correct for both the environment and the OPC used in that environment.

The particle size reported by an OPC is defined as an equivalent optical diameter; that is, the particle actually being measured has produced a signal equivalent to that from a calibration latex sphere of known size suspended in a specific fluid; the measured particle is reported as having that size. Latex spheres have a refractive index of 1.6 − 0i and are totally transparent to visible light. The first term is referred to as the "real" part of the refractive index and is related to the change in path of a beam of light when it is passing through the particle. The second term, referred to as the "imaginary coefficient," is related to the absorption of light passing through the particle.

When the OPC is used in the real world, however, the particles are seldom spherical and they do not have the refractive index of the calibration base material. Most atmospheric particles have a real refractive index close to 1.5 and absorb some light. The imaginary coefficient is usually in the range of 0.01 to 0.1. The particle shapes vary from spherical through a variety of crystal structures, or they may be fibrous. In some cases, the particle being detected may actually be an agglomerate of several smaller particles collected into a convoluted clump or chain. Close examination of such particles shows that their structure can often be described well by using fractal dimensions. This is particularly true for particles whose fine structure complexity is shown to increase with magnification. The use of fractals is not discussed here because it is not applicable to the subject of this discussion. However, an excellent introduction to fractals has been presented by Kaye (1989) and the reader is referred to that informative and interesting book for better understanding of the subject.

The measured particle is still reported by the OPC as a sphere of stated diameter. If different OPCs observe the same air sample, small differences in optical design or sensor resolution result in different data from each OPC, even though both were calibrated with the same batches of standard latex calibration material. This is due to the difference in optical properties of the calibration material and the particles in the air sample.

Essentially the same materials can be used to calibrate all OPCs. Primary requirements for these materials are that they are close to monodisperse, they are well characterized in their particle size distribution, they are capable of being easily dispersed in the fluid for which the OPC will be used, and they are stable over a reasonable time period during storage and in the calibration fluid. In addition, it is desirable that batches of calibration material are similar. The calibration material supplier should verify materials by defined test methods or by traceability to primary standard reference materials.

Most primary OPC calibration materials are monodisperse latex spheres supplied in the form of suspensions in liquid or as dry powders. This material is widely used for OPC calibration because it is in the form of isotropic, spherical monodisperse materials, it is available in a wide range of sizes, it is easily dispersible in water, it is stable over long storage periods with a few simple precautions, and a long history of use exists for this material. Suitable latexes can be procured from several suppliers in the United States and abroad. This material is usually very well characterized in terms of composition, median particle size, and standard deviation. A number of monodisperse latex materials are available from the U.S. National Institute of Science and Technology (NIST; previously the U.S. Bureau of Standards) in normal sizes of 0.1 μm, 0.3 μm, 1 μm, 3 μm, and 10 μm. An overview of NIST particle size standard reference materials and the means of characterizing these materials was recently presented by Hartman (1987). The monosized particles are typically supplied with relative standard deviations shown to be no more than 1%. The particles have been sized by several methods that agree quite well; the actual particle sizes are quite close to the nominal sizes stated for these materials.

A detailed description of the certification procedures used for 0.3-μm particles is available from NIST (Lettieri and Hembree 1988). Reputable calibration particle suppliers can define traceability of their products and of their measurement methods to NIST. The calibration particle sizes are defined on the basis of measurement procedures that are widely accepted. Near-monodisperse spheres are available from several suppliers in particle sizes ranging from less than 0.1 μm to greater than 1 mm. The relative standard deviation is seldom more than 10%, even for the largest particle sizes, with some materials having relative standard deviations well under 1%. The quality of many of these products is at least as good as that of the standard reference materials distributed by NIST.

Even though the relative standard deviation of the particles in most polystyrene latex (PSL) calibration materials is quite small, there is still some variability from batch to batch. Figure 26-1 shows a pulse height histogram for a batch of 0.102-μm particles, and Figure 26-2 shows a histogram for a batch of 0.109-μm particles. Both sets of data were obtained with the same high-resolution OPC. The response curve for this instrument is shown in Figure 26-3. Response is seen to be linear and monotonic for particles smaller than approximately 0.18 μm. Polytonic

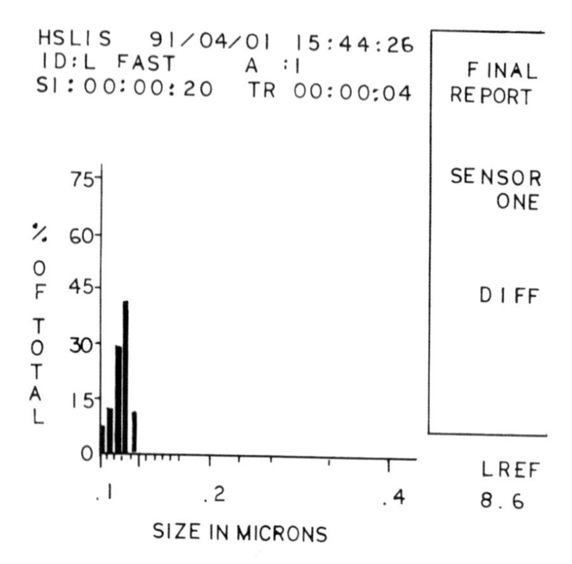

FIGURE 26-1. Pulse height histogram for 0.102-μm PSL particles. The histogram shows many smaller particles that may be due to either the presence of debris in the liquid or to the breadth of the PSL particle size distribution.

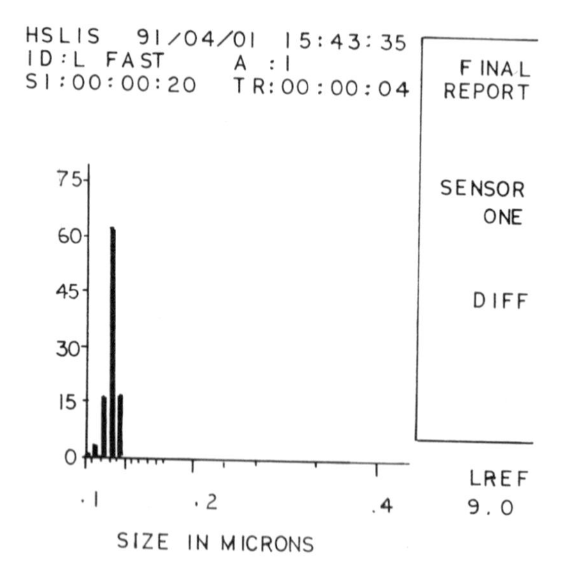

FIGURE 26-2. Pulse height histogram for 0.109-μm PSL particles. The histogram shows only a few oversize and undersize pulses; these may be ascribed to the resolution limits of the OPC. This batch of PSL appears to have a very small relative standard deviation and to be suspended in clean liquid.

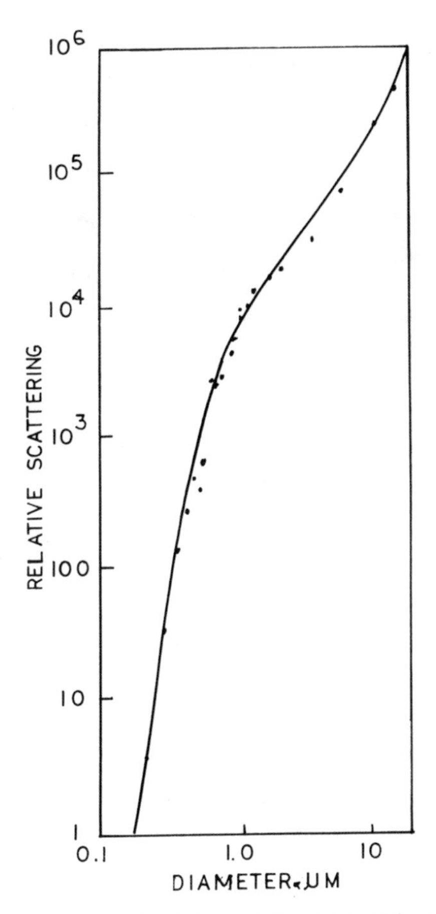

FIGURE 26-3. Response curve for high-resolution OPC. This curve was also shown as Figure 25-4.

response from a light scattering OPC does not appear serious for particles smaller than approximately 0.3 μm. Even so, it is obvious that the 0.109-μm particles are much more uniform than are the 0.102-μm particles.

Figure 26-4 shows a deposit of monosized latex spheres (PSL) used for liquidborne OPC calibration. These are 10-μm spheres with a relative standard deviation of approximately 1%. Figure 26-5 is a photomicrograph of Air Cleaner Fine Test Dust (ACFTD). Dispersions of this material are mainly used for testing particle penetration of coarse filters as used for either air and liquid. ACFTD has also been used for calibration of liquidborne OPCs, particularly for those OPCs used for measurements of parti-

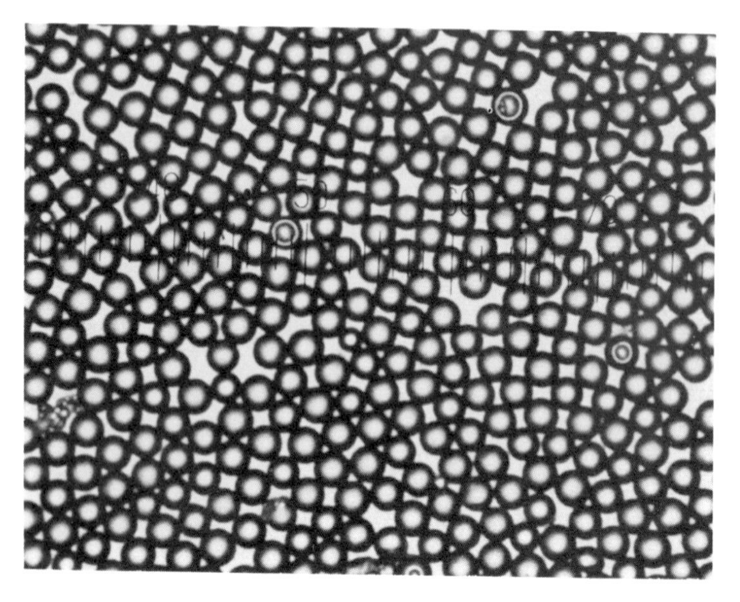

FIGURE 26-4. Near-monodisperse polystyrene latex calibration spheres. The median diameter is 9.0 μm, and the coefficient of variance is 6.7%. (Courtesy Duke Scientific Corp.)

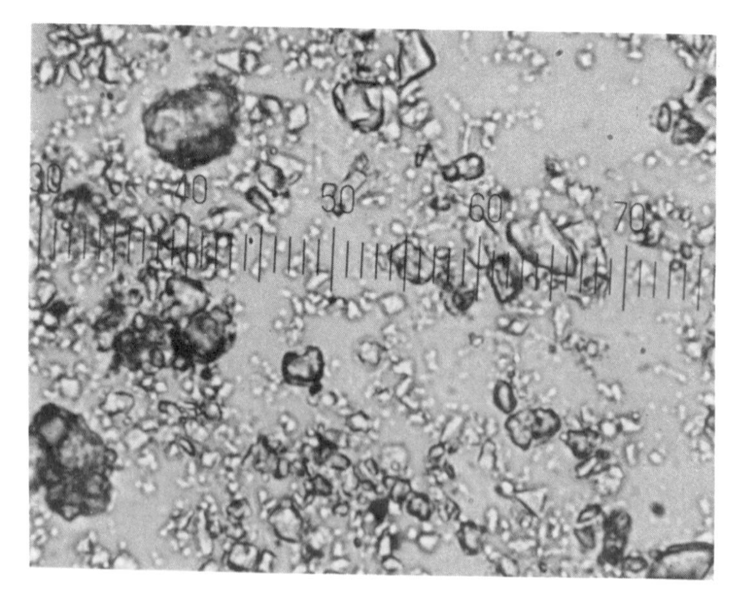

FIGURE 26-5. Air Cleaner Fine Test Dust. The scale is 2 μm per division. This material is not uniform in terms of shape or composition. (Courtesy Duke Scientific Corp.)

cles in lubricating oils and in the hydraulic fluid power industry. PSL calibration material is normally supplied as a 1 to 10% dispersion in water containing a small quantity of surfactant and bactericide. Mean sphere diameter and standard deviation values are usually stated for each batch. ACFTD is supplied as a dry powder with upper and/or lower size limits (with some tolerances) stated. These values are on a mass fraction base and are normally based on data produced by an elutriation process.

The differences between these materials is obvious in terms of shape. The latex particles are spherical and uniform, whereas the ACFTD is irregular in shape, as well as optically varied. There is also a great deal of difference in terms of chemical properties. ACFTD is composed of comminuted granite powder, whereas PSL is a relatively soft latex polymer. The composition and shape factor of ACFTD is seen to vary with particle size, and that of the PSL is constant. The PSL can be dispersed in liquids relatively easily, while the ACFTD requires high shearing force for dispersion. Air suspensions of PSL can be produced by atomization of dilute water suspensions; when the water evaporates, individual PSL particles remain. In addition to the PSL particles, some very small surfactant residue particles are also usually present. Air suspensions containing individual particles of ACFTD cannot be prepared easily from the dry powder because of electrostatic agglomeration; spraying dilute suspensions of ACFTD is not practical because the polydisperse nature and very broad particle size distribution of the ACFTD makes it impossible to prepare a droplet spray of reasonable concentration with single particles in each droplet over a working particle size range. Even so, an aerosol cloud from dry ACFTD can be dispersed in air by high-pressure jets. The cloud of powder is used to test for particle removal on a basis for relatively low-efficiency air filters. Most agglomerates larger than approximately 5 μm in diameter are removed by the dispersion process, and agglomerated particles smaller than this size are not a concern for the filters being tested in this application.

For calibration of surface scanning OPCs, monodisperse latex particles are deposited on blank wafers. Deposition is accomplished by exposing the wafer to a very dilute aerosol of latex particles prepared from a water suspension spray. Very clean wafers are used because a particle concentration of no more than a few hundred particles per wafer may be desired. For convenience, a calibration wafer may be prepared with several segments upon which defined quantities of particles of different sizes have been deposited.

There are several standard methods in use for OPC calibration. ASTM F328 (ASTM 1980a) and F649 (ASTM 1980b) are used for calibration of airborne OPCs. ASTM F658 (ASTM 1987) and F1226 (ASTM 1989) are used to calibrate liquidborne OPCs. F328 describes one procedure for determining response of an airborne OPC to latex aerosols of specific particle size and one to compare particle count data from the OPC to that from a referee method. Membrane filter collection, followed by microscopic observation, is shown as one potential referee method. Others can be used, including comparison to other previously calibrated OPCs. F649 describes a procedure to adjust the indicated particle count response of one OPC to that of a "master" OPC for particle concentration data when both are observing atmospheric aerosols; the test OPC gain controls are adjusted until the two instruments produce nearly identical concentration data. In this way, effects of differences in optical properties of the calibration material and the particles in the atmosphere being tested are minimized. F658 describes procedures to determine response of a liquidborne OPC to latex particle suspensions in sizes of 2 μm and larger. Response is defined in terms of reported particle size and sizing resolution, as well as for particle concentration. F1226 is used to adjust the response of a liquidborne OPC for submicrometer particles, again using latex spheres as a calibration base.

As mentioned previously, ACFTD is also used for calibrating liquidborne particle counters in specific industries. The standard methods presently used are ANSI B93.28 (ANSI 1972), primarily for hydraulic fluids, and ARP 1192B (SAE 1972), mainly for lubricating oils. These methods are based on response of the OPC to an ACFTD suspension of specific mass concentration. The OPC particle concentration at specific size ranges is adjusted to match the particle concentrations specified in the standard methods. Both methods were in process of revision in 1992 because the reproducibility and stability of ACFTD suspensions are difficult to maintain and there is no NIST traceability to the tabulated ACFTD particle size distribution. The revisions are based on the procedures of ASTM F658, with addition of some procedures to ensure stability of the latex particles in oil suspensions (Verdegan, Thibodeau, and Stinson 1990). It was shown that the revised method may produce improved correlation in the response of several OPCs to aliquots of contaminated hydraulic oil over that found when calibrated to an ACFTD suspension.

At this time, there is no documented standard method available for calibration of OPCs used for measurement of deposited particles. F328 states a procedure for verifying the counting accuracy of an airborne OPC for polystyrene latex spheres. However, there is no documented procedure for producing a reference calibration particle suspension of known concentration that can be used as a standard test material. Therefore, counting accuracy is defined by comparing the OPC particle concentration data with data obtained from a reference method observing samples from the same aerosol suspension. F649 is a procedure to aid in correlation of particle concentration data obtained when sampling from an atmosphere containing a normal polydisperse suspension. Passing an aerosol through an OPC and through a system that is known to count 100% of all particles at least as small as the OPC being investigated can count has been used to determine a counting accuracy curve for that OPC (Wen and Kasper 1986; Caldow and Blesener 1989). Figure 26-6 illustrates the shape of the counting accuracy curve for a typical OPC.

F658 describes methods for size calibration and determination of counting accuracy and sensor resolution for a liquidborne particle counter using monodisperse latex spheres suspended in water. The procedures for size calibration of the OPC and for defining sensor resolution are straightforward and relatively easy and fast. The procedure for counting accuracy requires comparison of data produced by the OPC and that from a referee method. The labor involved in carrying out the counting accuracy determination is arduous, and the accuracy is questionable. For those reasons, liquidborne OPC counting accuracy verification by the F658 method is seldom used. In 1989, a supply of suspensions of small monosized latex particles in sizes close to 1-μm diameter and with concentration specifica-

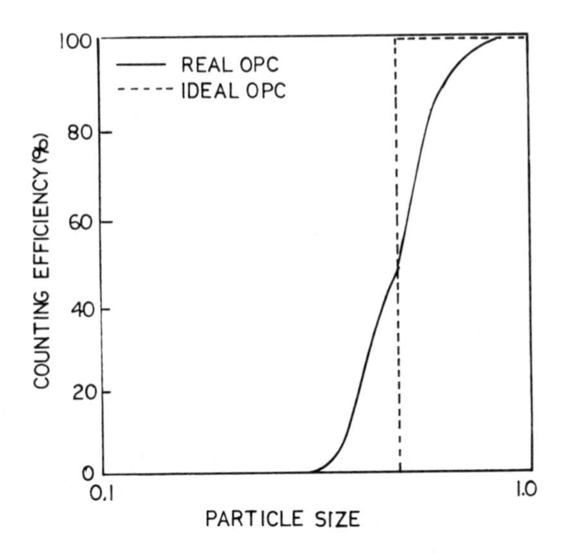

FIGURE 26-6. OPC counting accuracy versus particle size. Ideally, the counting accuracy should be zero for all particles smaller than the set threshold size and 100% for all larger particles. Resolution limits of a real OPC produce a counting efficiency curve with some slope. The OPC manufacturers set system operation so that the efficiency curve passes through the 50% point at the design particle threshold size. (From Caldow, R., & Blesener, J., 1989. A Procedure to Verify the Lower Counting Limit of Optical Particle Counters. *Journal of Parenteral Science and Technology* 43(4):174–179.)

tions as well as size data became available from Japan. Experimental testing has shown that the indicated values are accurate to within 10% or better, but the cost of the material is very high, and documented measurement procedures are not available.

In calibrating and correlating OPCs, the accuracy and the degree to which two OPCs can be matched are limited by a number of factors, including the nature of the OPC design and performance and the nature of the material that is being sampled. Other factors are particle sizing accuracy, sensor resolution, counting accuracy, sample flow rate control, counter concentration limits, and counter electronics dead time. The effects of these factors on response of the OPC are discussed elsewhere in some detail; the discussion here is limited to methods of defining their values.

Particle sizing accuracy is defined in terms of the range over which a particle of known size is reported by the OPC; for example, if a particle of 1 μm is passed through the OPC and electronic or optical component drift produces a signal that should arise from a particle of 0.98 or 1.02 μm, then a sizing error of 2% can be reported for that OPC. This parameter can be defined by noting the particle size range over which 95% (2σ) of the pulses from a monosized particle suspension are measured. Poor sizing accuracy causes variations in reported particle concentration data, especially at the smallest size range, for OPCs that report particle data in a number of size classes. Even if the total reported particle concentration is reasonably accurate, the reported concentration in any individual size range may be incorrect.

Resolution is defined as the ability to differentiate between particles of nearly the same size. It is quantified by measuring the increase in the breadth of the particle size distribution reported by the OPC for a monosized particle suspension. Widths of the actual and the reported size distribution are measured in terms of the particle sizes encompassing the median size and the size range plus and minus one standard deviation from that median. As the OPC sensor resolution varies, sensors with poor resolution tend to count more particles at the smallest measurable size ranges than does a sensor with good resolution. Because there are many more small particles than large ones, the former counter tends to oversize more particles than the latter counter. So, when counting at the smallest size range, more particles are counted by the poorer-resolution OPC. Because there are no particle concentration standards, care is needed in calibrating or correlating OPCs so that the maximum concentration capability of the sensor is not exceeded. If this happens, then the number of particles in the sensing zone becomes greater than one, and the OPC reports multiple particles as single ones.

Particle counting accuracy can be defined in terms of a percentage counting accuracy. This is 1 minus the ratio of the indicated particle concentration to the true concentration. At this time no documented means are available for generating standard aerosol suspensions with known concentration levels; thus it is necessary to pass one aliquot of a test particle suspension through the OPC to be calibrated for counting accuracy and then measure another aliquot from the same suspension by an accepted reference method. Surface-deposit OPCs can observe standard test wafers with either deposited particles or defined etch patterns that are protected from accumulation of atmospheric debris. For OPCs used for measurement of liquidborne particles, the procedures are fairly simple. One portion of the suspension is passed through the OPC, and another is passed through a membrane filter via a clean filter funnel. The concentration of particles in the liquid that has passed through the filter is determined by optical or electron microscope measurement. Particle concentration data are compared from the two methods. For OPCs used for airborne particles, the procedures are similar to those for liquid systems, but the problems of sampling inlet efficiency and of particle losses during sampling must be solved before the measurements are meaningful. Except for particles at the minimum size threshold, where the size threshold is set at the median signal level for those particles so that 50% are counted, OPC counting accuracy is close to 100% unless there is optical misalignment that prevents illumination of the entire sampled stream. This problem can occur more easily in airborne particle counters than in those for liquidborne particles. The orientation of the physically uncontained sample stream in the former OPC is more difficult to maintain or to control than that of the latter OPC.

Sample flow rate control is necessary, particularly for some older OPCs. As flow rate varies, then the data pulse duration and rise time also vary. Many older OPC electronic systems use voltage comparators to define pulse amplitude. For accurate characterization of pulse amplitude, a zero reference level must be established for a finite time interval, and the pulse amplitude and rise time must not vary from the values used for calibration. This design assures that any pulse that is sized by the electronic system will be sized after the signal level has dropped to the required zero reference level. Older OPCs can have dead times that range up to 90 μsec. Modern digital pulse processing systems decrease dead times to no more than 1 to 2 μsec. The short dead time allows the OPC to count and size particles in more concentrated suspensions at larger flow rates. Dead time values are defined by noting electronic component values. At high concentrations, the older OPCs may undercount because the electronic system recovery time may be longer than the interval between some of the particles if the flow varies through the OPC. Flow rate control is usually maintained by including a suitable volumetric or mass flow meter in the OPC. Even a relatively inexpensive 5% flow meter is satisfactory in terms of adequate control of pulse duration and rise time. However, reported concentration errors of 5% may not be acceptable; in that case a more accurate flow meter is required.

An OPC can be considered as a device observing discrete events that are random in terms of both temporal and spatial separation. The OPC observation zone is small enough so that the probability of more than one particle being in that zone at any time is satisfactorily low. Most OPC manufacturers specify a maximum recommended concentration at a level where physical coincidence of particles occurs less than 10% of the time; that is, 10% of the particle count data represents two or more particles being indicated as a single particle. The concentration value where this situation occurs can be determined by sequential dilution. As a suspension of high concentration is sequentially diluted, the indicated particle concentration changes. Dilution of an extremely high concentration can produce a reported increase in particle count if the concentration is decreased to a point where coincidence error is smaller. In this case, the initial saturation of the viewing volume is relieved to the point where individual particles can be counted. When two sequential indicated concentrations vary in the same ratio and in the same direction as the dilution ratio, then the coincidence error is at an acceptable level.

Once the OPC has been calibrated and assurance of correlation with other OPCs has been achieved, then operation can begin. There are certain requirements for trouble-free and reliable cleanroom operation over and above the directions in the operating manual. These include such simple precautions as making sure that the correct particle size ranges are selected (if that option is available), that the last calibration is still valid, that the OPC flow control is correctly set, that the printer paper supply (when used) is sufficient, and that the OPC case is clean enough for the specific cleanroom. The technician should be familiar enough with operation of the OPC to know when questionable data are being reported.

Present cleanroom systems use OPCs in several areas. The most common uses are for measurement of ambient air cleanliness or emissions from an installed filter. The importance of clean process fluids in all cleanroom operations has resulted in development of OPCs that can be used for characterizing cleanliness of compressed gases or of process liquids. The other important usage area for OPCs is observation of deposited particles. Although cleanroom operations have long been monitored for local large-particle deposition by collection on witness plates and subsequent visual particle measurement, the importance of verifying product freedom from even small-particle deposition has led to use of automated surface particle OPCs for observation of witness plate deposition. The same type of surface-analyzing OPC that has been used in the semiconductor industry to monitor submicrometer particle loading of wafers is being used for some witness plate measurements. Each of the OPC application areas has some operational requirements that differ from the others and they all still involve a number of common OPC operational needs.

Many OPCs are used at ambient pressure within an open area. Ambient air pressure OPCs are used mainly for determination of the particle content of cleanroom air or verification of installed filter performance. Discussion at this time is limited to operation within the cleanroom in general. Filter testing, whether for penetration determination or for in-place filter integrity definition, is discussed later.

The OPC may be carted or hand-carried to specific sample point locations within the cleanroom, or it may be located external to the specific sample joint location with sample air from one or more locations transported to the OPC by sample lines. In any case, the first requirements for good operation are verification of the OPC calibration and assurance that the OPC particle size ranges are set correctly for the conditions in the area to be measured. Next the operator should make sure that representative samples are procured from the area of concern. The operator must select the sample collection points for system monitoring or lay out the sampling point grid in accordance with the cleanroom verification requirements of Federal Standard 209D (FS209D). The operator should then select the particle size ranges to be measured to satisfy the cleanroom class requirements or to characterize the area that is being monitored. The OPC operating control format (sampling time, number of replicates per sample, sample frequency, particle size selection, alarms, etc.) should be selected; either the OPC is manually controlled by the operator, or the OPC operation and data output operations are handled by a computer. Figure 26-7 illustrates one OPC operating control format display. For some OPCs, more than one page of display may be necessary to lay out all the necessary parameter selections. The OPC is operated at the sample point, or samples are transported to the OPC by suitable tubing. In either case, sample inlet collection efficiency and sample transport line losses are important. The basics of particle suspension sampling are discussed elsewhere and are not repeated here. However, it is still important to remember that samples acquired in a unidirectionally flowing airstream should be acquired isokinetically if at all possible. Even if the particles examined are in the 0.1 to 1 μm size range where anisokinetic sampling errors are insignificant, it is

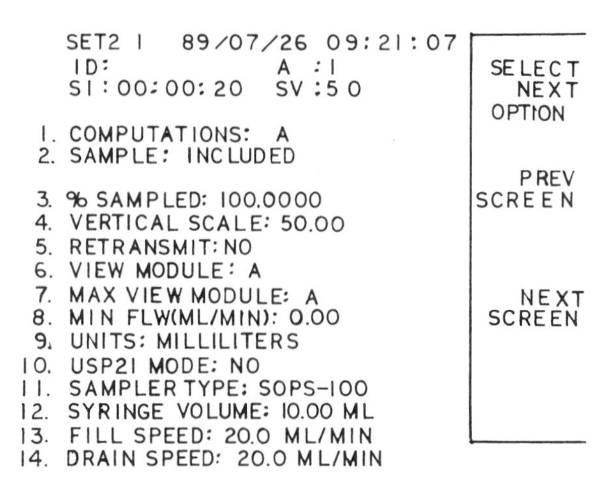

FIGURE 26-7. OPC operating control format display. An example of one manufacturer's selection of operating control choices for a liquidborne particle counter operation. Sample mode, flow control, and sample size selections are shown here. (Courtesy Particle Measuring Systems, Inc., Boulder, CO.)

still important to sample isokinetically so that any particles that are seen can be traced back to their source by following air streamlines from the point of measurement to that source area.

When cleanroom verification operations are carried on, either the operator should have a standard form sheet available to fill in, or the OPC is controlled by a microprocessor that ensures that the information required for verification is procured and entered. In either case, the sample point identification data for clean area verification are recorded first. Next, the OPC status in terms of sample flow rate, particle sizes selected for the cleanroom class to be verified, and the date of last calibration should be entered. The operator should then verify that the OPC is producing acceptable zero count information. The operator should indicate the inlet filter used to verify the zero count data, the sample size used, and the particle counts for at least two zero count measurements. If more than the recommended maximum particle counts of ASTM F50 or Appendix B of FS209D are recorded, then the OPC should not be used before repair or adjustment for acceptable zero count data has been carried out. For many high-sensitivity OPCs that use a gas laser active cavity illumination source, problems may be solved by carefully cleaning the internal cavity mirror or the Brewster's angle window. Next the OPC should be used to collect data from the various sample points selected for verification. Again, isokinetic sampling should be sought even if inlet sample losses are negligible at the particle sizes of concern in many areas. Depending upon the cleanroom operating condition when sampling, the operator should be careful that normal personnel activity is neither interfered with nor allowed to bias the OPC verification data.

Essentially the same sample acquisition procedures must be used for cleanroom monitoring as were used for verifying the cleanroom class. The major differences in OPC operation for monitoring purposes are usually the more permanent OPC placement and sample point locations and the sampling schedule for monitoring. Sample point locations for routine monitoring are usually chosen so that their data are representative of conditions in a large portion of the overall cleanroom. Data procured during verification can be used as a basis for selecting optimum sample point locations. The sample collection may be carried out on a continuous or scheduled basis for general cleanroom monitoring purposes. When some potentially critical procedures occur in the cleanroom, air monitoring may be required at specific times and locations for these situations. In addition, more detailed particle size data may then be required. Figure 26-8 shows some particle size distribution data obtained in an operating downflow unidirectional-flow cleanroom. The change in slope of the size distribution curve indicates that more than one particle source was present in the area. Careful observation of such changes in both particle size distribution and concentration can aid in showing particle sources in specific areas of the cleanroom.

OPCs used for monitoring may be in operation almost continuously for long time periods. Most OPCs require maintenance at intervals of a year or less. If an OPC has been operating continuously for several months since its last maintenance, then recalibration may be required. Following recalibration and verification of OPC gain settings, the area may show a significant change from its last condition. If so, then both the cleanroom and the OPC operation should be verified to determine whether the change in indicated clean area contamination is real or an OPC artifact. If the latter occurs, then the OPC calibration must be examined. Some monitoring and filter testing measurements may use several OPCs to procure simultaneous samples from several locations. It is then necessary that the response of all OPCs allow correlation of data from each OPC in use. Normally this can be accomplished by selecting one OPC as a "master" instrument and "calibrating" each of the others so that particle concentration data from the cleanroom air corresponds to the data from the master OPC. The procedures of ASTM F649 can be used to satisfy this requirement. Once the OPC(s) has been calibrated and any required inter- and intra-OPC correlation has been obtained, then operation can begin.

There are requirements to ensure trouble-free operation over and above

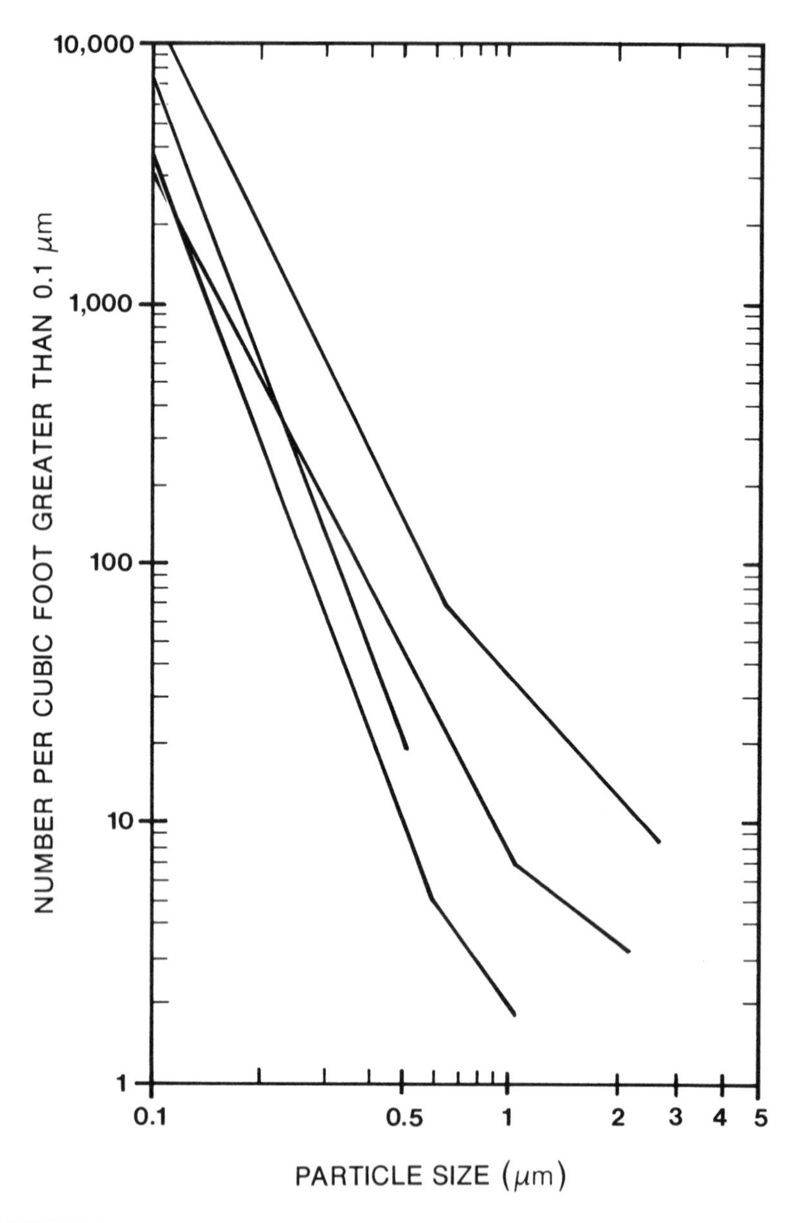

FIGURE 26-8. Particle count data in an operating cleanroom. Measurements down to 0.1 μm are shown. Particle generation within the room appears to be occurring for particles larger than approximately 0.5 μm.

the directions in the manufacturer's manual. These include making sure that correct particle size ranges are selected for measurement, that the last instrument calibration is still valid, that the OPC flow control is correctly set, that adequate printer paper supply (when used) is on hand, and that the OPC case is clean enough for the cleanroom. The operating technician, as well as the cleanroom supervisor, should know enough about the correct mechanical and pneumatic operation of the OPC to know when questionable data are being collected. In summary, the operator should verify that the OPC is in calibration, turn it on for an adequate warm-up period before measurements are made (most manufacturer's recommendations are extremely conservative), and verify that the OPC will produce zero false count data from clean air. Only then can reliable cleanroom data be obtained.

Zero false count data can be obtained easily by using a 0.2 or 0.45 μm pore size cartridge filter large enough so that the OPC pump draws the specified air sample flow easily. The filter is connected with a 10-cm or so section of clean tubing to the OPC inlet. Draw air through the filter and note the time needed for the OPC to report zero counts for that clean air. If more than a few minutes are required for the count to decrease to nearly zero, then check the OPC, the filter, and tubing for leaks or malfunctions.

Extremely clean compressed gases are used mainly in electronic and pharmaceutical manufacturing operations. In these industries, most compressed gas is stored, moved, and used at pressures from 2.7 to 10 bars (40 to 150 psig). Some gases are supplied in standard metal K bottles at 150 bars (2200 psi). Specifications for some critical semiconductor manufacturing gases may allow no more than 20 particles $\geq$ 0.02 μm per cubic foot of gas. At this time, this level of cleanliness is required for inert gases such as nitrogen and helium. Although specifications for compressed reactive process gases are not as severe, they must also be quite clean. In the pharmaceutical industry, the major compressed gas types are compressed air or nitrogen, usually used for drying clean containers. This air is usually specified as class 100 (FS209) cleanliness. Concentrations are usually defined in terms of particles per unit volume of atmospheric pressure gas. Measurements are usually made at pressure, and the data are then converted. Three methods are used to determine particle content in compressed gas: The gas can be expanded and passed through an atmospheric pressure OPC; a side stream sample can be passed through an OPC operating at pressure; or a section of tubing through which the compressed gas is flowing can be fitted with windows and an in situ OPC can be used.

Use of an atmospheric pressure OPC allows direct reading of the particle concentration at atmospheric pressure. The same OPC can be used for other measurements. An atmospheric pressure OPC can be used to count and size particles as small as 0.05 μm in diameter. Expansion of compressed gas to atmospheric pressure may result in adiabatic expansion resulting in condensation of any condensible vapors to form ''false'' particles (Kasper and Wen 1988). For this reason, this method is seldom used for gases at pressures greater than 10 bars. The expansion process is usually carried out by passing the gas through a capillary or an orifice and drawing a portion of the gas into the OPC. This process usually vents excess gas into the atmosphere. For this reason it should not be used with toxic or flammable gases. The expansion process usually results in loss of most particles larger than approximately 2 to 3 μm. If compressed gas pressure is greater than approximately 4 bars, the sound level becomes excessive during expansion.

Passing a side stream sample through an OPC that is designed to operate at line pressure reduces particle losses and sample-handling problems. The problems are not eliminated completely, but violent changes in gas flow rates do not occur. The system designer must keep in mind that the diverted side stream sample must eventually be disposed of. Disposal of inert gases can be accomplished by venting to atmosphere, taking care that excessive noise is not caused when the pressurized gas is released. Disposal of hazardous gases is best accomplished by returning these materials to the process line after measurement. If the hazardous gas cannot be returned to the process line, then scrubbing or adsorption must be used for disposal. Because most operating compressed gas line pressures are never constant, isokinetic sample acquisition is not possible; at best an isoaxial sample collection system can be used. Transport of the sample from the compressed gas line to the OPC results in some additional losses, even for very short transit lines. For this reason, measurement in the particle size range above approximately 2 μm is very questionable. The combination of diffusional losses of very small particles in the transit line to the OPC and the effect of reduced OPC sensitivity has resulted in a lower particle size sensitivity of 0.1 μm for measurement at 10 bars at a sample flow rate of 0.1 cubic feet per minute at standard conditions (scfm). Modification of one OPC sensor originally designed for liquid systems has allowed measurement at 150 bars of 0.5 μm particles at a flow rate of 0.003 cfm at that pressure, which corresponds to a flow rate of nearly 0.5 scfm. Sensitivity is reduced primarily because of the increased molecular density at pressure. This results in increased background scattering signal and a decrease in signal-to-noise ratio for the OPC.

Most surface-deposited particle OPCs are capable of detecting individual particles as small as 0.1 to 0.3 μm on polished surfaces up to 200 mm in diameter. Because these devices are used widely in defining cleanliness of wafers in semiconductor manufacturing, the need for identifying the location as well as the size and number of contaminant particles is important. In order to carry out these processes without interference from background contamination, these OPCs should be operated in a very clean environment.

References

American National Standards Institute, 1972. *Method for Calibration of Liquid Automatic Particle Counters Using "AC" Fine Test Dust,* ANSI B93.28–1972. New York: American National Standards Institute.

American Society for Testing and Materials, 1980a. *Standard Practice for Determining Counting and Sizing Accuracy of an Airborne Particle Counter Using Near-Monodisperse Spherical Particulate Materials,* ASTM F328-80. Philadelphia: American Society for Testing and Materials.

American Society for Testing and Materials, 1980b. *Standard Practice for Secondary Calibration of Airborne Particle Counter Using Comparison Procedures,* ASTM F649-80. Philadelphia: American Society for Testing and Materials.

American Society for Testing and Materials, 1987. *Defining Size Calibration, Resolution and Counting Accuracy of a Liquidborne Particle Counter Using Near-Monodisperse Spherical Particulate Material,* ASTM F658-87. Philadelphia: American Society for Testing and Materials.

American Society for Testing and Materials, 1989. *Test Method for Calibration of Liquid-Borne Particle Counters for Submicrometer Particle Sizing,* ASTM F1226-89. Philadelphia: American Society for Testing and Materials.

Caldow, R., & Blesener, J., 1989. A Procedure to Verify the Lower Counting Limit of Optical Particle Counters. *Journal of Parenteral Science and Technology* 43(4):174–178.

Hartman, A. W., 1987. Standards for Particle Size. Proceedings of the Parenteral Drug Association International Conference on Liquidborne Particle Inspection and Metrology, May 1987, Arlington, VA.

Kasper, G., & Wen, H. Y. On-Line Identification of Particle Sources in Process Gases. Proceedings of the 34th Institute of Environmental Science Annual Technical Meeting, pp. 485–490, April 1988, King of Prussia, PA.

Kaye, B. H., 1989. *A Random Walk through Fractal Dimensions.* New York: VCH Publishers.

Lettieri, T. R., & Hembree, G. G., 1988. *Certification of NBS SRM 1691: 0.3 μm-Diameter Polystyrene Spheres,* NBSIR 88-3730. Gaithersburg, MD: U.S. National Bureau of Standards.

Society of Automotive Engineering, 1978. *Procedure for Calibration of and Verification of Liquid-Borne Particle Counter,* Aerospace Recommended Procedure 1192B. Warrendale, PA: Society of Automotive Engineering.

Verdegan, B. M., Thibodeau, L., & Stinson, J. A., 1990. Using Monodispersed Latex Spheres Dispersed in Nonpolar Liquids to Calibrate Particle Counters. *Microcontamination* 8(2):35–39, 64–65.

Wen, H. Y., & Kasper, G., 1986. Counting Efficiency of Six Commercial Particle Counters. *Journal of Aerosol Science* 17(6):947–961.

27

Optical Particle Counter Errors and Problem Areas

Although particle counters are capable of rapidly counting and sizing very small particles in extremely low concentrations, they are also capable of producing erroneous data with no indication to the operator when this situation arises. Particle sizing errors and/or particle counting errors are produced. Particle sizing errors happen when the particles passing through the OPC are defined as being either larger or smaller than their actual size. All OPCs define particle size in terms of the diameter of a sphere whose projected area is equivalent to that of the particle being observed. Very few real particles are spherical, and the reported dimension varies with the orientation of the irregular particle as it passes through the OPC viewing zone. As indicated previously, the OPC response varies with the refractive index of the particle as well as with its shape. OPCs are calibrated with transparent latex spheres of refractive index 1.6. Any particle measured by the OPC is defined as having the same size as the latex sphere that would produce the same scattered light flux as did that particle. The error types that might be found in an optical particle counter can be summarized (Makynen et al. 1982). It was shown that counting efficiency could be affected by OPC resolution, sample-handling system losses, electronic system bandwidth effects on pulse amplitudes, and internal flow system recirculating of particles. Because scattered light from particles does not disperse equally at all angles, OPCs with different optical systems may respond differently to the same particles, as shown in Figure 27-1. The data shown here were obtained by two OPCs with different optical systems; both were just previously calibrated with PSL spheres. Both OPCs were then used to sample from an ambient airstream. Not only is particle concentration at several size ranges reported differently by the two OPCs

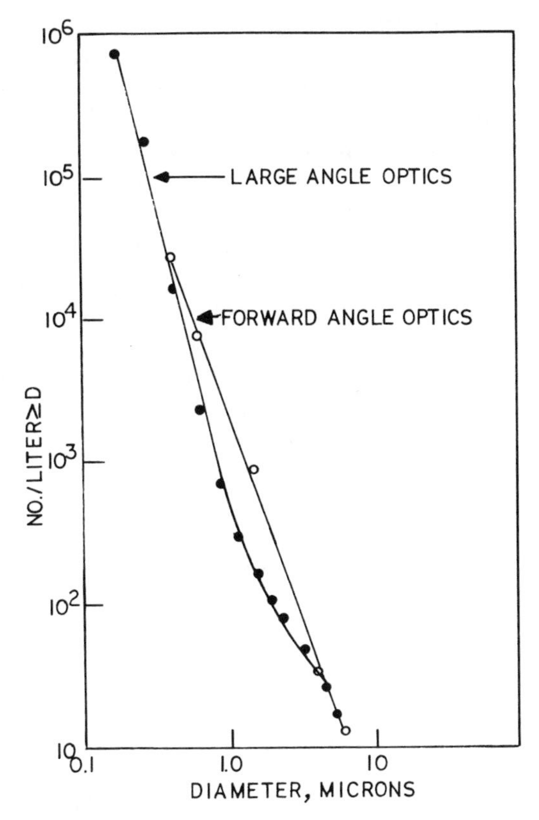

FIGURE 27-1. Varying optical system effects on particle data. Differences in calibration material and actual measured particle optical natures produces different data from two OPCs with different optical systems. (From Lieberman, A., 1980. Laboratory Comparison of Forward and Wide Scattering Angle Optical Particle Counters. *Optical Engineering* 19(6):870–872.)

but also the indicated size distribution function is different. To minimize this error effect, either the OPC to be used is calibrated over its working size range with particles with similar properties to the particles to be measured, or the OPC response is varied so that the signal from the analyzed particles matches that of the calibration particles. A comparison has been made of actual measurements of particles with theoretical calculations for particle counters, one type with narrow-angle collection optics and one with wide-angle forward scattering collection optics (Szymanski and Liu 1986). It was concluded that response for transparent particles is not monotonic for the narrow-angle system, whereas for the wide-angle system, output is attenuated for light-absorbing particles. Care in interpreting OPC response is always required, especially when the optical nature of the particles being measured is not unequivocally identifiable.

Counting errors occur when the OPC reports either more or fewer particles than are actually present in the air sample. These errors can occur either from incorrect OPC performance or from incorrect sample handling. In the former situation, the OPC optical system design or construction can produce varying illumination levels for particles passing through different portions of the OPC sensitive volume. The light scattered from particles passing through the low light level area is not sufficient to produce a measurable signal; these particles are not registered. The effects of such sizing errors on indicated concentration for a normal particle size distribution are shown in Figure 27-2. The concentration of particles in many distributions varies inversely with the square or cube of particle radius. A small error in particle sizing is magnified by the relationships of that power law distribution to a much larger error in particle concentration.

In addition to concentration errors caused by incorrect sizing as a result of nonuniform illumination problems, some OPCs can produce correct

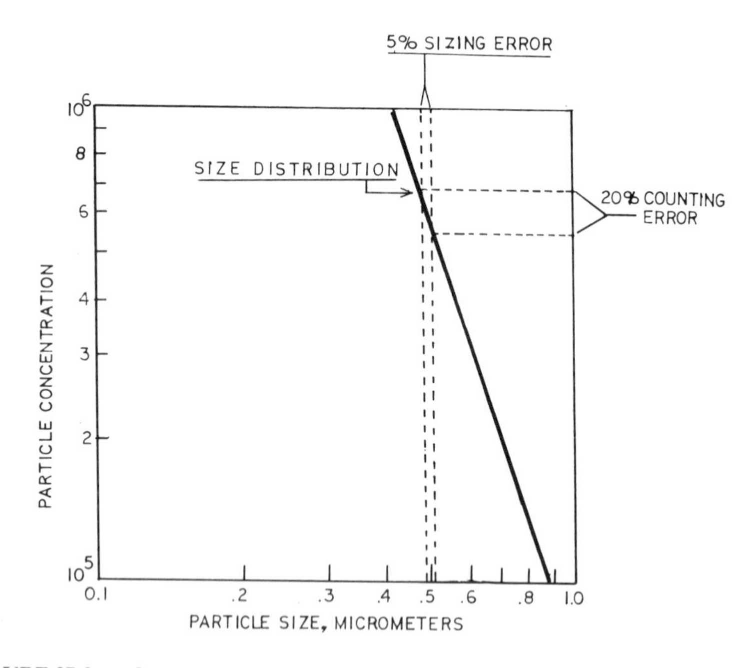

FIGURE 27-2. Counting error as a function of sizing error. The particle population usually varies inversely with a power of the particle size. This example is based upon a particle size distribution in which concentration varies with the negative 3.5 power of particle size. Many commercial OPCs are produced to operate with sizing errors of 5%.

particle size data but report incorrect concentration data. Particles are lost or otherwise affected in transport from the OPC sample tube inlet to the point of measurement. Significant inlet section losses can also occur, particularly for particles larger than approximately 1 μm in diameter. Some of these can be ascribed to irregularities in the OPC sample transport assembly that cause particle impaction during transit across an area where very high turbulence occurs (Wen and Kasper 1986).

With some OPCs, the background noise level at the smallest active particle size threshold is so high that it produces pulses large enough to cause the OPC to report excessive numbers of particles, even though these signals result from OPC background noise. This problem is particularly serious when an OPC is used to count and size particles in electronics-grade process gases. Where the average number per cubic foot of particles ≥ 0.1 μm in diameter is less than 10, the background noise count from the OPC should be no more than 0.1 counts per cubic foot in order to provide data that can be expressed with a 95% confidence level (Van Slooten 1986).

In many cleanroom applications, several OPCs are used. They may be operated in the same area or in different areas at the same or different times. For example, two OPCs may be used to observe operation of a filter. Obviously it is necessary that any difference in particles reported at the inlet and outlet of the filter should be the result of filter operation, not of differences between the two OPCs. A common problem is that different OPCs may not report the same concentration or size distribution when they are sampling from a common source. When the OPCs are operating correctly and have been calibrated to a common base, concentration data correlation to within 10% can be expected (Dahneke and Johnson 1986). Even when OPCs with differing optical designs are used, correlation to this level can be expected, particularly for measurement of 0.5-μm particles. However, optical design differences can cause differences in reported concentration data as great as a factor of 10 at other particle sizes (Lieberman 1980; Liu, Szymanski, and Ahn 1986).

Careful calibration reduces correlation errors between different OPCs. The correct calibration procedure for airborne particle counters is based upon F649-80, a standard ASTM calibration method (ASTM 1980). The procedure requires operation of a system to generate standard calibration aerosols with control of both particle size and concentration, along with a reference instrument for accurate measurement of the aerosol standards. Following calibration, the OPCs should not be used to characterize particulate material that differs very much in composition or size range from the calibration material. Even with these precautions, close correlation between different OPCs over their entire operating range will seldom be achieved.

Semiconductor manufacturing cleanroom air should not contain significant numbers of particles much larger than 1 μm. Even so, the operation of many semiconductor processing tools and the fluids used in those areas may release some particles much larger than that size. If that situation occurs, then it is important that the particle counter be able to detect those anomalous large particles while monitoring airflow around a tool. Particle counters should be able to detect both small and large particles for this reason. Light scattering from particles in the size range below ≈ 0.3 μm varies with the sixth power of the particle radius, whereas light scattered from particles larger than ≈ 0.7 μm varies with the second power of the radius. The linear dynamic range of a fast electronic pulse height analyzer system is limited unless special circuitry is used in the system. Therefore, standard particle counters capable of detecting very small particles have a smaller dynamic range than those that are not so sensitive. For most cleanroom operations, the added cost of special systems is seldom justified. A standard 0.05-μm counter sizes particles with good resolution not much greater than 1 μm in diameter, a 0.1-μm counter sizes particles well up to 5 μm, and a 0.5-μm counter sizes particles well up to 20 μm. These limits do not mean that larger particles are not counted at all. They are simply defined as being present and equal to or larger than the largest bin size for the particle counter.

Particle counter sizing resolution defines how well a particle counter is able to differentiate between two particles that are nearly the same size. This capability had long been considered a minor concern. Unfortunately, it also affects particle-counting accuracy, particularly in the smallest size range. Counters with poor resolution identify a broader range of particles both above and below the threshold size as being *at* that threshold size than do counters with better resolution. This results in the counter with poor resolution indicating a larger number of particles at the minimum counter size threshold because of the steep slope of the power law particle size distribution curve. When the cleanroom operator attempts to correlate measurements made in different areas with different particle counters of varying resolution, widely varying data are obtained. Present-day particle counter resolution is stated as the increase in width of the size distribution reported for near-monodisperse particles reported by the particle counter. A counter with acceptable resolution increases the reported relative standard deviation for monosize calibration particles by no more than 10%. Values of 3 to 5% are common for newer OPCs with sample flow rates of no more than a few liters per minute. For higher sample flow rates, it is necessary to increase the illuminated zone area in the direction orthogonal to the flow path. Otherwise, the sample velocity must be increased to the point where the sample pump cannot move air fast enough or the particle residence time in the sensitive volume is so short that only very sophisticated electronic systems can count and size the pulses correctly. Even when acceptably uniform illumination can be obtained over a larger area, the particle position within the larger area results in variations in the scattering angle for particles at different locations within that area. Figure 27-3 shows scattering angle variations within an older 1-cfm forward scattering angle OPC. It is emphasized that OPCs with high sample flow rates are inherently incapable of particle sizing with maximum resolution. High flow rate OPCs are particle counters, not particle sizing spectrometers. For precise particle size data, the OPC must have a uniform intensity illumination source, a signal-to-peak noise ratio of at least 4 at the smallest particle size, and a flow rate not much greater than 50 cc per second.

Particles in any sample are usually distributed randomly rather than uniformly. At any point within an air parcel, the instantaneous concentration can vary widely about the average value. If the average concentration becomes too high, then two phenomena occur and can cause errors: particle coincidence within the viewing volume and electronic saturation of the data processing system. *Coincidence* is defined as the simultaneous occurrence of more than one particle within the OPC sensitive volume. Coincidence can cause both counting and sizing errors. *Saturation* is the

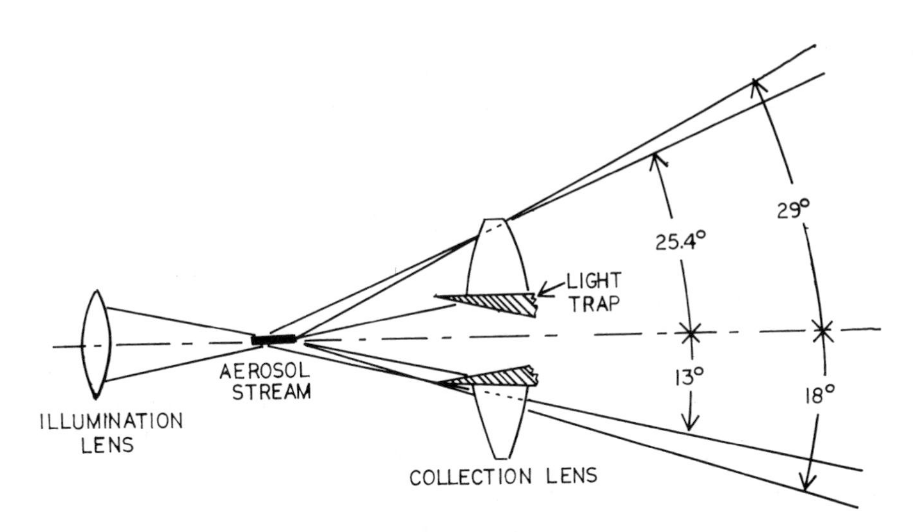

FIGURE 27-3. Finite sample stream size effect on scattering angles employed by an OPC. This illustrates a commercial forward scattering optical system with aerosol stream dimensions in actual use. Depending on whether a particle is at one end or the other of the rectangular aerosol stream, the solid scattering angle can vary sufficiently so that the OPC sizing resolution is poor.

inability of the electronic pulse processing system to detect and size individual pulses when the pulse rate exceeds the capability of that system to differentiate between successive pulses. The indicated concentration is lower than the true value, and the reported particle size is shifted toward larger sizes because the summed pulses from two or more smaller particles are reported as a single pulse from a larger particle (Raasch and Umhauer 1984).

The true particle concentration can be calculated from an equation derived from statistical probability theory. The equation describes the probability that zero, one, or more than one particle can be present within a fixed volume of space if the particle distribution is Poisson in nature, and the particle dimension is insignificantly small compared to that of the reference volume. It is assumed that the particle population is distributed in a random manner, and the probability of one or more particles being present at any time within the view volume is calculated on this basis. An acceptable reduced version of the equation states that the ratio of indicated to true particle concentration is equal to 1 minus half of the product of the true concentration and the sensitive volume. This product reflects the long-term average sensitive volume population. This reduced version of the exponential equation is valid to the point where that population is 0.1. If the particle concentration increases to the level where the sensitive volume population begins to exceed 0.1, then the indicated concentration values eventually begin to decrease. Eventually, the OPC indicates a zero population at extremely high true population values. This occurs when one or more particles are present in the sensitive volume at all times and the scattered light level never decreases to the point where the electronic system signal drops to the reset level. This situation is shown in Figure 27-4. At that point, the OPC data are produced by random variations in the concentration of particles in the sensitive volume rather than from individual particles. In other words, the background optical noise level varies, with occasional levels exceeding the threshold level that would be produced by the smallest particle that the OPC should be capable of detecting.

A summary of OPC specifications that directly affect the possibility of errors is given in Table 27-1 (Lieberman 1989). This table shows some OPC performance parameters that are common to essentially all OPCs. It summarizes the pertinent performance specifications that are important for data integrity from any OPC. Because the selection of operating conditions depends upon these OPC performance capabilities, they are discussed in some detail here. Even though these parameters have been mentioned in preceding sections that discuss the basis for OPC operation, more detail is provided here because thorough understanding of the effects of variations in operating parameters is so important in selecting the operating levels for

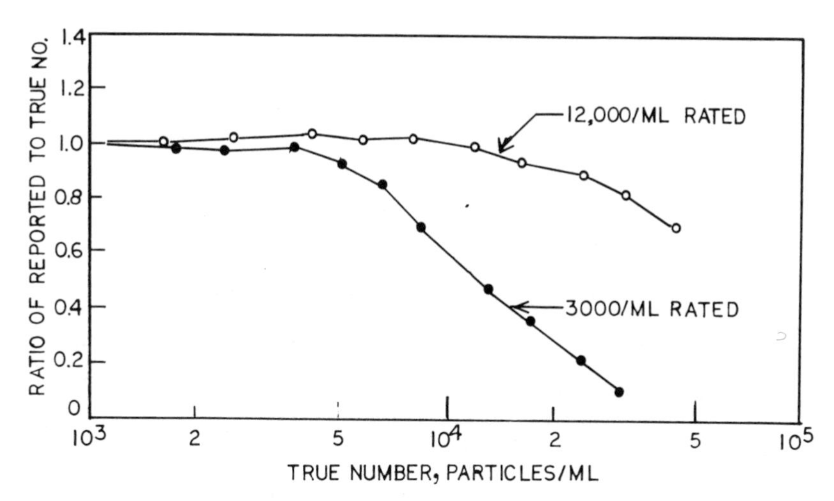

FIGURE 27-4. Coincidence effect on concentration measurement. Data are shown for two sensor systems, one with a maximum rated concentration of 12,000 particles per milliliter and one rated for 3,000 particles per milliliter. As the concentration of particles increases to the point where more than one particle can be in the sensing zone at any time, the OPC indicates more than one particle as a single, albeit larger, particle. The total reported concentration is less than the actual concentration, even though the number of reported large particles is greater than the actual number.

any OPC. Many OPCs are procured and used on the basis of terminology that may not clearly define some of the requirements for good OPC performance. A better understanding of how the performance specifications affect the data produced by the OPC should help in understanding the true value of the data reported by that OPC. Although there are differences in the OPCs for each of the application areas, there are some common performance requirements for good OPC operation in all areas. The OPC must define the number and size of particles that may be randomly and sparsely distributed in a very clean sample. The particle size distribution is usually described with a power law, but concentrations may be so low that sufficient particle population data for a statistically valid definition of that function, or even of the true concentration value for the entire population that the measured sample should represent, may be difficult to obtain in a reasonable sample size.

GOOD SIGNAL-TO-NOISE RATIO

The data produced by an OPC must be derived solely from particles. Most specifications for defining particle populations call for a numerical value of the sample concentration mean with a specific confidence level. The con-

TABLE 27-1 Particle Counter Operating Requirements

- Good Signal-to-Noise Ratio
- Acceptable Particle Sizing Sensitivity
- Maximum Particle Sizing Accuracy
- Acceptable Particle Sizing Resolution
- Adequate Particle Counting Accuracy
- Maximum Sample Size or Volumetric Flow Rate
- Maximum Particle Concentration Capability
- Minimum Data Processing Time
- Correlation with Other Instruments
- Adequate Particle Size Definition Range
- Compatibility with Fluids and Environment
- Documented Calibration Record
- Reliable Operation and Ease of Maintenance

Source: (From Lieberman, A., 1989. Performance Definitions Effects on Submicrometer Particle Measurement. Proceedings of the Fine Particle Society Meeting, Boston.)

fidence level states how certain one can be that the true concentration in the population from which the samples were procured will lie between the upper and lower confidence limits. For a measured sample average, knowledge of the standard deviation (the range of the raw data values) compared to the average value allows one to state the limits within which the true average will fall with a specific confidence level. Therefore, the more data that can be procured, the more reliable it will be. The quantity of data from an OPC can be increased by either increasing sample size or increasing OPC sensitivity so that more particles will be recorded. In either case, it is also necessary to be sure that the data quality is not degraded. For most OPCs, the data quality is degraded when significant portions of data are produced by OPC noise, rather than by particles. Because most OPCs are used at their greatest sensitivity, the noise pulse generation rate should be insignificant, compared to the expected rate of particle rate data acquisition. At a signal to root-mean-square noise ratio of

10, measurable random noise pulses (at a level equal to the pulse generated by the smallest particle that the OPC detects) will not be present in significant number.

ADEQUATE PARTICLE-SIZING SENSITIVITY

The need for adequate particle size sensitivity arises from the fact that clean sample particle concentrations are very low, but the population usually increases inversely with approximately the third power of particle size. If the OPC can detect the smallest possible particle, then more data can be procured with better size sensitivity. Extreme size sensitivity is not always required because many OPCs are operated in areas where a sufficient "large" particle population is present. Where high sensitivity is sought, the OPC may be operating with particle signals that are not much greater than internal noise levels. It is suggested that acceptable size sensitivity can be defined as that level where a sufficient number of particle count events are identified by the OPC in an acceptable sampling time period. For example, in extremely clean compressed gases or on the surface of carefully controlled virgin wafers, sample measurement periods of several hours may be required even for an OPC with sensitivity well below 0.1 μm. In other ambient air measurement situations, sampling periods may be limited to no more than 1 or 2 minutes. In either case, the OPC sensitivity must be adequate to produce sufficient data in the accepted time period from the particle size distribution present in the measured sample.

MAXIMUM PARTICLE-SIZING ACCURACY

Particle-sizing accuracy is needed because most desired contamination level information is expressed in terms of particle concentration at and above a specified particle size. For the power law distributions frequently seen in real particulate contaminants, any error in particle size measurement results in a concentration error that increases in magnitude with the power function associated with the particle size distribution. Figure 27-2 shows how a sizing error of 5% can produce a concentration error of 16% (1.05^3) for a third-power particle size distribution. Even for the particle size "distribution" used for FS209D, with an assumed slope of 2.2, the concentration error for a 5% sizing error would be nearly 12%.

ACCEPTABLE PARTICLE-SIZING RESOLUTION

Particle-sizing resolution is related to sizing accuracy. *Resolution* is defined as the OPC's ability to differentiate between particles that are nearly the same size. The OPC references particle size by the amplitude of the pulse produced by that particle; therefore, if identical particles produce pulses of widely varying amplitude, then that OPC has poor sizing resolution. Further, if the OPC is set to define particles in the smallest size range, then some of the particles close to the smallest size threshold may or may not be recorded at all. Resolution can be quantified as the increase in standard deviation caused by the OPC pulse broadening to the true standard deviation of a group of monosized particles. Reported particle population data for polydisperse particles may be significantly greater from an OPC with poor resolution than those data from an OPC with better resolution. This effect occurs because the power function size distribution results in the presence of many more particles in an interval just below the smallest detection level than in the same size interval just above that level. The number of indicated "subcountable" particles (Whitby and Vomela 1967) is much greater than the number of uncounted particles that are just over the threshold level. Figure 27-5 shows how OPC sensors with poor resolution can broaden the range over which a particle size distribution is reported.

ADEQUATE PARTICLE-COUNTING ACCURACY

Accurate data are expected from any instrumental measurements. *Counting accuracy* for an OPC is defined as the ratio of the reported particle population to the true population in the measured sample. Even though it is normally assumed that any OPC counting efficiency is 100%, this assumption is not always justified. In some situations, an OPC optical system observing particles in fluid may be misaligned so that part of the sampled fluid bypasses the OPC sensing zone. This situation can usually be remedied by optical realignment. For OPCs used for airborne particles, poor inlet sampling efficiency and particle losses during transit to the sensing zone can result in serious undercounting of particles larger than a few micrometers in diameter. For liquidborne OPCs, the same effects can occur, but differences in the fluid physical properties result in errors not becoming important for particles smaller than 10 to 15 μm in diameter. Airborne OPC inlets should be constructed so that the inlet nozzle is not blunt, but it should not be so sharp as to cause a personnel hazard. The

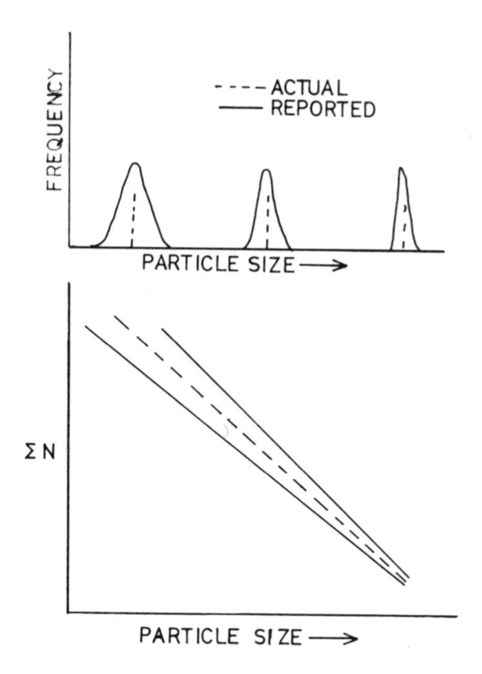

FIGURE 27-5. Resolution effects on size distribution data. The upper portion of the illustration shows the particle size distributions reported for monosize particles. An OPC with poor resolution reports a broader particle size distribution than one with better resolution. The result in terms of reported particle size distribution is shown in the lower portion of the illustration. For a true particle size distribution (*center line*), the OPC with poor resolution can report size distribution that can be anywhere within the bounds of the upper and lower lines.

inlet should be dimensioned to provide isokinetic sampling wherever possible. For typical cleanroom air moving at 90 to 100 feet per minute, an OPC sampling 1 cubic foot per minute should have an inlet nozzle with an area of 0.009 to 0.01 square feet (1.3–1.4 square inches). The transit line from the inlet nozzle to the OPC detection area of the OPC should be as short as possible for particles smaller than 0.1 μm and larger than 2 μm. If any curvature in the transit line is required, the radius of curvature should be at least 8 inches. With this precaution, centrifugal effects do not remove large particles.

When an OPC is calibrated, its resolution limits cause production of pulses with some amplitude distribution from a suspension of monosized calibration particles. The standard procedure is to establish the reference pulse amplitude level at the median of the pulse amplitude distribution produced by the calibration particles. Thus, half of the calibration particles are recorded in the size range for particles ≥ the median particle size, and half are recorded in the size range for particles ≤ the median particle size. If the calibration particle size is at the minimum measurable size sensitivity for the OPC, then the counting efficiency for calibration at the smallest size range is never more than 50%. When polydisperse particles are measured by the OPC, all particles that produce pulses with an amplitude larger than the smallest particle size threshold level should be counted with 100% efficiency. Therefore, it can be stated that the OPC operational counting efficiency is 100% for all particles with an equivalent optical size greater than the median size of the calibration particles at the minimum particle size sensitivity, even though the calibration counting efficiency is 50%. This relationship is shown in Figure 27-6. This portion of the discussion assumes that the OPC resolution is ideal; that is, the OPC does not produce any increase in the reported relative standard deviation for monosized particles.

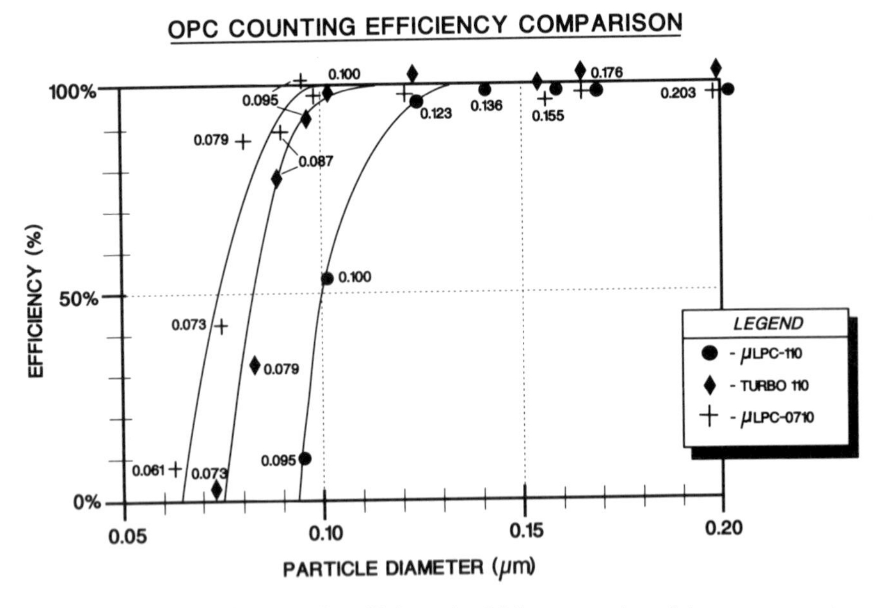

FIGURE 27-6. Particle-counting efficiency for OPC systems. Actual data are reported for several OPCs with varying reported maximum sensitivities. (Courtesy Particle Measuring Systems, Inc., Boulder, CO.)

MAXIMUM SAMPLE SIZE OR VOLUMETRIC FLOW RATE

For many process operations, very low particle levels are present, and very few data are produced by the OPC. For example, some proposed specifications for semiconductor process gases would allow no more than 10 particles $\geq$ 0.1 μm per cubic foot in acceptable gas. At this low concentration, the particle spatial distribution would certainly be nonuniform. This means that sufficient data must be acquired so that at least an indication of the average particle distribution can be stated. Otherwise, it would be very difficult to be reasonably sure that the actual particle concentration in the gas being sampled does not exceed the median value concentrations of a few samples. The proposed specification assumes an OPC operating with a flow rate of 0.1 cfm or less and requires that each sample is based on a measured volume of 8 cubic feet. Similarly, measurement of surface particle contamination deposited on wafers during semiconductor manufacture is aimed at defining deposition rates for 0.3-μm particles as low as 10 per 200-mm wafer. The advantage of using an OPC with maximum sampled gas flow rate or maximum sample size capability is apparent.

MAXIMUM PARTICLE CONCENTRATION CAPABILITY

Most of the preceding discussion concerns OPCs that are used for examination of clean samples. However, many times it is necessary to characterize samples with high particle concentrations. Cleanrooms in the 1980–1990 period vary over 6 orders of magnitude in particulate cleanliness levels. Some normally clean fluids may become inadvertently contaminated, and the operator must be aware of the problem as soon as possible; a process tool may malfunction, and a burst of particles may be present at a critical location. Contamination control operations cover a very wide range of situations. The same OPC should not be used for measurements in both very clean and in very contaminated environments. As pointed out previously, an OPC observes only the total amount of light emitted from the sensing zone and does not image individual particles. Therefore, if a single larger particle or several small particles are present in that sensing zone, the OPC signal can be identical. Because particles in any suspension are not uniformly distributed, there is a finite probability that local inhomogeneities in the particle spatial distribution will cause some coincident particles to be present in the sensing zone at some time, no matter how low the concentration or how small the sensing zone. The coincidence error can be calculated from basic probability considerations. Most OPC manufacturers specify a maximum recommended concentration that should result in coincident particles causing no more than 10% of the data. This level allows calculation of the coincidence error with an approximation that states that the ratio of indicated to true particle numbers in a measurement is equal to 1 minus half the true particle concentration in the sensing volume. Figure 27-7 illustrates how physical coincidence error increases with particle concentration. It becomes apparent that small sensing volumes permit measurement in samples where interparticle spacing is small, and thus high concentration systems can be handled best with an OPC with the smallest possible sensing volume. It must be kept in mind, however, that a small sensing volume does not allow measurement of large volume samples unless long sample measurement times are used. Note that these comments apply for measurement of particles whose volume is insignificant compared to that of the sensing volume. If a large particle passes

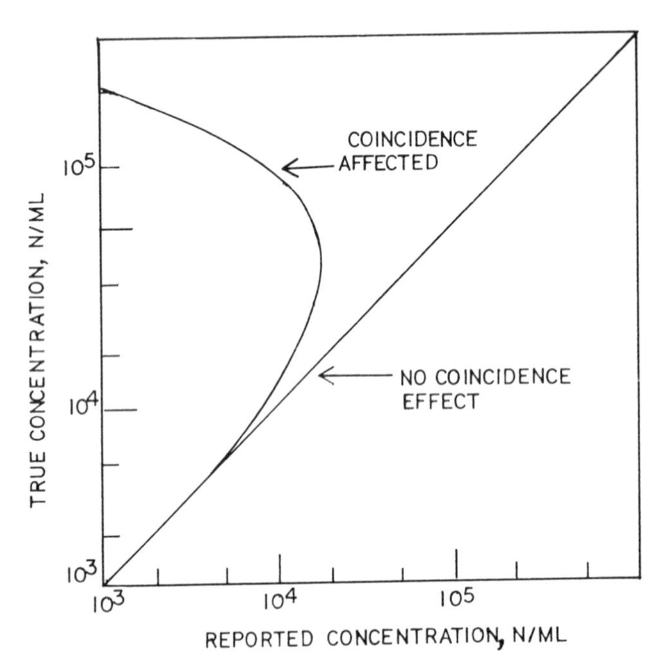

FIGURE 27-7. Coincidence error effects on concentration data. If actual concentration continues to increase, eventually the sensing zone is always occupied with one or more particles. In that situation, the OPC detector signal does not decrease to the point where the electronic system can reset. The indicated number of particles can decrease to very low levels.

through the sensing volume, its trailing edge signal can mask any small-particle signal.

MINIMUM DATA PROCESSING TIME

Most OPC data processing electronic systems are basically pulse height analyzers. The electronic system counts each pulse and defines its amplitude in relation to a reference level. The operating level must be reached between successive particle-derived pulses. Some finite residence time, or dead time, at that operating level is required to ensure that an accurate value for pulse amplitudes can be established. If another particle signal arrives from the sensing volume during the dead time interval, then that signal is not detected. Older analog comparator circuit pulse counting systems may require as long as 90 μsec dead time to recover from pulses produced from large particles. When this time interval is added to the 10- to 30-μsec pulse duration for most particle passage through the sensing volume, then a maximum pulse counting rate of some 10,000 per second can be accommodated for uniformly spaced pulses. Because the particles are randomly spaced, the pulse counting system must be able to handle the short-term pulse rate as well as the average rate. A rule of thumb that has been used states that the short-term rate is seldom more than 1 order of magnitude greater than the long-term average. Thus, a 10-kHz counter should not be used for measuring systems where the long-term average pulse rate will be greater than 1 kHz. Particle data losses arising from the electronic system dead time interval and physical coincidence are interactive and additive.

CORRELATION WITH OTHER INSTRUMENTS

The ideal OPC should produce particle size and concentration data from any particle sample that is equivalent to that produced by another OPC or even a different particle characterization device. Because OPC data are affected by particle properties that may differ from those that may affect other devices, then direct correlation to any device without a means of converting data to match the controlling particle properties is not possible. At best, the slopes of the particle size distribution functions from the OPC and from the other device correspond (Johnston and Swanson 1982). An empirically derived conversion factor may then be developed to aid in correlating particle size data from the two devices. In this situation, experience has shown that the particle size distribution data may be correlated acceptably well over 80% or so of the particle size range, but the tails of the distribution vary significantly. Therefore, correlation with other devices is possible, but verification is needed for the specific instrument type and particle composition.

For cleanroom operations, interfacility and intrafacility particle data correlation is required, particularly for verification of particle content in process fluids as supplied from vendors and as received or in use on a process line. For this application, most vendors and their customers use OPCs with similar sensitivity and general design features. The major variation in these OPCs may be in optical design, with some variations in illumination wavelength. For these factors, response changes can be predicted from physical theory well enough so that correlation can be accomplished between two different OPCs (Knollenberg 1986) by theoretical means. It is, of course, necessary that the OPCs are in good operating condition and have been calibrated accurately. In general, concentration correlation can be accomplished to within 5% for OPCs of the same optical type and to within 20 to 30% for OPCs of different optical designs if they are calibrated carefully. Airborne OPCs should be calibrated to ASTM F649 and liquidborne OPCs to NFPA T.2.9.6R1 (NFPA 1989) or to ASTM F658 or ASTM F1226 for better correlation.

ADEQUATE PARTICLE SIZE DEFINITION RANGE

As convenient as it is to characterize particle size and concentration in great detail, most cleanroom particle measurements made with an OPC are aimed at defining trends rather than providing detailed data. For this purpose, an OPC that states only the population of particles equal to and larger than a single size threshold may be adequate for almost all routine measurements. Even so, there will be occasions when detailed particle data are required. An OPC that can characterize particles over a broad size range is convenient to aid in defining particle sources where the particle size distribution varies with the source. Capability for counting and sizing particles over a broad size range is very useful in this respect. However, some requirements must be considered. If one wishes to define a broad range of particle sizes in a sample, then a very large range of signal levels must be measured accurately. This requirement increases the cost of the measurement system. For this reason, the cleanroom measurement should be carried out with an OPC that has adequate, rather than maximum, particle size measurement range. The term *adequate* means that the OPC has sufficient size range capability to characterize those particles that are assumed to occur in significant quantities. For example, assume a particle size distribution with a −3 power relation between population and particle size. If 10,000 > 0.1-μm particles are present, then 10 > 1 μm-particles and no more than one or two > 2-μm particles are present. A 20:1 size measurement dynamic range allows size definition from 0.1 to 2 μm. Knowing that any particles larger than 2 μm are present is sufficient; their exact size need not be known.

OTHER OPC SPECIFICATIONS

Operation of an OPC in an environment where OPC components may be corroded or otherwise damaged appears to be an obviously erroneous procedure. However, problems can occur if an OPC used for measurement of liquids is exposed to a series of reactive materials that may produce deposits of adherent reaction products upon critical optical elements. Similarly, long-term operation of an OPC used for measurements of gas cleanliness with an acid gas may not cause a problem unless excessive quantities of water vapor are inadvertently added to the system; in this situation, acid deposition can degrade the OPC operation. Operation at extreme temperatures or pressures can cause problems for standard OPCs. Excessively high temperatures may cause standard electronic components to fail; operation at excessive pressures may cause standard low-pressure seals or O-rings to fail. Operation of high-sensitivity surface-analyzing OPCs in normal atmospheres usually results in deposition of atmospheric particles on otherwise clean surfaces being observed; erroneous data are produced as a result of these artifacts from the atmosphere.

Documented calibration records are strongly recommended to ensure that the OPC performance has not drifted so that any data produced by the OPC are not grossly erroneous or produced by internal OPC noise. Further, the operator should know if the specific application for which he is using the OPC is compatible with the calibration data used to set up the OPC particle size ranges. If an OPC is used where the anticipated contaminant particles are mainly opaque, then the calibration data produced from transparent latex particles may require modification to compensate for the added imaginary coefficient of the measured particle refractive index.

A requirement for reliable operation is apparent. It includes uncomplicated start-up and normal operating procedures, such as establishment of operating levels for sample size, particle size ranges, sample duration, data format, and data recording. Easily interpreted warning indications should be produced if erroneous operation occurs. Routine maintenance may be required. On-board printer tape must be replaced regularly, and airborne particle-counting OPCs that use active cavity laser systems have mirrors that must be kept clean. This type of maintenance should be easily carried out without requiring metrology laboratory or manufacturer's service personnel.

References

American Society for Testing and Materials, 1980. *Standard Practice for Secondary Calibration of Airborne Particle Counters using Comparison Procedures,* ASTM F649-80. Philadelphia: American Society for Testing and Materials.

Dahneke, B., & Johnson, B., 1986. A Method for Cross Calibrating Particle Counters. *Journal of Environmental Science* 29(5):31–36.

Johnston, P. R., & Swanson, R., 1982. A Correlation between the Results of Different Instruments Used to Determine the Particle Size Distribution in AC Fine Test Dust. *Powder Technology* 32(2):119–124.

Knollenberg, R. G., 1986. The Importance of Media Refractive Index in Evaluating Liquid and Surface Microcontamination Measurements. Proceedings of the 32d Institute of Environmental Science Annual Technical Meeting, May 1986, Dallas, TX.

Lieberman, A., 1980. Laboratory Comparison of Forward and Wide Scattering Angle Optical Particle Counters. *Optical Engineering* 19(6):870–872.

Lieberman, A., 1989. Performance Definitions Effects on Submicrometer Particle Measurement. Proceedings of the 1989 Fine Particle Society Meeting, Boston.

Liu, B. Y. H., Szymanski, W. L., & Ahn, K. H., 1986. On Aerosol Size Distribution Measurement by Laser and White Light Optical Particle Counters. *Journal of Environmental Science* 28(3):19–24.

Makynen, J., et al., 1982. Optical Particle Counters: Response, Resolution and Counting Efficiency. *Journal of Aerosol Science* 13(6):529–535.

National Fluid Power Association, 1990. *Hydraulic Fluid Power: Calibration Method for Liquid Automatic Particle Counters Using Latex Spheres,* NFPA T2.9.6-R1 1990. Milwaukee: National Fluid Power Association.

Raasch, J., & Umhauer, H., 1984. Errors in the Determination of Particle Size Distribution Caused by Coincidences in Optical Particle Counters. *Particle Characterization* 1(1):53–58.

Szymanski, W. W., & Liu, B. Y. H., 1986. On the Sizing Accuracy of Laser Optical Particle Counters. *Particle Characterization* 3(1):1–7.

Van Slooten, R. A., 1986. Statistical Treatment of Particle Counts in Clean Gases. *Microcontamination* 4(2):32–38.

Wen, H. Y., & Kasper, G., 1986. Counting Efficiencies of Six Commercial Particle Counters. *Journal of Aerosol Science* 17(6):947–961.

Whitby, K. T., & Vomela, R. A., 1967. Response of Single Particle Optical Counters to Nonideal Particles. *Environmental Science and Technology* 1:801–814.

28

Gas Filter Testing Methods

Gas filtration of air in the cleanroom is carried out with HEPA (high-efficiency particulate air) or ULPA (ultralow penetration air) filters for ambient pressure air cleaning or with high-efficiency cartridge filters for compressed gas cleaning. Filter testing is required for both types not only to verify the performance of the filter element or filter medium proper but also to ensure the integrity of the installed filter. The ambient air filters for the cleanroom are relatively fragile and require great care in handling to avoid damage to the filter medium and the resultant leaks. The cartridge filters are not so delicate but still require care in installation to ensure that seals and O-rings are correctly installed when the cartridge is changed.

Filter testing consists of a performance evaluation and a method of ensuring that filter media integrity has not been compromised. Testing is carried out on filters to be installed in the cleanroom to make sure that the basic filter element integrity is acceptable. In addition, testing after installation is carried out to make sure that the filters have been installed correctly and that no leaks are present. Thus, the preinstallation tests provide a quantitative definition of performance of individual filters. The in-place tests verify the performance of the entire filter system in terms of acceptable air quality; these tests include verification of the integrity of installation. Filter testing in the cleanroom is normally limited to in-place testing or to leak testing for installed HEPA or ULPA filter elements. Preinstallation tests are normally carried out by the filter manufacturer, especially for HEPA or ULPA filters.

There are a number of documented standard methods and specifications for preinstallation testing of HEPA and ULPA filters. The procedures used for testing compressed gas cartridge filters are reported in the technical literature, but there are no documented standard methods available for these devices. The manufacturer's testing is frequently carried out by challenging the cartridge with an aerosol of specific size latex or bacterial particles in known concentration and then measuring the downstream aerosol concentration. Although these tests are quite reliable, the procedures are not documented as standard methods, and interlaboratory test protocol differences exist. It has also been found that testing with different filter challenge materials or test methodology may produce different performance indications for the same filter types.

Testing methodology for HEPA and ULPA filters are usually based upon challenging the filter with an acceptably defined liquid droplet aerosol of dioctyl phthalate (DOP) or similar material in a concentration of 10 to 100 micrograms per liter. The particles are primarily less than 1 micrometer in diameter. Particle penetration is determined by measuring upstream and downstream concentration with a photometer. In addition to this method, testing penetration of the DOP aerosol with a condensation nucleus counter has been carried out (Johnson et al. 1990), and the method correlates well with the photometer measurement procedure. Some well-known preinstallation tests for ambient pressure filters are summarized:

IES RP-001, *Recommended Practice for HEPA Filters,* in which polydisperse DOP particle penetration, pressure drop at rated airflow, and frame integrity are measured. This procedure is available from the Institute of Environmental Sciences, 940 E. Northwest Hwy, Mount Prospect, IL 60056, USA.

MIL-F-51068 (E), *Military Specification: Filter, Particulate, High Efficiency, Fire Resistant,* defines filter module design, materials of construction, and filtration efficiency for DOP particles of stated size. This procedure is available from the Naval Publication and Forms Center, 5801 Tabor Ave., Philadelphia, PA 19120, USA.

DOE NE F 3-43, *Quality Assurance Testing of HEPA Filters and Respirator Canisters,* covers requirements for the quality assurance inspection and filtration efficiency testing of HEPA filters, including test procedures and equipment for test. This procedure is available from the Nuclear Standards Management Center, Oak Ridge National Laboratory, Bldg. 9204.1, Rm. 321, MS 10, P. O. Box Y, Oak Ridge, TN 37830, USA.

In some cleanrooms, filters that have been tested with DOP as a challenge are not acceptable because some DOP vapor may be released from the filter after testing has been completed. For this reason, alternate challenge materials and testing methods are sometimes used to define filtration efficiency for HEPA filters. Figure 28-1 shows a test setup for HEPA filters using ambient air particles as a challenge (Gutacker 1985). Particle size and concentration data for this system are determined with an optical particle counter. Anisokinetic sampling at a location upstream of the filter is used to define the challenge level, whereas isokinetic sampling with a rectangular probe across the filter face is used to define penetration and to locate any leaks. Use of the OPC to determine particle concentration and size distribution allows definition of fractional penetration for ambient air particles because the filter will be used to control these materials, rather than an artificial challenge material. The use of the rectangular probe (Fig. 28-2) for isokinetic sampling accelerates the leak-testing procedure.

Some in-place filter test procedures in frequent use are summarized here. The first method is aimed at defining whether leaks exist in any portion of the filter frame or filter medium after installation. The second and third methods are used to determine penetration efficiency for filter systems.

IES RP-2, *Recommended Practice for Laminar Flow Clean Air Devices,* covers procedures for evaluating performance and major requirements of laminar flow clean air devices, including clean benches and other devices

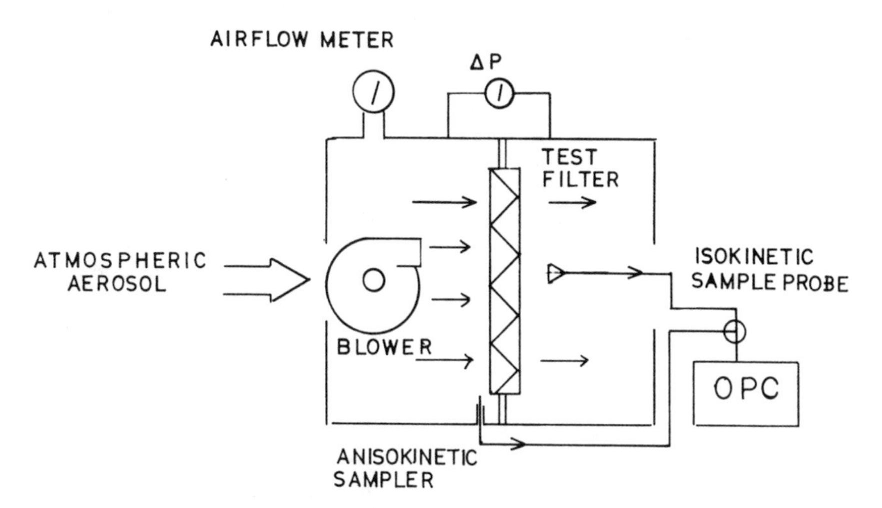

FIGURE 28-1. HEPA filter penetration test system with ambient air challenge. Where high-efficiency filters are tested, the ambient air challenge concentration must be large enough so that the downstream particle count data are sufficient for good statistical definition. The integrity of the system does not allow any ambient air to leak into the downstream measurement area. (Courtesy A. Gutacker, ARGOT, Inc.)

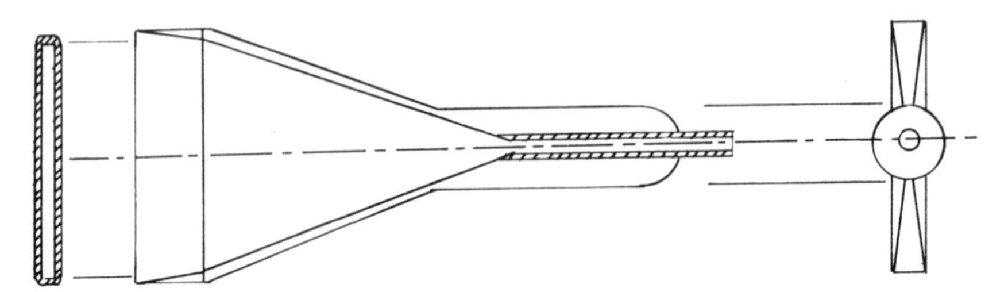

FIGURE 28-2. Rectangular isokinetic probe for in-place filter penetration testing. Sample probes need not be circular. As long as the sampled material population is sufficient for valid data, the rectangular probe can be used to scan filters more rapidly than can be done with a circular probe.

with self-contained blowers. The leak test procedures use polydisperse DOP smoke and a photometric determination that no more than 0.01% of the challenge DOP smoke concentration has passed through the filter at any point. Airflow velocity, induction leaks or backstreaming effects, and blower performance are measured. This procedure is available from the Institute of Environmental Sciences, 940 E. Northwest Hwy, Mount Prospect, IL 60056, USA.

ANSI N101.1, *Efficiency Testing of Air-Cleaning Systems Containing Devices for Removal of Particulates,* presents a method for in-place testing filtration efficiency by use of polydisperse DOP smoke and measuring DOP concentration before and after the system. This procedure is available from the American National Standards Institute, 1430 Broadway, New York City, NY 10018, USA.

DOE NE F 3-41, *In-Place Testing of HEPA Filter Systems by the Single-Particle, Particle-Size Spectrometer Method,* is designed for ULPA or tandem HEPA systems. It uses polydisperse DOP challenge and measures upstream and downstream concentrations and size distributions with an OPC sensitive to 0.1-μm particles. This procedure is available from the Nuclear Standards Management Center, Oak Ridge National Laboratory, Bldg. 9204.1, Rm. 321, MS 10, P. O. Box Y, Oak Ridge, TN 37830, USA.

ASTM subcommittee F21 on filtration is preparing an ASTM method based on the DOE procedure. Development of the method has been shown practical by Rivers and Engleman (1984), who studied operation of both a dual-laser particle spectrometer system and a dual-probe, single-particle spectrometer system to define fractional penetration of polydisperse DOP challenge aerosol through ULPA filters. Further work on development of the method was carried out by Gogins et al. (1987), who showed the importance of optimizing system hardware and applying statistical procedures for rapid evaluation of HEPA and ULPA filters where fractional penetrations may be as low as 10^{-7} for particles as small as 0.1 μm.

It is very important that installed filter systems be tested to ensure that no damage that could cause leakage has occurred during installation. The test procedure, based on IES RP-2, used an upstream DOP challenge aerosol with a concentration of 10 to 100 micrograms of aerosol per liter of air. The entire downstream surface of the filter and the filter edge seals are scanned with an isokinetic probe connected to a photometric analyzer or particle counter to search for aerosol leaks, defined as a concentration of 0.01% or more of the upstream concentration. The probe is scanned across the filter face at 10 feet per minute. Some differences in performance of the particle counter and the photometer are reported (Greiner 1990). The differences arise from the fact that the filter challenge material concentration and nature can vary with the two methods and that the sensitivity of the two devices also varies. Because of these and other similar factors, no meaningful statements can be made as to whether one instrument is superior to the other. However, differences in performance, particularly for leak testing, appear to be present.

It is apparent that leak testing of large filter banks can require large amounts of personnel time. For that reason, some automated filter scanning systems have been produced for use in large filter systems. A system with some 3,600 filters that required scanning was tested with a computer-controlled tracking and sampling system that moved 16 OPC inlet nozzles simultaneously in rectangular patterns under a filter ceiling (Ziemer 1988). The computer moved the sampling inlet nozzles across the face of the filter system at the required velocity in the required pattern and recorded particle concentration for each filter. The system was developed because manual leak testing of the filter system would have required an estimated 600 to 800 person-hours. There were also some questions as to the reliability of that quantity of data when collected by manual measurements.

Air filtration by a HEPA filter in good condition normally reduces the concentration of particles in the 1-micrometer size range by a factor of 10^6 or more. In this situation, viable microorganisms in the air are reduced to the point where aseptic conditions exist. For this reason, the pharmaceutical industry has used HEPA filters to provide aseptic packaging environments. Filters that have been tested for particle penetration and are in use are verified periodically for inert particle penetration and for leaks. For this reason, great care is used in testing work stations and clean benches to minimize any possible bacterial penetration of filters. Primary emphasis is placed on assurance of satisfactory airflow within the clean bench and of assurance that no leaks exist in the clean bench filter. Many pharmaceutical clean bench testing procedures simply consist of defining total airflow and airflow uniformity at the bench work access opening, in addition to leak testing of the installed filter. The favored leak test method is to use a DOP challenge aerosol at a concentration of 10 to 100 μgm per liter of air and test for leaks with a photometer (La Rocca 1985).

In almost all of the previously mentioned test documents, DOP is the challenge aerosol. It is generated by spraying liquid from a submerged orifice pneumatic Laskin nozzle (Laskin 1948). This nozzle produces a liquid droplet aerosol with a mean diameter of approximately 0.25 μm with a geometric standard deviation of 1.53 when liquid DOP is sprayed using gas at a pressure from 1 to 2 bars (Hinds, Macher, and First 1983). These data were obtained using an OPC, so the size information is on the basis of an equivalent optical diameter. Acceptably high particle concentrations can be produced by a Laskin nozzle, as shown in Table 28-1.

DOP is not acceptable in all areas. It is an organic liquid commonly used as a plasticizer for polyvinyl chloride. Although quite low, the vapor pressure is finite. After filters have been tested for penetration with this material, there is frequently a noticeable odor. A question of possible

TABLE 28-1 DOP Challenge Aerosol Number Concentrations versus Mass Loadings Downstream of Test Filter

Mass Conc'n µgrams/Liter	Equiv. Eff.	Particles per cubic foot at unit density >0.5 µm	>0.3 µm
Monosized Particles			
100	0%	4.32×10^{10}	2.01×10^{11}
10	90%	4.32×10^{9}	2.01×10^{10}
1	99%	4.32×10^{8}	2.01×10^{9}
0.1	99.9%	4.32×10^{7}	2.01×10^{8}
0.01	99.99%	4.32×10^{6}	2.01×10^{7}
0.001	99.999%	4.32×10^{5}	2.01×10^{6}
0.0001	99.9999%	4.32×10^{4}	2.01×10^{5}
ANSI N101.1 Particle Size Distribution			
100	0%	8.5×10^{9}	2.5×10^{10}
10	90%	8.5×10^{8}	2.5×10^{9}
1	99%	8.5×10^{7}	2.5×10^{8}
0.1	99.9%	8.5×10^{6}	2.5×10^{7}
0.01	99.99%	8.5×10^{5}	2.5×10^{6}
0.001	99.999%	8.5×10^{4}	2.5×10^{5}
0.0001	99.9999%	8.5×10^{3}	2.5×10^{4}

When using an OPC to test filter penetration or leaks, the operator must be careful that the maximum recommended concentration capabilities of the OPC are not exceeded by the DOP concentration being measured.

carcinogenicity has also been raised. Tests have been carried out to find a substitute challenge material with the same physical properties as DOP but without some of the problem areas. Dioctyl sebacate appears to be a suitable substitute material (Gilmore, McIntyre, and Petersen 1982). It is compatible with commonly used filter testing equipment, it exhibits similar operational behavior to DOP, it effectively discriminates between penetration values of less than 0.03%, and it replicates test results obtained with DOP. Even so, organic material with even low vapor pressure may not be allowed near some products. The possibility of vapor emission from tested filters with possible subsequent deposition on sensitive surfaces is a concern. This is especially true where very-high-quality optical systems are assembled or for some semiconductor manufacturing areas, where any organic vapor collecting on surfaces can degrade the product.

For these reasons, other alternate filter test materials have been sought. Materials that have been investigated include fluorescent-dyed silica powder, standard air cleaner test dust, sodium chloride particles, ferric oxide powders, and ambient air. Each of these presents some problems. A satisfactory generation procedure for producing well-dispersed reproducible challenge aerosols from dry powders does not exist. Potential corrosion from deposition of sodium chloride with subsequent water absorption is a concern. Complete deagglomeration of the silica or ferric oxide powders is essentially impossible. Commercially available systems for generation of the silica challenge aerosols have been used (Bishop 1988), but some questions exist as to reproducibility of operation of the aerosol generation system. Standard air cleaner test dust particle size distribution is most suitable for challenging coarse filters rather than HEPA filters. Many of these materials may not be acceptable in some cleanroom areas.

A major problem occurs when ULPA or tandem HEPA filters are being tested where there is difficulty in gaining access to the duct after the prefilter for satisfactory challenge aerosol sample acquisition. The problem here is that the concentration of particles after the final filter is extremely low. In this situation, the challenge concentration must be very high in order to acquire sufficient downstream data in a reasonable time period. With such high challenge concentrations, dilution is necessary for measurement by most single-particle analyzing devices. Satisfactory dilution devices reduce the overall concentration sufficiently that small-particle concentrations can be defined, but the upstream concentration of particles much larger than 1 μm is reduced to the point where meaningful measurement is extremely difficult.

As an alternate challenge material, monosized polystyrene latex particles have been used. These are satisfactory for penetration testing of relatively small-area flat-sheet media. Concern has been expressed regarding cost for large quantities of PSL as would be required for penetration measurement or in-place testing of very large flow systems. For this reason, large element testing is usually carried out by using a polydisperse challenge and determining fractional penetration by use of an OPC system with size discrimination capability. The challenge aerosol is usually polydisperse DOP or atmospheric dust. With these materials, suitable access locations in the HVAC system must be available for injecting the challenge material. It is also desirable that the HVAC system configuration allows sufficient space for the test aerosol to disperse uniformly across the plenum or duct to the filter being tested and accessibility for sample line and test device installation.

Pharmaceutical processing is normally carried out in a sterile environment. Because most microbial particles are larger than a micrometer in size and are frequently deposited upon larger particles, HEPA filters in the air supply system should remove most viable aerosols, even those that consist of single unattached bacteria. Some air filter testing has been carried out with bacterial aerosol challenges. For example, endospores of *Bacillus subtilis* var. *niger* NCIB 8056 have been used as a test aerosol (Sinclair and Tallentire 1982). This material was used to generate uniform dispersions of elliptical spores of some 0.06 μm^3 in volume. Testing with these particles showed that filter penetration varied significantly with airflow rate.

Several problem areas are involved in testing filters. Tests before installation should not present any serious physical problems. The testing is carried out in a test stand that can accept almost any filter size that is manufactured. However, care is still required to minimize any possible errors in filter penetration measurement. An evaluation of error sources in testing ULPA filters (Schmitz and Fissan 1988) indicates that particle losses in the test system and measurement instrument noise data are the major error sources. A suggestion is made to use Poisson statistics to aid in differentiating downstream particle data from electronic noise effects or unsteady particle emissions from the filter or from other downstream test equipment; particle occurrences can be expected to occur on a random basis, whereas noise effects are more or less continuous and uniform.

Testing filters that have been installed in an existing HVAC system usually presents several problems. Perhaps the one most frequently seen is that of access to the upstream area in order to present the challenge aerosol in a controllable and uniform manner. Many filter systems are installed in an existing facility where available space is at a premium. This is particularly true for filter banks installed in ducts, where space for installation of flow-straightening vanes does not exist. In some installations, the space between the prefilter bank and the final filter is measured in inches, at best.

The problem of installing a line to insert a challenge aerosol and of providing even minimal concentration uniformity across the filter face is extremely difficult. When the testing protocol includes a requirement to determine the filter challenge level, then the problems become almost insoluble. No universally satisfactory solution has been found to this problem. The testing personnel must understand the nature of each individual problem and either use remedial measures in measurement or data interpretation or simply report the problem and point out the fact that the reported data may be of low reliability because of such problems.

Once the filters have been installed, penetration efficiency testing is seldom required. This test is usually carried out on individual filters in the manufacturer's facility before delivery. Where cleanroom operations cannot allow any possibility of organic vapor emission, as may occur after standard DOP test aerosol is used, the cleanroom operator may require an in-place penetration test with a special solid challenge aerosol. This material must be selected so that no problems can occur if any penetrating material deposits upon products. Alternately, it is possible to install untested filters from a batch of filters on the basis of test data from sample filters that are not installed in the critical cleanroom environment. However, in-place leak testing still must be carried out with assurance of adequate challenge for each filter being tested. This test must be carried out after installation even of tested filters because possible damage to the filter can always occur during installation. Provision of adequate challenge is a difficult problem with ULPA filters because leak definition requires a high challenge level in order to verify their performance. Because of the high efficiency of this filter type, the downstream data may be too sparse for development of statistically valid penetration data unless testing is carried out for a very long time. If this situation occurs, then the filter data may be limited to a maximum penetration statement.

Normal replacement schedules for HEPA and ULPA filters in cleanrooms are based on a pressure increase as the filter becomes loaded. New filters usually operate with a differential pressure of 0.1 inch of water or less. When the pressure drop increases to approximately 1 inch of water, the energy needs for air movement suggest that filter replacement costs may be more economical than the increase in fan power costs. Note that the small increase in pressure drop through the loaded filter does not result in additional particle penetration. In fact, as the filter loads with solid particles, the typical dendrite formation of collected particles upon the fibers may increase the filter performance significantly. In general, a cleanroom operated with 20% makeup air in a temperate urban environment may require prefilter replacement at intervals of 1 to 3 years and final filter replacement at intervals of 5 to 10 years. Figure 28-3 (Gutacker 1991)

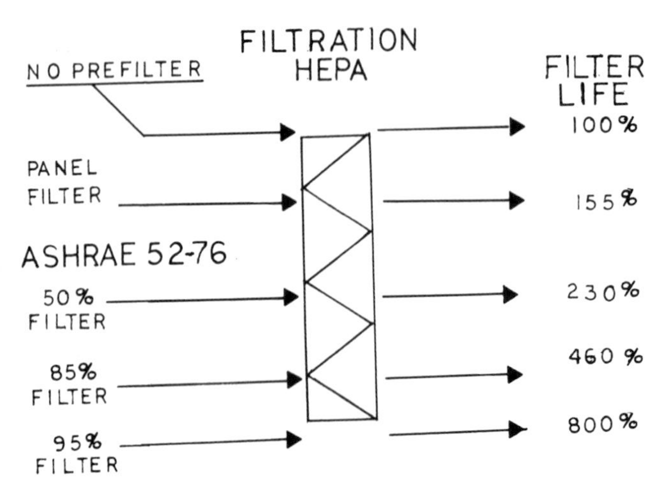

FIGURE 28-3. Prefilter effect on HEPA filter life. Changing prefilters is less costly in terms of both material and time than changing HEPA filters. (Courtesy A. Gutacker, ARGOT, Inc.)

illustrates how prefilters can extend HEPA filter life. As indicated by use of the American Society of Heating, Refrigeration, and Air Conditioning Engineers (ASHRAE) ratings, even prefilters with relatively high penetration are very effective in protecting and prolonging the life of the cleanroom final filters. Considering the cost of replacing HEPA filters, particularly for a cleanroom ceiling or wall, compared to costs of replacing prefilters in a noncritical area, the advantages of a scheduled prefilter replacement plan become obvious. This plan can be based on a differential pressure change base or on a regular time schedule.

Compressed gas filters are usually made with very-high-efficiency membrane or stacked disk types that operate at face velocities normally less than those for ambient air filter systems. The filter media may be permanently installed in cases that are replaced when necessary, or the media may be replaced in permanently installed cartridges. These filters operate under sufficient pressure so that the filter medium must be enclosed in a suitable leakproof container. Direct access to the filter medium is impossible during use. Penetration determination is rarely, if ever, carried out by the filter user. This test procedure is usually done by the manufacturer using special equipment to allow generation and measurement of well-characterized challenge aerosols. Because the compressed gas system in use must present gas with essentially zero particle content to the process line, the compressed gas filter must neither allow any gasborne particle to penetrate nor generate or release any particle from the medium or from the filter container itself. The testing procedure requires measurement of large quantities of gas, usually at pressures that can range from a few to several hundred bars, to ensure that the compressed gas is as clean as required. Where particle content must be controlled at size ranges of 0.01 μm or so, measurement with a condensation nucleus counter after expansion to ambient pressure may be required. In that situation, the required reduction in pressure to ambient level must be carried out in such a way that no artifact generation occurs.

In addition to ensuring removal of particulate material from the compressed gas supply, it is necessary that no contaminating particles, gases, or vapors be added to the supply by emission or outgassing from any element in the gas system. When new components, especially filters with their extended surface areas, are installed, some outgassing can be expected. Adsorbed or entrapped atmospheric gases may be released from the filter system, or some organic materials may be released from polymeric components such as seals. A test method has been proposed (Gotlinsky et al. 1989) in which filtered clean gas is presented to the filter and long-term testing is carried out to determine any particles or undesired vapors that may be released from the filter. Because the test is so sensitive, care is required to make sure that the test system is not the source of problem materials.

References

Bishop, D. E., 1988. Dri-Test™: The New Way to Challenge HEPA Filters. Proceedings of the 9th International Committee of Contamination Control Societies Conference, pp. 179–182, September 26, 1988, Los Angeles.

Gilmore, R. D., McIntyre, J. A., & Petersen, G. R., 1982. Operational Experience Using Diethylhexylsebacate (DEHS) as a Challenge Test Material in Filter Testing. Proceedings of the 17th DOE/NRC Nuclear Air Cleaning Conference, pp. 821–835, July 1982, Boston.

Gogins, M., et al., 1987. Design and Operation of Optimized High Efficiency Filter Element Test Systems. Proceedings of the 33d Institute of Environmental Science Annual Technical Meeting, May 1987, San Jose, CA.

Gotlinsky, B., et al., 1989. Outgassing of All Metal Filters for Use in the Semiconductor Industry. Proceedings of the 35th Institute of Environmental Science Annual Technical Meeting, pp. 401–405, April 1989, Anaheim, CA.

Greiner, J., 1990. HEPA Filter Leak Testing Using the Particle Counter Scan Method. *CleanRooms* 4(9):36–39.

Gutacker, A. R., 1985. *Permanent Atmospheric Background Testing for High Efficiency (HEPA) Filters and Filter Systems,* Document LCCD-155. Webster, NY: ARGOT, Inc.

Gutacker, A. R., 1991. Ambient Air Filtration. In *Contamination Control Technologist Handbook,* ed. A. Gutacker, Chapter 28. Webster, NY: ARGOT, Inc.

Hinds, W. C., Macher, J. M., & First, M. W., 1983. Size Distribution of Aerosols Produced by the Laskin Aerosol Generator Using Substitute Materials for DOP. *American Industrial Hygiene Association Journal* 44(7):495–500.

Johnson, E. M., et al., 1990. A New CNC Based Automated Filter Tester for Fast Penetration Testing of HEPA and ULPA Filters and Filter Media. Proceedings of the 36th Institute of Environmental Science Annual Technical Meeting, pp. 250–256, April 1990, New Orleans.

La Rocca, P. T., 1985. Testing Requirements for HEPA Filters and Clean Work Stations. *Pharmaceutical Manufacturing* 2(3):47–51.

Laskin, S., 1948. *Submerged Aerosol Unit,* AEC Project Quarterly Report UR-38. Rochester, NY: University of Rochester.

Rivers, R. D., & Engleman, D. S., 1984. Evaluation of a Laser Spectrometer ULPA Filter Test System. *Journal of Environmental Science* 27(5):31–36.

Schmitz, W., & Fissan, H., 1988. Error Analysis of Penetration Ratio Measurements of ULPA Filters. *Journal of Aerosol Science* 19(7):1421–1424.

Sinclair, C. S., & Tallentire, A., 1982. Microbiological Evaluation of Papers: The Influence of Dispersion Flow Rate on Penetration. *P & MC Industry* September/October: 52–57.

Ziemer, W., 1988. In-Situ Test of ULPA Filters. Proceedings of the 9th International Committee of Contamination Control Societies Conference, pp. 115–122, September 26, 1988, Los Angeles.

29

Liquids: A Cleaning Overview

Once the cleanroom operation is defined and in good order, airborne particulate contaminants are usually reduced to minor problems. The major contamination problems are then the emissions from personnel and processing tool actions (discussed previously) and the contamination transferred to products from process fluids. The particle concentration in most reactive process liquids is normally orders of magnitude greater than that in the ambient air or in a compressed gas supply. Because so many high-technology products either are produced using liquid processing materials or are cleaned by liquids, contamination transfer from the supposedly clean liquid to the product must be controlled. Production of a modern finished integrated-circuit component may require the use of as much as 150 gallons of water for all of the processing steps (Iscoff 1986). A study of the mechanisms involved in transfer of particles from liquids to solid surfaces has shown some of the important processes that control deposition and retention of particles from liquids to solid surfaces (Michaels et al. 1988). These processes are similar in nature to those in deposition from air suspension, but particle transfer across the liquid-solid and liquid-air interfaces is also important.

Table 29-1 shows some possible cleanliness classes for clean liquid supplies. A number of specification classes have been proposed for process water. The ASTM specifications for process water cleanliness for the electronics industry has been in a continuous state of revision (ASTM 1989) and may never be in a final, generally accepted form. A model has been proposed for classification levels that may be usable for semiconductor process liquids (Sielaff and Harder 1986). It assumes a particle size distribution slope of 2 and simply sets up class level steps at concentration

TABLE 29-1 Some Liquid Cleanliness Classifications for Clean Fluids

Semiconductor Process Liquid Particle Levels, No/100 ml

Class	>0.5 μm	>1 μm	>2 μm	>5 μm	>10 μm
0	10	1	0	0	0
1	300	10	1	0	0
2	700	50	5	1	0
3	1500	100	10	1	0
4	5000	300	30	2	1
5	20000	1000	80	5	1
6	100000	3500	250	15	5

levels that decrease by a factor of 2 for each step. Table 29-2 shows another proposed listing of semiconductor pure water guidelines (Balazs 1989). These levels were obtained after measurements of water samples from several working semiconductor fab areas. Some actual contaminant measurements (Carmody, Lindahl, and Martyak 1990) in an operating semiconductor facility indicate that particle content in deionized process water in 1989 were at levels of some 300 particles per liter for particles larger than 0.1 μm, whereas goals for the 1990 period were to reduce this concentration to less than 1 per liter. Even clean water, at 100 particles per 100 ml, contains as many particles per unit volume as the air in a class 28,000 cleanroom. Most reactive process liquids are orders of magnitude dirtier than water in terms of particle concentration alone.

The contaminants usually seen in liquids are particulate debris or dissolved materials of all types. Figure 29-1 illustrates the configuration of contaminant particles often present in some hydraulic liquids. These materials are accumulated from environmental air and are not greatly dissimilar from particles that might be transported into a cleanroom. Particles in liquids are seen in size ranges from submicrometer to several millimeters in size and can have any of a large range of shapes, including regular crystals, spheres, flakes, and long fibers. Data on particle levels in English semiconductor deionized water systems were obtained in 1984; these showed particle concentrations in size ranges greater than 0.5 μm (Hall 1984). The cleanest water, three to four particles per ml in sizes ≥ 0.5 μm, was seen just after a point-of-use filter; concentrations an order of magnitude greater, including particles larger than 5 μm, were found in some portions of the deionized water lines. Particularly in process fluids, liquid-borne particles can be present in concentrations of several thousand per milliliter! Contamination levels in sulfuric acid cleaning baths have been measured, and as many as 10,000 particles per ml in sizes ≥ 0.3 μm were seen in some acids (Eisenmann and Ebel 1988).

TABLE 29-2 Deionized Water Cleanliness Specifications and Guidelines

		SPECIFICATIONS			GUIDELINES			
			256K DRAM		1M DRAM		4M DRAM	
Item		Detection Limit**	1985 Specs Attainable	1985 Specs Acceptable	1988 Specs Attainable	1988 Specs Acceptable	<1μ VLSI	ULSI Target
Resistivity @ 25°C		18.2 max.	18.2	17.9	18.2	18.0	18.2	18.2
TOC (ppb)		5	<20	<50	<10	<30	<10	5
THM (ppb)		<1	–	–	–		<3	–
Particle / L by SEM	0.1-0.2μ		–	–	–	–	<1500	<1000
	0.2-0.3μ		–	–		<2000	<800	<500
	0.3-0.5μ				<200	<200	<50	<10
	>0.5μ				<1	<1	<1	<1
Particle / L by on-line laser	0.3-0.5μ	<1	–	–	–	–	<50	<10
	>0.5μ	<1	–	–	–	<100	<1	<1
Bacteria / 100mL								
by culture		<1	0	<6	0	<6	0	0
by SEM			–	–	<1	<10	<5	0
by EPI			–	–	<5	<50	<10	<1
Silica-dissolved (ppb)		0.25	<3	<5	<0.4	4	3	1
Boron (ppb)		0.05	–	–	<0.05	2.0	0.005	*
Ions (ppb)								
Na^+		0.05	0.05	0.2	<0.05	0.1	0.025	
K^+		0.1	0.1	0.3	<0.1	0.1	0.05	
Cl^-		0.05	0.05	0.2	<0.05	0.1	0.025	
Br^-		0.1	<0.1	0.1	<0.1	0.1	0.05	
NO_3^-		0.1	<0.1	0.1	<0.1	0.1	0.05	
$SO_4^{=}$		0.1	0.1	0.3	0.05	0.2	<0.05	
Ions total		0.5	<0.5	1.2	<0.5	<0.7	<0.2	
Residue (ppm)		<0.1	<0.1	<0.3	<0.1	0.1	<.05	*

Source: (Courtesy M. Balazs, Balazs Laboratories.)

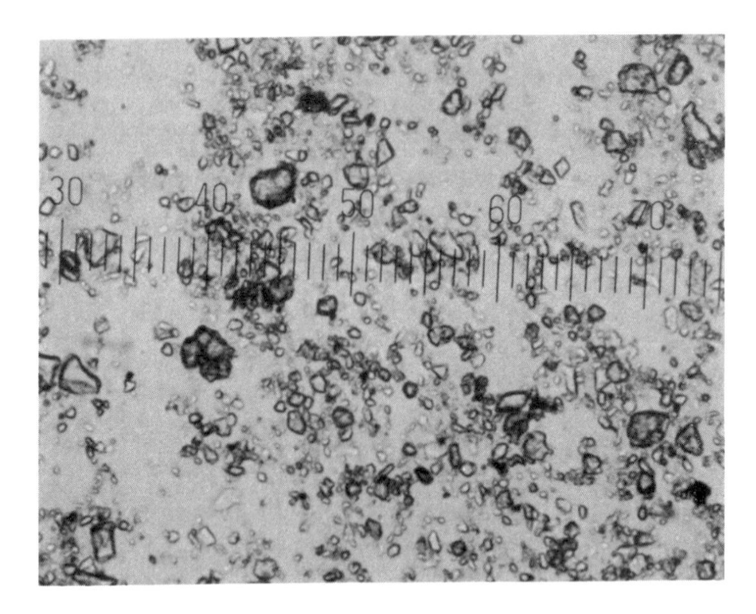

FIGURE 29-1. Contaminant particles as frequently seen in hydraulic fluids. The scale shown is 4 μm per division. ACFTD has long been used as a surrogate contaminant material for hydraulic fluid power system testing.

The particles may be organic or inorganic, depending on their source. Crystalline materials are usually accumulated during use or preparation or are ingested from the air during liquid storage or processing. They can also crystallize or precipitate from high concentrations of dissolved materials during temperature changes. They can be wear particles from operation of liquid flow control devices. For example, tests were carried out on a perfluoralkoxy (PFA) water supply system. It was found that particles were generated by abrasion of moving parts in valves and fittings and were washed out of crevices and gaps in components (Warnecke and Herz 1988). Spherical particles in liquids are usually either air pollutants, such as plant spores, bacteria, or fly ash particles, or droplets of other immiscible liquids. Bubbles produced from chemical reactions or air ingression are also spherical; defining these materials as particles may stretch the meaning of the word. Flake-shaped particles are worn from laminated surfaces or thin coatings. They may also be sections of plant- or animal-derived particles ingested from the air. Fibers are generated from defective filter media, from clothing, or from plant or animal sources. They can have diameters from less than 1 μm to 20 μm and aspect ratios of 5 to 10 or more. Long fibers are usually curled and are generally found to be interwoven with other fibers.

Bacteria and pyrogen particles can be found in almost any liquid. Bacteria have been found in water, reactive chemicals, fuels, and lubricants. Deionized water systems are fertile breeding grounds for bacteria unless care is taken to keep the system sterile and to avoid areas where the water can remain stagnant. Bacteria are usually in the 1- to 2-μm size range but can agglomerate to each other or to inert particles to form much larger clumps. Pyrogen fragments are usually smaller than the original bacteria. These materials are a potential production problem in semiconductor and optical system production, but the regulatory limitations imposed on the pharmaceutical manufacturing industry require extreme care in controlling microbiological contamination (Lieberman 1990). The control technology in these industries begins with planning for production areas and for definition of validation of necessary components and services; it continues through operating procedures for the entire process. Facility validation is required at reasonable intervals, and detailed documentation for both the method of controlling contamination and the observed data on contaminant levels is required.

Dissolved contamination in liquids can result from the original manufacturing process or from the process for which the liquid is being used. The type and quantity of material that can be found as a dissolved contaminant is limited only by the reactivity of the liquid. Removal of dissolved contaminants from liquids is usually quite difficult. The contamination varies with the liquid type, the liquid temperature, and the operating conditions. Contamination can be ionic, organic, or biological. A number of purification methods have been considered for cleaning microbiological contaminants from liquids. One of these methods, which is borrowed from municipal water treatment, is to oxidize organic and biological contaminants with ozone. It is claimed that the ozone will oxidize many organic and biological materials to "innocuous" carbon dioxide and that the ozone can be removed from the water before use by irradiation with ultraviolet light (Nebel and Nezgod 1984). Where the water will be used for cleaning silicon wafers during processing, care is required to ensure that dissolved oxygen levels are not so high as to cause excess oxide growth on the wafer surface after removal of other contaminants by the water. It has been shown (Ohmi et al. 1990) that oxide growth rates on wafer surfaces will increase if the wafer has been processed with water containing higher levels of dissolved oxygen.

Particulate contaminant removal from liquids is relatively simple, at least in principle. Filtration is the primary method of cleaning liquids, even though maintenance of cleanliness is extremely difficult. A variety of filtration mechanisms, including adsorption and sieving, are effective (Grant et al. 1989). Where adsorption mechanisms are involved, the particle enters into a pore and may be removed by adsorption on the pore wall. Retention depends on a number of factors, including the liquid's pH, its polarity and viscosity, the liquid velocity, and the particle charge. Where sieving is the major mechanism, all particles larger than the specific pore to which a particle is directed are captured. Figure 29-2 is a photomicrograph of a coarse filter used for liquid cleaning, where sieving is the major particle removal mechanism. In addition to this type of fibrous filter, membrane filters are frequently used for liquid filtration. Both collection mechanisms are effective with these media. Effective cleaning for process liquids may require use of a recirculating filtration system to remove submicrometer particles effectively (Gruver, Silverman, and Kehley 1990).

Membrane filters can be produced by several processes. A common method produces membranes by deposition of polymer solution films. Removal of solvent allows polymer precipitation in the form of a wet gel, which is dried to form a porous membrane. Microporous polytetrafluoroethylene (PTFE) membranes can also be formed by calendering PTFE resins into a membrane, which is then stretched to form a mat of elongated resin fibers. Pores of this material are in the form of slots. "Track-etched" polyester membrane filters are produced by exposing a coherent film to heavy radioactive fission particles. When these pass through the film, they leave a track of radiation-damaged material some 4 nm in diameter. If the film is exposed to caustic solution, the radiation-damaged tracks are dissolved, leaving essentially cylindrical pores through the film. Longer exposure time to the caustic solution results in larger etched openings.

Membrane filters can be modified to accentuate electrostatic effects. Charge modification can be accomplished by incorporating either positively or negatively charged compounds to aid in attracting and retaining oppositely charged particles. Positively charged membranes usually are produced with addition of quaternary amine groups. These are particularly effective when bacteria (usually with negative charge) are of concern. Negatively charged membranes usually incorporate carboxylic acid groups. Charge-modified filters contain a limited number of charge sites. When these are occupied, then filtration efficiency can decrease significantly (Tolliver and Schroeder 1983). With the exception of the track-etched filters, membrane filters always have a range of pore morphologies and size distributions that can affect the filtration performance of any individual filter from a lot (Meltzer 1988).

Other particle control methods, such as reverse osmosis or high-gradient magnetic filtration, are used for special applications. Dissolved contaminant removal is most difficult. Ion exchange or reverse osmosis are the favored methods, with the former actually being an exchange of

FIGURE 29-2. Coarse filter medium photomicrograph. Even with coarse filters, the pore dimension at a plane surface of the filter does not determine the size of the particle that will be retained. Particle deposition upon fibers and entrapment within the filter body are effective collection mechanisms.

one dissolved material with another. Bacterial particle removal is more difficult because the bacteria are self-replicating and can grow through filters (Simonetti and Schroeder 1984). For this reason, ultraviolet radiation is frequently used to break bacteria and large oxidizable particles into smaller fragments. Caution is required in system design when ultraviolet systems are used for this purpose. When the bacteria are killed by ultraviolet radiation, fragmented bacterial materials may be reduced in size to the point where penetration through the filter in use may increase (Shadman, Governal, and Bonner 1990). For this reason, ultraviolet radiation may be more effective if it is used after filtration. Total oxidizable carbon levels in the water are then reduced.

Operation of any liquid-cleaning system requires a great deal of care. Even if the liquid-cleaning system is quite efficient, artifact formation and release of particles from component surfaces can keep the contaminants in liquids at an unacceptably high level. It is necessary to make sure that the cleaning system design allows for easy replacement of components such as filters or heavily used valves or nozzles and that flow problems are avoided. If a section of a liquid loop will be used only occasionally, then it is usually worthwhile to recirculate liquid in that section to avoid dead leg formation where bacterial growth can most easily occur. Line sections where intermittent flow occurs downstream of a point-of-use filter must be minimized. When a filter or other component is replaced, filter seal installation and/or O-ring integrity must be assured.

A variety of methods exists for testing liquid-cleaning systems. In the same way as air filters are tested, liquid filters are tested for resistance to penetration by hard particles of various sizes as well as for leakage through filter media and filter systems with potential loss of integrity. Because the liquid filters are exposed to much greater shearing forces than are air filters, the relationships between particle penetration and filtration mechanisms shift with particle size from those seen in air filter media studies. In addition, the electrical charge mechanisms that are effective when passing nonpolar liquids through a filter are modified markedly when a polar liquid is being cleaned. For many pharmaceutical applications, freedom from bacteria for filtered material is of great importance; in addition, passage of a single organism from a challenge of up to 10^{12} is detectable by common microbiological assay methods (Krygier 1986). For some liquid filters that may be exposed to varying liquid flows, concern over possible filter medium distortion as a result of flow changes has indicated the need to test under conditions of both steady and pulsating flow rates. Pulsating flow effects on penetration efficiency were found to change with both filter material and construction types (Gotlinsky 1988).

For these reasons, the testing methods for liquid filters vary appreciably from those used for air filtration. Even though the collection mechanisms for particles in the liquid being filtered may be similar to those for particles in gases being filtered, the shearing force that can be applied to collected particles by fluid flow is much greater in liquids than in gases. In addition, electrical properties of liquids can vary greatly. For these reasons, the fluid conditions are specified more carefully. The environmental conditions should be carefully defined for the test procedure. Even so, rating membrane filters in terms of pore size is not simple (Johnston and Meltzer 1987). Pharmaceutical filter systems are rated differently when challenged by microbial or latex suspensions. In some cases, filter performance is measured when the fluid is being recirculated (ANSI 1973). This method is particularly applicable to definition of filters for hydraulic fluid power systems or applications where the liquid is also used as a coolant or lubricant in a recirculating system.

Filtration is frequently used for "sterilization" of pharmaceutical materials that are chemically or physically not suited for terminal sterilization processes that require excessive heat or radiation. For this reason, the integrity of sterilizing filters must be validated to be sure that any bacteria penetration is not significant. In many cases, the validation process involves definition of the pore size distribution of the filter medium (Hardwidge 1984; Chrai 1988; Emory 1989) by using nondestructive filter integrity testing. Nondestructive liquid filter integrity testing is based on the fact that wetted filters do not allow gas to penetrate a filter until sufficient pressure is applied to force liquid from the filter pores. That pressure is directly related to the liquid surface tension and inversely related to the pore size. Several automated integrity testers are commercially available (Olson, Gatlin, and Kern 1983). Care is required to ensure that any automated integrity test instruments are reliable and accurate, particularly for use in the pharmaceutical industry. The relationship between pore size defined by bubble point testing and by spherical particle penetration has been examined recently (Roberts, Velasquez, and Stofer 1990); the conclusion was that many membrane filter pore size ratings defined by the bubble point method are approximately half the dimension of the spherical particles that penetrate the filter. Even if the filter integrity is verified, the filter housing must be sterile to avoid release of bacteria into the filtered liquid. Procedures are available to sterilize components in place by steam, so it is not necessary to remove the filter housing from the system, transport it to and from an autoclave, and then reconnect to the system (Chrai 1989).

Possible release of particles that may be present on the filter is tested because the filter may be exposed to very high liquid shear that can strip retained particles from filter medium surfaces. The particles may be composed of debris from manufacturing operations, fragments of the filter or of sealing compounds, or previously collected particles that are driven from the filter. Several test methods have been described for defining particle emission from filters. The methods are based on recirculating very clean liquid through the test filter system and analyzing the effluent for emitted particles. The particles may be characterized by number and size alone (Goldsmith, Barski, and Grundelman 1984; Grant et al. 1986), or morphological information can be used to identify the emitted particle source. Measurements have indicated that shedding rates vary greatly from one filter material to another (Peacock et al. 1986) and that filter-to-filter variability can be as great as that between different filter materials (Meltzer 1987).

References

American National Standards Institute, 1973. *Multi-Pass Method for Evaluating the Filtration Performance of a Fine Hydraulic Fluid Power Filter Element,* ANSI B93.31. New York: American National Standards Institute.

American Society for Testing and Materials, 1989. *Draft Standard D19.02.03.03.* Philadelphia: American Society for Testing and Materials.

Balazs, M., 1989. *Semiconductor Pure Water Specifications and Guidelines.* Sunnyvale, CA: Balazs Laboratories.

Carmody, J. C., Lindahl, A. R., & Martyak, J. E., 1990. Meeting the Challenges of RO/DI System Contamination in the 1990s. *Microcontamination* 8(2):29–32, 62–63.

Chrai, S. S., 1988. Integrity Tests for Filter Systems. *Pharmaceutical Technology* 12(10):62–71.

Chrai, S. S., 1989. Validation of Filtration Systems: Considerations for Selecting Filter Housings. *Pharmaceutical Technology* 13(9):84–96.

Eisenmann, D. E., & Ebel, C. J., 1988. Sulfuric Acid and DI Water Point of Use Particle Counts and Resultant Silicon Wafer FM Levels. Proceedings of the 9th International Committee of Contamination Control Societies Conference, pp. 547–559, September 26, 1988, Los Angeles.

Emory, S. F., 1989. Principles of Integrity-Testing Hydrophilic Microporous Membrane Filters, Part II. *Pharmaceutical Technology* 13(10):36–46.

Goldsmith, S. H., Barski, J. P., & Grundelman, G. P., 1984. A Method for Measuring Particle Shedding from Microporous Membrane Filter Cartridges in Liquid Streams. *Microcontamination* 2(3):47–52.

Gotlinsky, B., 1988. *Steady and Pulsed Flow Particle Counting of Filters for High Purity Water,* Scientific & Technical Report PUF04BB, Glen Cove, Long Island, NY: Pall Filter Corp.

Grant, D. C., et al., 1986. A Comparison of Particle Shedding Characteristics of High Purity Water Filtration Cartridges. Proceedings of the 5th Semiconductor Pure Water Conference, January 17, 1986, San Francisco.

Grant, D. C., et al., 1989. Particle Capture Mechanisms in Gases and Liquids: An Analysis of Operative Mechanisms in Membrane/Fibrous Filters. *Journal of Environmental Science* 32(4):43–51.

Gruver, R., Silverman, R., & Kehley, J., 1990. Correlation of Particulates in Process Liquids and Wafer Contamination. Proceedings of the 36th Institute of Environmental Science Annual Technical Meeting, pp. 312–315, April 1990, New Orleans.

Hall, D., 1984. Contamination Control of High Purity Water. *Semiconductor International* 7(5):182–186.

Hardwidge, E. A., 1984. Validation of Filtration Processes Used for Sterilization of Liquids. *Journal of Parenteral Science and Technology* 38(1):37–43.

Iscoff, R., 1986. The Challenge for Ultrapure Water. *Semiconductor International* 9(2):74–82.

Johnston, P. R., & Meltzer, T. H., 1987. Microbial Control: The Rating of Membrane Filter Pore Sizes. *Ultrapure Water* 4(8):14–17.

Krygier, V., 1986. Rating of Fine Membrane Filters Used in the Semiconductor Industry. *Microcontamination* 4(12):20–26.

Lieberman, A., 1990. Contamination of Parenteral Products. *Pharmaceutical Manufacturing International* 1990, pp. 227–228.

Meltzer, T. H., 1987. An Investigation of Membrane Cartridge Shedding: I, A Quantitative Comparison of Four Competitive Filters. Proceedings of the 6th Semiconductor Pure Water Conference, January 15, 1987, Santa Clara, CA.

Meltzer, T. H., 1988. Microfiltration: The Pore Structures of Microporous Filters. *Ultrapure Water* 5(5):49–55.

Michaels, L. D., et al., 1988. Particle Deposition at the Solid-Liquid Interface. Proceedings of the 34th Institute of Environmental Science Annual Technical Meeting, pp. 439–442, April 1988, Anaheim, CA.

Nebel, C., & Nezgod, W. W., 1984. Purification of Deionized Water by Oxidation with Ozone. *Solid State Technology* 27(10):185–193.

Ohmi, T., et al., 1990. Examining Performance of Ultra-High Purity Gas, Water and Chemical Delivery Subsystems. *Microcontamination* 8(3):27–33, 60–63.

Olson, W. P., Gatlin, L. A., & Kern, C. R., 1983. Diffusion and Bubble Point Testing of Microporous Cartridge Filters: Electromechanical Methods. *Journal of Parenteral Science and Technology* 37(4):117–124.

Peacock, S. L., et al., 1986. A Comparison of Particle Shedding from Different Chemical Filtration Products. Proceedings of the Microcontamination Conference, May 19, 1986, Santa Monica, CA.

Roberts, K. L., Velasquez, D. J., & Stofer, D. M., 1990. Dispelling the Rating Myths of Microporous Membranes. Proceedings of the 9th Semiconductor Pure Water Conference, January 17, 1990, Santa Clara, CA.

Shadman, F., Governal, R., & Bonner, A., 1990. Interactions Between UV and Membrane Filters during Removal of Bacteria and TOC from DI Water. Proceedings of the 36th Institute of Environmental Science Annual Technical Meeting, pp. 221–223, April 1990, New Orleans.

Sielaff, G., & Harder, N., 1986. A Classification Model for Liquidborne Particles in Semiconductor Process Chemicals. *Microcontamination* 4(1):43–48.

Simonetti, J. A., & Schroeder, H. G., 1984. Evaluation of Bacterial Grow Through. Proceedings of the 30th Institute of Environmental Science Annual Technical Meeting, pp. 141–151, May 1984, Orlando, FL.

Tolliver, D. L., & Schroeder, H. G., 1983. Particle Control in Semiconductor Process Streams. *Microcontamination* 1(1):34–43.

Warnecke, H. J., & Herz, R., 1988. Investigation of Particle Generation in Liquid Supply Systems. *Solid State Technology* 31(10):1–5.

30

Particulate Contamination Measurement in Liquids

Definition of contamination level in liquids requires that a representative sample of the liquid be obtained for measurement and that a measurement method be used that is compatible with both the liquid and the contaminant of concern. Contaminants in liquids that are of concern include dissolved ionic inorganic materials and nonpolar organic materials, dissolved gaseous compounds, and particulate contaminants, both inert and viable. This chapter is directed primarily toward measurement of the particulate contaminants in cleanroom liquids. Procedures for sampling, storing, and measuring dissolved contaminants are outside the scope of this chapter.

Particulate contaminants in a liquid are usually distributed randomly rather than uniformly throughout the liquid. Therefore, any sample must be large enough to be truly representative of the entire liquid from which it was taken, and a sufficient quantity of data must be developed to state an acceptable confidence limit. *Compatibility* refers to physical and chemical compatibility between the liquid, the contaminant, the measurement instrument, and the material requirements of the method. Wetted parts of an instrument to be used for measurement of contaminants in a corrosive liquid should be made of materials that cannot be affected by that liquid.

Because particle spatial distribution in liquids is seldom uniform, a sample-acquisition system should be employed that can handle a representative sample of the liquid with no material losses or artifact generation during the liquid sample handling process. Liquids in a cleanroom can be flowing in a line or be within a container or a reaction vessel. The liquids can be extremely reactive. Pure water is a *strong* solvent for many materials. The quantity of liquid of concern can range from milliliters to kiloliters. The contaminant levels are usually well below the ppm level. At this level, particle number concentration can be more than 10,000 per milliliter for particles in the 1-μm range and much greater when smaller particles are considered.

The methods for sample handling and measurement of particulate contamination in liquids usually depend on the level and type of contamination that must be defined. For example, the methods for measurement of submicrometer particles are significantly different from those that are normally used for measurement of larger particles. The differences range from requirements for cleaning of sample containers to the instrumentation for measuring the particle content of the liquid. When using optical single-particle counters, light extinction devices are normally used for measurement of particles larger than 1 to 2 μm, whereas light scattering devices are used for submicrometer particle measurement. It is very difficult to clean sample containers completely for liquids whose content of particles in the 0.1 to 0.5 μm range must be determined. A tare background correction from the container is frequently accepted if extremely clean liquids are to be measured.

Several standard methods are used for definition of liquid cleanliness. Definition of adequate deionized water cleanliness for semiconductor flushing requires measurement of many components, including reactive silica at the 1 part per billion (ppb) level, sodium ion at the 0.1 ppb level, total oxidizable carbon at the 10 ppb level, a variety of inorganic and organic materials and amino acids and metal ions at the part per trillion (ppt) level, and particulate materials at very low size and population levels (Wotruba, Coulter, and Thomas 1989). Deionized water for flushing very large scale integration (VLSI) semiconductor products in the 1990 era should be essentially free of particles larger than 0.1 μm in diameter (Kearney 1989). With a wide variety of measurement methods and the requirements for extremely clean processing liquids, development of particle levels standards in semiconductor process liquids has been a continuing effort for many years (Dillenbeck 1985). Semiconductor Equipment and Materials International has prepared a variety of standard documents for many semiconductor process liquids, and industry is showing a growing acceptance of these documents.

Standards for maximum particle content in pharmaceutical liquids are distributed by governmental agencies. In the United States, control is required only of inert particles larger than 10 μm and 25 μm for parenteral liquids. Allowable concentrations depend on the container size. Large-volume parenteral liquid containers ($>$ 100 ml) should contain no more than 100 particles per ml; small-volume containers should contain no more than 10,000 particles per container. Standards in Europe and Australia require control of inert particles at the 2-μm level. No viable microorganisms are allowed in these liquids in any of the standards. Sterile and aseptic filling procedures are required for these materials, and equally strict controls are imposed for packaging of medical devices. The other areas where controls are imposed on contamination levels in liquids are those of lubricant or hydraulic fluid power systems. For the systems now in use and those planned for the near future, control is required only for particles larger than approximately 5 μm. Depending on the application area, most systems now in use can operate with hydraulic fluid containing up to several thousand particles per ml $\geq$ 5 μm. Most manufacturers and users of hydraulic systems routinely observe particle content at the 2-μm level to control against silt accumulation.

Whenever liquid samples are to be examined, it is necessary to take into consideration the process that may affect the liquid that is to be sampled. The operator should know if the process involves a flowing stream or takes place in a quiescent vessel or a stirred tank. For flowing streams, the operator should know if the flow rate is large or small in order to aid in selecting sample size. The sample should be large enough that statistically valid data are obtained, but not so large that the process material is changed by removal of the sample. The operator should know if the process is continuous and uniform or intermittent so that sample times and/or intervals can be chosen to represent actual process conditions. If the process is uniform and any variations are cyclic with known frequency, a constant sample rate and sample interval can be established on that basis. If the process is not uniform and variations in the nature of the process can be associated with specific operations, then sample acquisition times should be selected on the basis of those specific operations. When sampling from lines where valve operations occur, the operator should be aware that those operations frequently generate bursts of particles from the valve components. Sampling procedures should take this phenomenon into consideration. If the process involves sudden changes in liquid flow system dimensions when liquid is flowing under pressure, then cavitation may generate stable bubbles in the liquid that may be reported as particles. Therefore, no sudden changes in line dimensions should occur close to sample acquisition locations. Design of valves, gasket fittings, O-ring seals, retainers, and the like should be examined carefully to make sure that cavities are not present that may generate eddies in the liquid flow pattern. Particles can be deposited in these cavities by reverse eddy flows and then be reentrained to contaminate later samples.

Sample point location selection depends on the nature of the system being sampled. Experience aids in selecting locations in some possible liquid systems. For example, if a liquid system includes a cleaning filter, sampling shortly after the filter location can provide information about the integrity of that filter system. If the liquid system involves mixture, addition, or reaction because of chemical or thermal effects with materials, a sample point can be located shortly after the pertinent process has been completed. It may be desirable to install a sampling point at a reservoir inlet or outlet line location to verify the condition of as-received or in-use process liquids. Other obvious sample points can be located by judicious observations. In addition, convenience and physical constraints affect sampler design. In many locations, existing line configurations may make it impossible to install ideal sample acquisition systems. Undesirable sample locations may be imposed by physical constraints, line configurations, or equipment locations. If these situations exist, the data record should reflect the limitations imposed by the necessity for such choices.

Sampling of flowing streams can be carried out by an in-line or in situ method, where the analytical device observes a portion of the liquid flow within a line. Sample observation is carried out by nonintrusive energy transfer devices that illuminate a portion of the flowing stream and by observing the change in energy level when particles are present in that portion of the stream. A problem exists in that it is necessary to know the liquid flow rate in the sample line at the time of measurement and to define the exact portion of the total flow that is being observed. The liquid flow rate can be measured by an external flow meter to provide real-time data if the flow rate is varying during the sample period. If access for such a device is not available, then it may be necessary to observe pulse duration from the particle passage through the measurement system of known dimension to estimate fluid velocity. This technique should be reserved for flow systems known to be operating in the turbulent flow regime. Otherwise, the parabolic velocity profile for laminar flow conditions may cause large differences between average flow velocity and the velocity at the measurement point. Figure 30-1 shows the velocity profile for laminar and turbulent pipe flows. The velocity variation across the pipe radius is appreciably less for turbulent flow. For that reason, the precise location of the sample line inlet is not so critical if turbulent flow can be assured.

In setting up an in situ measurement system, the sensor element should be located at a suitable measurement point. That point should be located as close as possible to the actual point of use in determining whether a liquid is acceptable for a particular process. The in situ observation sample volume should be large enough that a statistically valid sample can be obtained in a reasonable time. A particle count of 1,000 or more is desired for each measurement, but no more than 1 hour of sampling should be needed, even for very clean liquids. This means that the sensing volume should handle liquid flows ranging between 0.1 and 10 ml per minute, with the smaller flow rates for liquids with relatively high contamination levels.

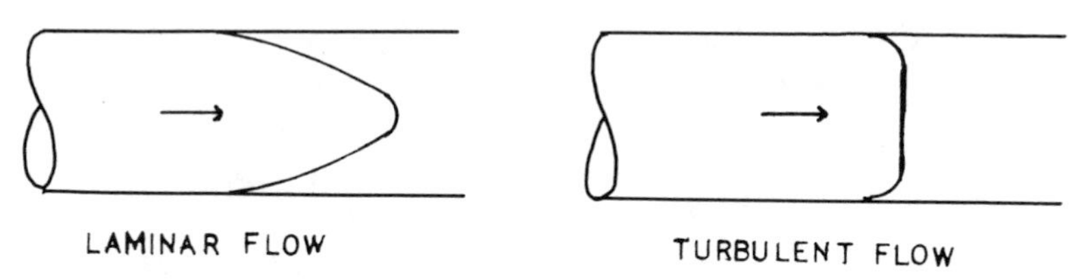

FIGURE 30-1. Velocity profiles for pipe flow. Turbulent flow is preferred within the sensing zone of an OPC because the electronic response can be markedly affected by residence time and pulse rise rate as particles pass through the sensing zone. The parabolic velocity profile peculiar to laminar flow can result in a range of particle velocity through the various portions of the sensing zone.

If liquids are very heavily contaminated, as may occur when filter penetration is tested, then it may be necessary to observe the challenge liquid in a side stream line to which diluent may be added as required. When single-particle counting systems are used, and the liquid being measured may be so heavily contaminated that the sensitive volume may contain more than one particle at any time, side stream sampling with dilution may be required. Sampling can be carried out with a side stream method, where part of the flowing stream is diverted through the analytical device and either returned to the line or to waste after analysis. Figure 30-2 shows recommended side stream sampling inlet arrangements. The sampler configuration choice is controlled mainly by the particle transport mechanism constraints discussed previously.

In designing and installing side stream sample probes, a fundamental decision must be made between permanently installed sample probes and removable types that can be inserted into the liquid system through suitable valves and packing glands. The removable probe can be easily inspected and cleaned when required, but well-designed fittings and seals must be present at each sample point. For liquid systems designed to operate in clean conditions, the permanently installed probes are recommended. Even though they may collect some particles from the liquid, they provide less opportunity for ingression of environmental contaminants into the liquid during probe removal or installation.

Connections and fittings to the sampled liquid system, to the sample line, and to the measuring device should be as smooth as possible. Polished metal or ceramic components are recommended. If sample transport lines longer than a few meters are used, no data should be recorded until sample flow in the line has stabilized, until the portion of line from the sample acquisition point to the measurement point has been flushed of any previous sample liquid, and until any gas that may be in the line has also been removed. Most sample lines that have been in use for any period of time

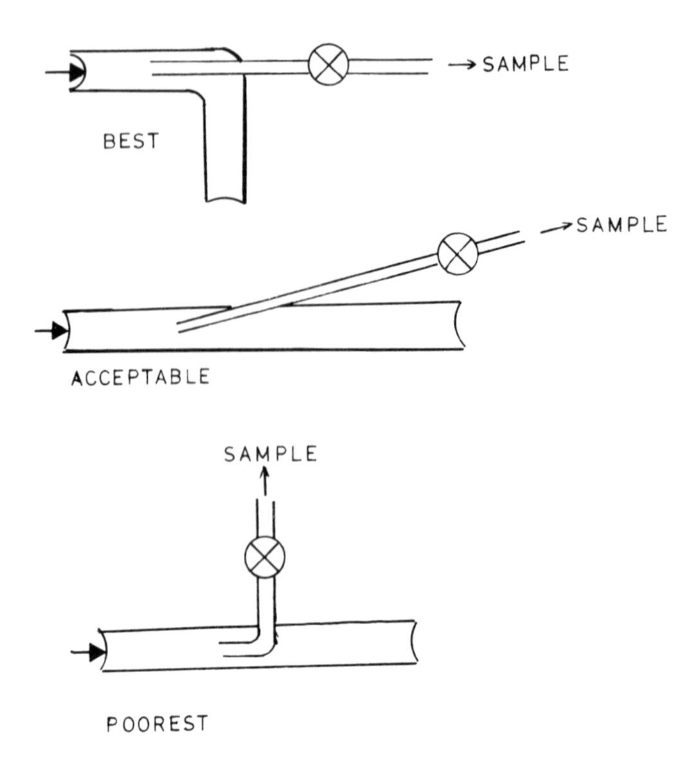

FIGURE 30-2. Recommended side stream sample inlet lines. The objective is to minimize particle losses caused by inertial deposition at sample line bends in the sample transit line. Many sample line installations cannot be ideal, but some attempt should be made to minimize sampling errors for all systems.

have some particles deposited on their inner wall surfaces. If the lines are flexible, then any motion releases particles that were previously deposited and may have become agglomerated. Care in line layout is required to avoid flexing or vibration that can cause such particle release.

The sample probe inlet in the sampled line should face into the liquid flow and be operated in isokinetic conditions whenever possible. If isokinetic sampling cannot be attained because of flow variations or any other reason, then isoaxial sampling at a greater than isokinetic rate should be sought. During sample transport to the measurement or observation point, the sample transport line should be as short and free from constrictions and sharp bends as possible.

After the side stream sample has been observed, the liquid must be disposed of. If inert, innocuous liquids are being measured, then disposition to waste is a minor problem. If hazardous materials are being sampled, however, then the measured material must be disposed of safely. Cost considerations are important but secondary to safety needs. If the operator is sure that the measurement process has not modified the liquid in any way, then the side stream can be returned to the main process line. If any change in properties of the side stream may result from the measurement process, then a disposition problem may arise. Discussions with local environmental protection and occupational safety agencies may be required. The same problem may also occur when batch samples are handled and cannot be used again in a process.

An in situ system may be incorporated into a flowing line by adding a window section to allow measurement of particle content within the line. Either acoustic or optical methods can be used for detection. Acoustic response systems are capable of measuring single 10-μm particles by an ultrasonic pulse echo response system. Optical light scattering systems are capable of measuring single 0.2-μm particles by observing scattered laser light from a small portion of a stream up to several inches in diameter. Windows are used for illumination and scattered light measurement from a defined portion of the line. The same method can be used to measure 0.05-μm particles in liquid from a portion of a stream approximately 1 cm in diameter that is passed through a sensor. The window section may use transparent materials that are inert to almost any liquid (Montgomery 1987). The in situ methods have the advantages that nonintrusive measurements are made, sampling extraction and liquid-handling errors are eliminated, and no sample disposition problem is present.

Sampling from containers is usually carried out as a batch process. A batch sample or a series of aliquot samples of the liquid is removed from a container and passed through the analytical device. Some preliminary planning may be required to avoid errors. For example, clean liquids may be stored in a reservoir from which material is drawn as needed. Samples have to be removed from the reservoir in order to characterize the material used from it. In the same way as occurs for in-line sampling, if a valve control or sample line is used to transport liquid from the reservoir, then artifact effects introduced by operation of the valve or by flexing of the line at the beginning of the withdrawal process can be ignored by discarding the data from the sample obtained during the initial flow. In sampling from large reservoirs, it may not be necessary to be concerned about sample uniformity within the reservoir because the sample-withdrawal process can simulate actual process liquid withdrawal. If a number of smaller containers are to be used and samples are selected from the group, then it is necessary to assure thorough mixing of the material in the sample container because essentially all material from the small container will be removed.

Before a batch container is opened for sample removal or use, the exterior surfaces should be cleaned as well as possible, especially around the container closure. Removal of material from the container is usually done by either aspirating or feeding by pressure through a clean sample line to a secondary batch sample holder or directly to the contaminant-measuring device. Removal of materials from small containers (up to 5 liters) may be done by pouring a few hundred ml into the smaller batch sample holder. When pouring, the technician should take care to minimize bubble generation and splashing that can be caused by pouring too rapidly. Pouring should not be done at so slow a rate so that a liquid film can adhere to the exterior upper portion of the container. This can allow debris on the container exterior to be washed into the smaller sample vessel. As always, if the liquid being handled is hazardous in any way, the laboratory supervisor and the technician should be cognizant of correct handling procedures and safety precautions for that particular liquid.

Batch samples can usually be measured adequately if the sample volume is in the range of 20 to 500 ml, depending on the particle concentration in the liquid. The choice between glass, plastic, or metal containers depends on the nature of the liquid being measured. Glass is preferred, if possible, because it can be cleaned easily and its cleanliness can be verified. If packaged correctly, it can be shipped under hostile temperature and vibration conditions. Corrosive liquids may require special plastic container materials. Metal containers are not normally recommended unless high-pressure operation is required for liquids that will not attack the metal. Batch sample container and closure cleaning procedures should include high-velocity liquid or manual scrubbing with surfactant solutions, rinsing with filtered solvents, and drying in clean gas, followed by cleanliness verification. The cleaning process is equally important for new and for reusable containers. If the containers are to be used again, a series of mutually miscible solvents must be used for the cleaning process. Otherwise, colloidal emulsion droplets may form by the mixing of immiscible liquids, adhere to the container surface, be released to subsequent samples, and be counted as contaminant particles.

Figure 30-3 shows the basis for a batch sampling system used for removal of liquid from a process vessel, and Figure 30-4 shows a batch sample handling system used for small-volume sampling from containers up to several liters capacity. When working with batch samples, particle settling or agglomeration caused by storage effects must be avoided before measurement. An equally serious problem with batch sample handling systems occurs when vacuum aspiration is used to move liquids from a container through the OPC sensor assembly. When liquids with high vapor pressure and/or high viscosity are aspirated through an OPC sensor cell of small dimension, a pressure drop may occur that is sufficient to cause gas

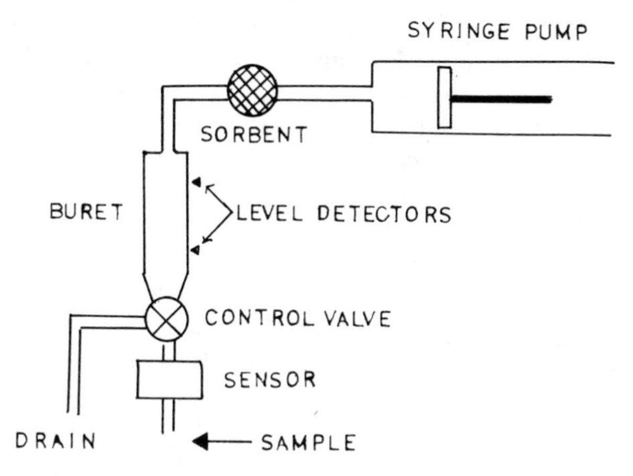

FIGURE 30-3. Aspiration sample feeder operation. This type of sample feeder is used to draw samples from a larger vessel. If the liquid being sampled may contain gases, measurement should not be made while the liquid is under reduced pressure. Bubble formation and growth may cause erroneous data.

or vapor bubble formation in the cell. The bubbles are erroneously reported as particles. Calculations have been made for a sensor cell with dimensions orthogonal to liquid flow of 0.5 mm × 1 mm and a flow length of 0.5 mm for a range of liquid flow rates (Fisher, Hupp, and Scaccia 1987). At 100 ml per minute, pressure drops ranging from 2 psi for an aqueous ammonia solution up to 12 psi for phosphoric acid were measured. These

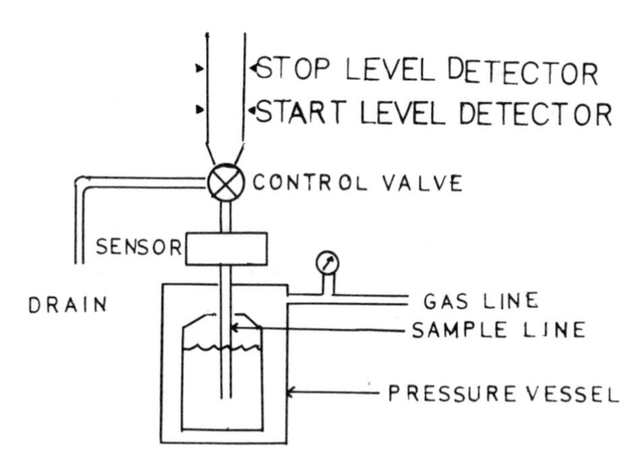

FIGURE 30-4. Pressure sample feeder operation. This type of sample feeder is used to feed samples from containers of up to 5 liters capacity. It is popular if the liquid may contain gases.

pressure drops are sufficient to induce either degassing or vapor bubble formation for many liquids. For this reason, samples should be fed under pressure for most electronic liquid particle measurements (Grant and Schmidt 1987; Dillenbeck 1987). For most systems, an adequate pressure is 1.5 to 2 bars. This is sufficient to prevent gas or vapor bubble formation after maximum pressure drop in the sensor cell.

In some situations, it is necessary to remove a liquid sample from an operating system and examine it for contaminant level in another location. In this situation, a means of removing the sample from the system is required, along with a method of suitable transport and/or storage prior to measurement. If the operating system operates at a pressure sufficient to force liquid from the system to a suitable container, then it is necessary only to open a sampling valve, allow sufficient liquid to pass through the valve to flush the valve surfaces while collecting the flushing liquid for suitable disposal, and then collect a sample in a suitable container for subsequent analysis. A procedure is defined for hydraulic fluid power systems that can be adapted to any pressurized systems with care to control toxic liquids (ANSI 1972). If a large number of sample points must be monitored, then the cost of installing individual particle-measuring systems at each point or of collecting a large number of samples, transporting to a measurement laboratory, and maintaining good records may become excessive. For this situation, computer control of a sequential, time-activated, multiport valving system may be used to deliver a series of samples to a single particle measurement system with control of data and identification of source (Vargason 1990). Careful design of the system is required to make sure that no artifact data are produced and that hazardous material handling is carried out safely both during and after measurement.

A variety of error sources exist for defining contaminants in liquid systems (Lieberman 1984). Understanding of error sources is important in reducing their effects. Comprehensive, meaningful standards and specifications are needed to define contamination levels in liquids. Sample-handling procedures are especially important in handling liquids to avoid artifact introduction and particle losses. Understanding the operation of the measurement devices is an absolute necessity. A need exists for better understanding of the statistical methods required for good measurement of sparse data. Skill in operation and interpretation of data is required. Otherwise, errors can occur in interpreting data (Barber 1988). Understanding the potential problems and developing procedures to eliminate the problem areas by using good laboratory practice can reduce the errors to insignificant levels (Coplen, Weaver, and Welker 1988).

The procedures used for measuring particles in liquids are well known.

The first method to be discussed is collection and observation, where a known quantity of liquid is passed through a membrane filter of pore size smaller than the particles of concern. The membrane is examined with either light or electron microscopy or bacterial culturing techniques to determine particle number and size or viable bacteria concentration. The method is effective for particles below 1 micrometer and has the advantage that samples are available for reexamination. In addition, the collected material is available for analysis to determine the composition of particles sometimes as small as 1 micrometer across. The method has the disadvantages that subjective processes for sample collection and observation are part and parcel of the method, that a long time may be required for data production, and that skilled personnel are needed for effective measurements. One method in use filters a water sample through a membrane filter, adds stain to allow subsequent discrimination for different types of particles, and observes particles as small as 0.2 μm using oil immersion light microscopy (Yang and Tolliver 1989). It is possible to determine size distribution and composition of particles smaller than 0.1 μm by using a scanning electron microscope, morphological definitions, and electron beam analytical methods (Balazs 1988). Scanning electron microscopy methods can be used on very clean water samples to detect both inert particles and bacteria at the 0.2-μm level with counting times of 30 to 45 minutes (Hango 1989). Sample volumes are in the range of 200 to 1,000 liters, requiring 1 to 5 days of sampling through track-etched membrane filters before the collected sample contains sufficient material for analysis. The method has the further advantage of allowing identification and sizing of specific bacteria.

A popular technique for enumerating bacteria in water involves epifluorescence microscopy. A positive correlation between direct counts using the method and bacterial endotoxin levels in water has been shown (Mittelman 1988). In this method, a sample of water is filtered through a track-etched membrane filter coated with acridine orange dye, which is specific for nucleic acids. After drying, the deposit is examined by using an epifluorescent microscope with a fluorescent light combination that is appropriate for the dye. Depending on their relative DNA/RNA concentrations, bacteria fluoresce in various shades of orange, green, and red. Bacterial cells as small as 0.1 to 0.3 μm can be resolved with this method. It is an especially useful method for characterizing bacterial levels in DI water because both live and dead cells fluoresce and a total count can be procured (Carpenter 1990).

Acoustic devices operate by passing a focused pulse of ultrasonic energy through a liquid. The echo from particles is detected. The method is effective for detection of single particles > 10 μm or so and for larger quantities of particles down to 0.8 μm. It is an in-place method that can easily be fitted to most flow lines. As with most energy interaction methods, it responds to more than the single parameter of particle size and results in some variability with liquid type. Bubbles and immiscible liquid droplets can cause problems. It is seldom used for very clean liquids but has been found effective for indication of overall debris levels or trends in liquids at low cleanliness levels.

Electrical resistance counters pass a liquid stream through a small orifice where electrical current is flowing. As particles pass through that orifice, the conductivity of the liquid path varies in accordance with particle volume. The method can define particles from ≈ 0.5 μm to much larger sizes with a 30 : 1 dynamic range for any single sensor. High counting speeds are used, and excellent sensor resolution is a result of the volume measurement base. The major disadvantages of the method arise from the need for conductive liquids and the susceptibility of the signal-carrying liquid path in picking up external electrical noise. Its primary contamination control application is for characterization of pharmaceutical liquids, mainly in European areas. The system is mainly used to define concentration of blood components.

Optical single-particle counting instruments are most widely used for particle measurement in cleanroom liquids. Light extinction counters are used for particles larger than 1 to 2 μm, and light scattering counters are used for particles in the size range 0.05 to 5 μm. A stream of liquid is passed through the sensing zone, and the signals caused by collected light level variation are measured and related to particle count and size for the liquid quantity measured. These devices have the advantage of consistent high-speed response and widespread use in the cleanroom area. They have the disadvantages of responding to more than just the size of the particles that are being observed, as with any other energy interaction method. Even so, it has been concluded that these instruments are still appropriate for accurate measurements of liquid cleanliness and high-efficiency filter performance for systems where concern exists for particles of 0.05 μm and larger (Grant 1990). Such particles can pass through some filters or may be generated in the fluid lines downstream of the filter.

A variation of the single-particle counting, light scattering system has been described (Kreikebaum 1990). This device is based on high-sensitivity nephelometry that uses light scattering optics. Testing was carried out that indicated sensitivity 2 orders of magnitude greater than the turbidimeters used for observation of particles in potable water. It was shown that the nephelometer was sensitive to a concentration of 5×10^4 per ml of 0.085-μm spheres. Sensitivity to larger particles can be expected in lower concentrations.

A method for detecting extremely low quantities of nonvolatile residues in liquids with vapor pressure at least equal to that of water has been described (Blackford, Sem, and Kerrick 1990). The liquid is atomized to form an aerosol of uniform droplets, and the droplets are vaporized in clean, dry air. The residual particles are passed through a condensation nucleus counter, and the quantity of particles can be used to define nonvolatile residue levels down to some 10 ppb. The method reports both particulate contaminants and any dissolved nonvolatile residues in the liquid.

Viable bacterial particles in liquids can be enumerated by several methods (Mittelman 1985), including viable count by transferring the liquid sample to a nutrient medium, culturing the sample, and observing bacterial colony formation; and direct count by scanning electron microscopy or by an epifluorescence method that is capable of detecting both viable bacteria and some pyrogens (ASTM 1988). In this last method, a liquid sample is passed through a membrane filter and stain is added to the filter surface. The bacterial particles respond to this stain and are counted. Some skill is required to interpret the images, but it is quickly learned. Biochemical detection depends upon measurable reaction to biochemical "markers" such as pyrogen-induced clot formation, bioluminescence measurement, or chromatographic analysis of bacterial phospholipid fatty acids.

References

American National Standards Institute, 1972. *Particulate Contamination Analysis: Extraction of Fluid Samples from Lines of an Operating System,* ANSI/B93.19M. New York: American National Standards Institute.

American Society of Testing and Materials, 1988. *Standard Test Method for Rapid Enumeration of Bacteria in Electronics-Grade Purified Water Systems by Direct-Count Epifluorescence Microscopy,* F1095-88. Philadelphia: American Society for Testing and Materials.

Balazs, M. K., 1988. Measuring and Identifying Particles in Ultrapure Water. *Microcontamination* 6(5):35–40.

Barber, T. A., 1988. Limitations of Light Obscuration Particle Counting as a Compendial Test for Parenteral Solutions. *Pharmaceutical Technology* 31(10):34–52.

Blackford, D. B., Sem, G. J., & Kerrick, T. A., 1990. Residue Analysis in High Purity Water Using a Condensation Nucleus Counter. Proceedings of the 2d PDA/IES International Conference on Particle Detection, Metrology and Control, Parenteral Drug Association, February 5, 1990, Arlington, VA.

Carpenter, S. E., 1990. Improving Epifluorescent Monitoring Methods in Estimating Numbers of Bacteria in Ultrapure Water. Proceedings of the 36th Institute of Environmental Science Annual Technical Meeting, pp. 212–216, April 1990, New Orleans.

Coplen, R., Weaver, R., & Welker, R., 1988. Correlation of ASTM F312 Microscope Counting with Liquidborne Optical Particle Counting. Proceedings of the 34th Institute of Environmental Science Annual Technical Meeting, pp. 390–394, May 1988, King of Prussia, PA.

Dillenbeck, K., 1985. Measuring Particulates in Processing Chemicals: The Need to Standardize. *Microcontamination* 3(11):21–30.

Dillenbeck, K., 1987. Advances in Particle-Counting Techniques for Semiconductor Process Chemicals. *Microcontamination* 5(2):31–39.

Fisher, D. H., Hupp, S. S., & Scaccia, C., 1987. Limitations of Vacuum Sampling Techniques for Counting Particles in Liquids. *Microcontamination* 5(2):14–20.

Grant, D. C., 1990. Measurement of Particle Concentrations in Central Chemical Delivery Systems. Proceedings of the 36th Institute of Environmental Science Annual Technical Meeting, April 1990, New Orleans.

Grant, D. C., & Schmidt, W. R., 1987. Improved Methodology for Determination of Submicron Particle Concentrations in Semiconductor Process Chemicals. *Journal of Environmental Science* 30(3):28–33.

Hango, R. A., 1989. DI Water-Quality Monitoring for Very Dense Electronic Component Manufacturing. *Ultrapure Water* 6(4):14–21.

Kearney, K. M., 1989. Ultrapure Water Requirements Squeeze into the Submicron Range. *Semiconductor International* 12(1):80–83.

Kreikebaum, G., 1990. Sub 0.1 Micron Particle Detection and Analysis Using High Sensitivity Nephelometry. *Journal of the IES* 33(4):19–24.

Lieberman, A., 1984. Problems Associated with Submicrometer Contaminant Measurement. In *Semiconductor Processing,* ASTM STP 850, ed. D. C. Gupta, pp. 172–182. Philadelphia: American Society for Testing and Materials.

Mittelman, M. W., 1985. Biological Fouling of Purified-Water Systems: II, Detection and Enumeration. *Microcontamination* 3(11):42–58.

Mittelman, M. W., 1988. Rapid Enumeration of Bacteria in Purified Waters by Direct-Count Epifluorescence Microscopy. Proceedings of the 9th International Committee of Contamination Control Societies Conference, pp. 531–536, September 26, 1988, Los Angeles.

Montgomery, C. N., 1987. In-Situ Particle Sizing in Liquids. Proceedings of the 33d Institute of Environmental Science Annual Technical Meeting, pp. 353–365, May 1987, San Jose, CA.

Vargason, R., 1990. Liquid Multiport System Provides Automatic Real-Time Monitoring of Wet-Process Station Liquids. *Microcontamination* 8(9):39–41.

Wotruba, W. F., Coulter, B. L., & Thomas, D. J., 1989. Instrumentation. *Ultrapure Water* 6(1):35–43.

Yang, M., & Tolliver, D. L., 1989. Ultrapure Water Particle Monitoring for Advanced Semiconductor Manufacturing. *Journal of Environmental Science* 32(4):35–42.

Index